THE SOCIETY FOR APPLIED BACTERIO
TECHNICAL SERIES NO. 10

MICROBIAL ULTRASTRUCTURE

The Use of the Electron Microscope

Edited by

R. FULLER
National Institute for Research in Dairying, Shinfield, Reading, England

AND

D. W. LOVELOCK
H. J. Heinz Co. Ltd., Hayes Park, Hayes, Middlesex, England

1976

ACADEMIC PRESS
LONDON · NEW YORK · SAN FRANCISCO
A Subsidiary of Harcourt Brace Jovanovich, Publishers

ACADEMIC PRESS INC. (LONDON) LTD
24/28 OVAL ROAD
LONDON NW1

U.S. Edition Published by
ACADEMIC PRESS INC.
111 FIFTH AVENUE
NEW YORK, NEW YORK 10003

Library of Congress Catalog Card Number : 76–1080
ISBN : 0–12–269450–3

Printed in Great Britain by
Cox & Wyman Ltd,
London, Fakenham and Reading

Contributors

D. Amici, *Institute of General Physiology, University of Camerino, Italy 62032*

Judith P. Armitage, *Department of Botany and Microbiology, University College London, Gower Street, London WC1E 6BT, England*

Barbara Bole, *Long Ashton Research Station, University of Bristol, England*

B. E. Brooker, *National Institute for Research in Dairying, Shinfield, Reading, England*

Roberta Bartlett, *Department of Bacteriology, University of Birmingham, Birmingham B15 2TJ, England*

R. R. Davenport, *University of Bristol, Department of Agriculture and Horticulture Research Station, Long Ashton, Bristol, England*

Susan C. Dickens, *Unilever Research Laboratory, Colworth House, Sharnbrook, Bedford, England*

John D. Dodge, *Department of Botany, Birkbeck College, Malet Street, London, England*

G. J. Dring, *Unilever Research Laboratory, Colwarth House, Sharnbrook, Bedford, England*

C. S. Dow, *Biological Sciences, University of Warwick, Coventry, CV4 7AL, England*

B. P. Eyden, *Zoology Department, Glasgow University*, Glasgow *G12 QQ, Scotland*

Madilyn Fletcher, *Marine Science Laboratories, University College of North Wales, Menai Bridge, Gwynedd, Wales*

G. D. Floodgate, *Marine Science Laboratories, University College of North Wales, Menai Bridge, Gwynedd, Wales*

R. Fuller, *National Institute for Research in Dairying, Shinfield, Reading, England*

Audrey M. Glauert, *Strangeways Research Laboratory, Cambridge, England*

G. W. Gould, *Unilever Research Laboratory, Colworth House, Sharnbrook, Bedford, England*

Peter J. Highton, *Department of Molecular Biology, Kings Buildings, Mayfield Road, Edinburgh EH9 3JR, Scotland*

W. Hodgkiss, *Terry Research Station, Aberdeen, Scotland*

M. H. Jeynes, *Department of Bacteriology, Medical School, University of Birmingham, England*

A. M. Lawn, *Houghton Poultry Research Station, Houghton, Huntingdon, England*

Beverley McLeod, *University of Bristol, Department of Agriculture and Horticulture, Research Station, Long Ashton, Bristol, England*

Stephen T. Moss, *Department of Biological Sciences, Portsmouth Polytechnic, Portsmouth, Hampshire, England*

H. N. Newman, *Department of Dental Medicine and Medical Research Council Dental Unit, University of Bristol Dental School, Lower Maudlin Street, Bristol BS1 2LY, England*

M. Paparelli, *Institute of General Physiology, University of Camerino, Italy 62032*

Elizabeth Parsons, *University of Bristol Department of Agriculture and Horticulture, Research Station, Long Ashton, Bristol, England*

R. Parton, *Department of Microbiology, University of Glasgow, Alexander Stone Building, Garscube Estate, Bearsden, Glasgow, Scotland*

Susan M. Passmore, *Long Ashton Research Station, University of Bristol, England*

Phyllis Pease, *Department of Bacteriology, University of Birmingham, Birmingham B15 2TJ, England*

Gillian Roper, *Welcome Research Laboratories, Langley Court, Beckenham, Kent, England*

C. R. Rowles, *174 Landor Road, Whitnash, Leamington Spa, England*

I. Santarelli, *Institute of General Physiology, University of Camerino, Italy 62032*

J. A. Short, *Wellcome Research Laboratories, Beckenham, Kent, England*

U. B. Sleytr, *Strangeways Research Laboratory, Cambridge, England*

D. G. Smith, *Department of Botany and Microbiology, University College, London, Gower Street, London WC1E 6BT, England*

J. M. Stubbs, *Unilever Research Laboratory, Colworth House, Sharnbrook, Bedford, England*

Janice Tallack, *Department of Bacteriology, University of Birmingham, Birmingham B15 21J, England*

G. G. Tedeschi, *Institute of General Physiology, University of Camerino, Italy 62032*

Kareen J. I. Thorne, *Strangeways Research Laboratory, Cambridge, England*

Margaret J. Thornley, *Immunology Division, Department of Pathology, University of Cambridge, England*

C. VITALI, *Institute of General Physiology, University of Camerino, Italy 62032*

P. D. WALKER, *Wellcome Research Laboratories, Langley Court, Beckenham, Kent, England*

D. WESTMACOTT, *Biological Sciences, University of Warwick, Coventry CV4 7AL, England*

R. WHITTENBURY, *Biological Sciences, University of Warwick, Coventry CV 7AL, England*

Preface

THIS volume, which is number 10 in the Technical Series, is based on the contributions made to the 1974 Autumn Demonstration Meeting of the Society for Applied Bacteriology.

The meeting covered the field of microbial structure as revealed by the electron microscope. The use of this instrument in studying micro-organisms has increased enormously in recent years and much information relating structure to function at a sub-cellular level has accrued. The book is by no means a comprehensive treatise on the use of the electron microscope in the study of microbial ultrastructure but does, we hope, illustrate the potential of the instrument by drawing together observations obtained by the application of varied techniques to a wide range of specimens including bacteria, fungi, algae and protozoa.

We would like to thank everyone who contributed demonstrations and subsequently converted them into manuscript. We are also indebted to Professor D. Lewis, Dr D. G. Smith and other members of the staff of the Botany and Microbiology Department of University College, London, for assistance with the staging of the demonstration.

R. FULLER

April 1976 D. W. LOVELOCK

Contents

Techniques for Transmission Electron Microscopy

J. A. Short and P. D. Walker

Wellcome Research Laboratories, Langley Court, Beckenham, Kent, England

Introduction

There now exists a great mass of excellent scientific literature giving details of preparative techniques employed in transmission electron microscopy for the examination of biological specimens. Indeed, recent textbooks are usually compiled in three or four volumes and the reader, for a knowledge in broad outline, can do no better than refer to one or more of these.

The notes, references and micrographs that follow are guidelines which it is hoped will encourage the reader to explore the subject in much greater depth.

In general, the methods employed for the visualization of ultrastructure in eukaryotic cells are applicable to the prokaryotes. One should, therefore, consider the disciplines of negative staining, shadowing, thin sectioning and freeze fracture and etching. Associated with these disciplines and subdividing them are techniques such as histochemistry, autoradiography and immunocytochemistry which should also merit attention.

It cannot be stated too strongly that although of necessity the microbiologist may become extremely proficient in one aspect of electron microscopy he should be familiar with all the techniques and thus be aware of their combined possibilities.

Negative Staining

This is probably the simplest and quickest technique used. It has been of particular value in the elucidation of the ultrastructure of particulate specimens, especially viruses. More recently, however, negative staining has become important in the study of surface structures of bacteria, particularly Gram-negative organisms. Good illustrations of the technique in use are given in later chapters of this book.

In principle the particles to be examined are mixed in a dilute solution of a heavy metal salt. A drop of this mixture is placed on the grid surface and excess removed with the edge of a filter paper. The thin film of

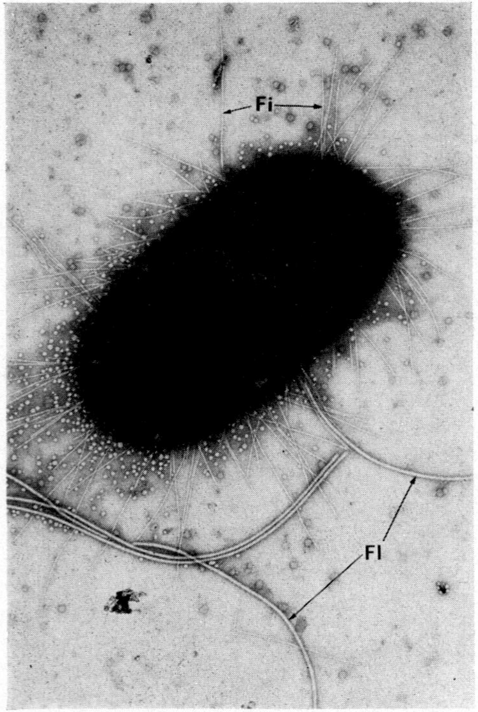

FIG. 1. *Salmonella dublin*. Negatively stained with 1·5% sodium phosphotungstate pH 7·2 showing flagella (Fl) and fimbriae (Fi) (× 51 000).

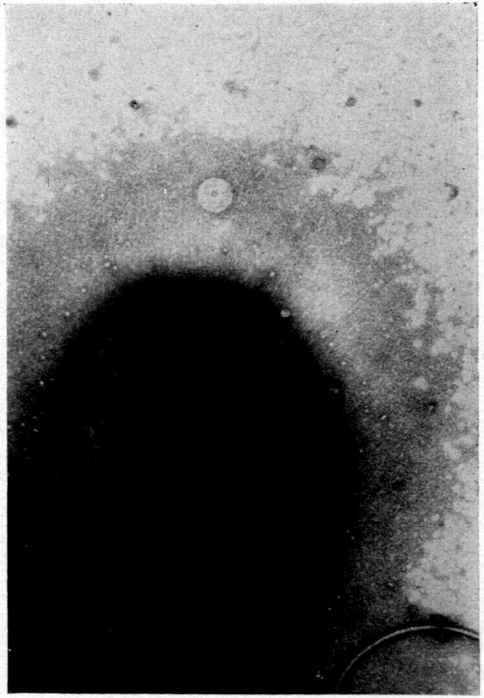

FIG. 2. *Escherichia coli*. (Strain Abbotstown). Negatively stained with 1% sodium phosphotungstate pH 7·2 showing K88 antigen (× 72 000).

mixture left is allowed to evaporate during which time the particles become "embedded" in the stain, primarily by surface tension effects. When viewed in the electron microscope the particles appear as light areas due to their low scattering power, surrounded by the electron dense stain (Figs 1 and 2). This is the opposite of positive staining (see references under "Thin Sectioning and Staining") which make particles visible by actually combining a heavy metal salt with them.

References for Negative Staining

BAHR, G. F. & ZEITLER, E. H. (1965). Quantitative Electron Microscopy. *Lab. Invest.*, **14,** 729.

BRENNER, S. & HORNE, R. W. (1959). A negative staining method for high resolution electron microscopy of viruses. *Biochem. Biophys. Acta.*, **34,** 103.

FINCH, J. T. & HOLMES, K. C. (1967). Structural studies of viruses. In *Methods in Virology*, Vol. 3, (Maramorosch, K. & Koprowski, H., eds) 351, New York: Academic Press.

HASCHEMEYER, R. H. & MYERS, R. J. (1972). Negative staining. In *Principles and Techniques of Electron Microscopy*, Vol. 2, (Hayat, M. A., ed.) 101. New York: Van Nostrand Reinhold Co.

HASCHEMEYER, R. H. (1970). Electron microscopy of enzymes. In *Advances in Enzymology*, Vol. 3, (Nord, P. F., ed.), 71. New York: John Wiley & Sons Inc.

HORNE, R. W. (1965). Negative staining methods. In *Techniques for Electron Microscopy*, (Kay, D. H., ed.), 328. Oxford: Blackwell Scientific Publications.

HORNE, R. W. (1967). Electron Microscopy of Isolated virus particles and their components. In *Methods in Virology*, Vol. 3, (Maramorosch, K. & Koprowski, H., eds), 521. New York: Academic Press.

HUXLEY, H. E. & ZUBAY, G. (1960). Electron microscope observations on the structure of microsomal particles from *Escherichia coli.*, *J. Mol. Biol.*, **2,** 10.

JOHNSON, M. W. & HORNE, R. W. (1970). Some observations on the relative dehydration rates of negative stains and biological objects. *J. Microscopy*, **91,** 197.

MELLEMA, J. E., VAN BRUGGEN, E. F. J. & GRUBER, M. (1967). An assessment of negative staining in the electron microscopy of low molecular weight proteins. *Biochem. Biophys. Act.*, **140,** 182.

MUSCATELLO, U. & HORNE, R. W. (1968). Effect of the tonicity of some negative staining solutions on the elementary structure of membrane bound systems. *J. Ultrastruct. Res.*, **25,** 73.

VALENTINE, R. C. & HORNE, R. W. (1962). An assessment of negative staining techniques for revealing ultrastructure. In *The interpretation of ultrastructure*, (Harris, R. J. C., ed.), 263. New York: Academic Press.

VAN BRUGGEN, E. F. J., WEIBENGER, E. H. & GRUBER, M. (1960). Negative staining electron microscopy of proteins at pH values below their isoelectric points. Its application to haemocyanin. *Biochem. Biophys. Acta.*, **42,** 171.

WILLIAMS, R. C. & FISHER, H. W. (1970). Electron microscopy of tobacco mosaic virus under conditions of minimal beam exposure. *J. Mol. Biol.*, **52,** 121.

Shadowing and Replication

These two techniques were amongst the first used for the electron microscopic examination of specimens. Shadow casting (Figs 3 and 4) has been

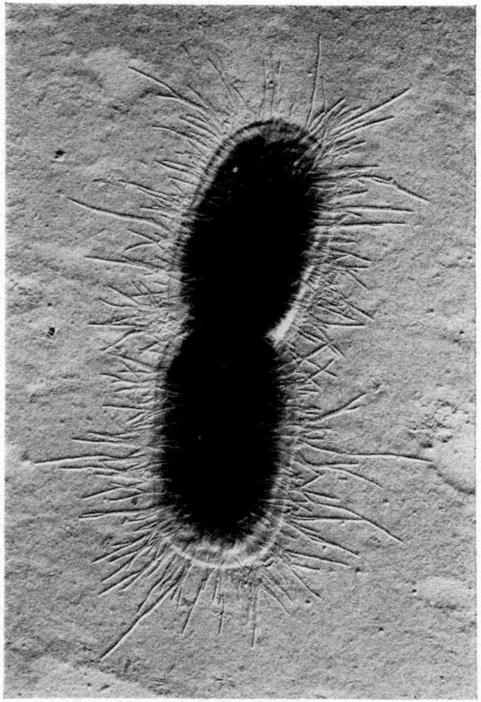

FIG. 3. A carbon-platinum shadowed preparation of a *Salmonella* species (× 45 000).

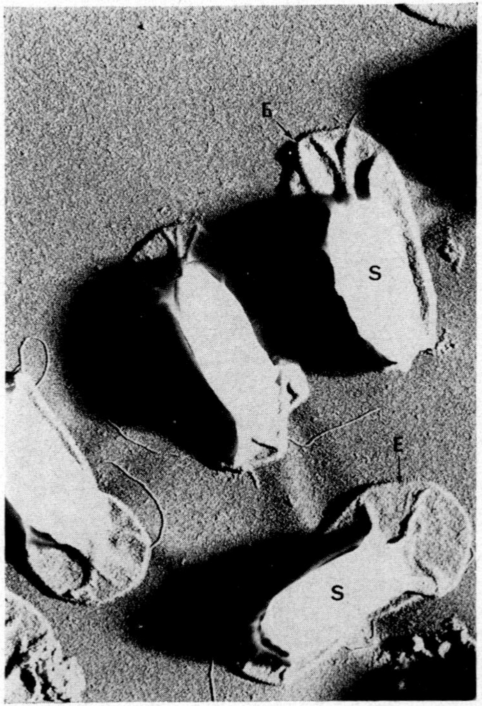

FIG. 4. A gold-palladium shadowed preparation of mature spores of *Bacillus cereus*. S, spore body; E, exosporium (× 51 000).

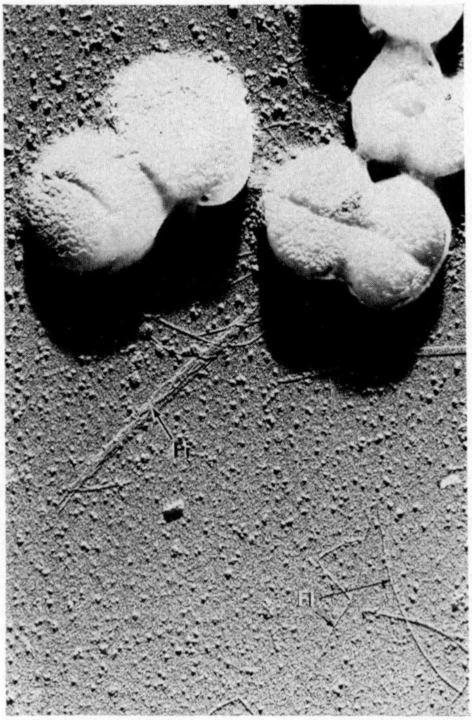

FIG. 5. A platinum–carbon replica of culture *Neisseria gonorrhoea*. Note rough bacterial surface and fimbriae (Fi) ($\times$ 40 000).

used in the estimation of size of particles, for example, viruses. It is also a vital part of the replication process giving increased contrast as well as enabling the third dimension of the specimen surface to be examined.

Replication (Fig. 5) is the process whereby a thin film of material is obtained which exactly corresponds to the surface topography of the specimen. Although replication was first applied to the study of non-biological material current interest in the properties of cell surfaces, host/pathogen relationships and the advent of freeze etching has revitalized replication as a technique capable of yielding worthwhile results in the biological field.

References for Shadowing and Replication

ABERMANN, R., SALPETER, M. M. & BACHMANN, L. High resolution shadowing. In *Principles and techniques of electron microscopy*, (Hayat, M. A., ed.), 1972. New York: Van Nostrand Reinhold Co.

BACHMANN, L. (1962). Shadow casting using very high melting metals. *Proc. 5th Inter. Cong. Electron Micros.*, **1,** FF3.

BACHMANN, L., ABERMANN, R. & ZINGSHEIM, H. P. (1969). Hochauflösende Gefrierätzung, *Histochemie*, **20,** 133.

BACHMANN, L., ORR, W. H., RHODIN, T. N. & SEIGEL, B. M. (1960). Determination of surface structure using ultra-high vacuum replication. *J. appl. Phys.*, **31,** 1458.

BASSETT, G. A., MENTER, J. W. & PASHLEY, D. W. (1959). The nucleation, growth and microstructure of thin films. In *Structure and properties of thin films* (Neugebauer, C. A., Newkirk, J. B. & Vermileyea, D. A., eds). New York: John Wiley & Sons Inc.

BRADLEY, D. E. (1954). Evaporated carbon film for use in electron microscopy. *Brit. J. appl. Phys.*, **5,** 65.

BRADLEY, D. E. (1956). Uses of carbon replicas in electron microscopy. *J. appl. Phys.*, **27,** 1399.

BRADLEY, D. E. (1959). High resolution shadowing-casting techniques for the electron microscope using the simultaneous evaporation of platinum and carbon. *Brit. J. appl. Phys.*, **10,** 198.

BRADLEY, D. E. (1967). Replica and shadowing techniques. In *Techniques for electron microscopy* (Kay, D. H., ed.), 96. Oxford: Blackwell Scientific Publications.

HENDERSON, W. J. (1969). A simple replication technique for the study of biological tissues by electron microscopy. *J. Microscopy*, **89,** 369.

HENDERSON, W. J. (1971). Application of an extraction replica technique for the study of biological materials. *Micron.*, **2,** 250.

HENDERSON, W. J. & GRIFFITHS, K. (1972). Shadow casting and replication. In *Principles and techniques of electron microscopy* (Hayat, M. A., ed.), 151. New York: Van Nostrand Reinhold Co.

KLEINSCHMIDT, A. K., LANG, D., JACKERTS, D. & ZAHN, R. K. (1962). Preparation and length measurements of the total desoxyribonucleic acid content of T2 bacteriophage. *Biochim. Biophys. Acta.*, **61,** 857.

MOOR, H. (1970). High resolution shadow casting by means of an electron gun. *Proc. 7th Inter. Cong. Electron Micros.*, **1,** 413.

POOLEY, F. D. & HENDERSON, W. J. (1966). PVA as a replication medium for electron microscopy. *Res. Techn. Instrum.*, **18,** 6.

REIMER, L. & SCHULTE, C. (1966). Elektronenmikroskopische Oberflächenabdrucke und ihr Auflösungsvermögen. *Naturwissenschaften*, **53,** 489.

Freeze Fracture and Etching

This technique is one of the newest procedures for the examination of biological material. Interest in the technique has increased dramatically since its introduction in 1961, primarily due to the great advantages it has over other techniques. The views it gives of the ultrastructure of membrane systems, if not unique, are certainly difficult to obtain using other methods.

The difference between the terms freeze fracture and freeze fracture and etching should be noted. The former denotes cleavage of a frozen specimen and replication of the fractured surface (Fig. 6). The latter

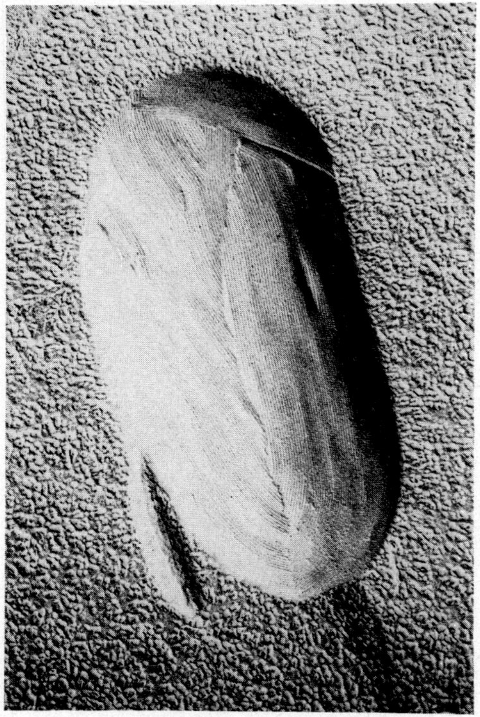

FIG. 6. A freeze fractured spore of *Bacillus subtilis*. Pretreated with 20% glycerol as a cryoprotective agent (× 90 000).

FIG. 7. A freeze fractured and etched preparation of *B. cereus*. E, exosporium; Sc, sporecoat composed of fibres and pits; C, cortex (× 67 000).

implies that after cleavage ice is allowed to sublime away from the fractured surface under vacuum conditions, thus exposing structures which might otherwise be hidden. This ice sublimation is known as etching (Fig. 7).

The steps which may be involved in the technique are: pretreatment of the specimen with cryoprotective agents, freezing, fracturing, etching, replication, cleaning of the replica. The technique, therefore, whilst having its own unique methodology draws heavily on that from other disciplines. For this reason references have been included that embrace freeze substitution and freeze drying and the reader should also be well acquainted with the principles of replication and shadowing.

References for Freeze Etching

ASAHINA, E., SHIMADA, K. & HISADA, J. (1970). A stable state of frozen protoplasm with invisible intracellular ice crystals obtained by rapid cooling. *Exp. Cell Res.*, **59,** 349.

BRANTON, D. (1966). Fracture faces of frozen membranes. *Proc. Nat. Acad. Sci.*, **55,** 1048.

BRANTON, D. (1969). Membrane structure. *Ann. Rev. Plant Physiol.*, **20,** 209.

BULLIVANT, S. & AMES, A. (1966). A simple freeze-fracture replication method for electron microscopy. *J. Cell Biol.*, **29,** 435.

BULLIVANT, S. (1970). Present status of freezing technique. In *Some biological techniques in electron microscopy* (Parson, D. F., ed.), 91. New York: Academic Press.

CHALCROFT, J. P. & BULLIVANT, S. (1970). An interpretation of liver cell membrane and junction structure based on observation of freeze fracture replicas of both sides of the fracture. *J. Cell Biol.*, **47,** 49.

CLARK, A. W. & BRANTON, D. (1968). Fracture faces in frozen outer segments from the guinea pig retina. *Z. Zellforsch*, **91,** 586.

KOEHLER, J. K. (1968). The Techniques and applications of freeze etching in ultrastructure research. In *Advances in Biol. Med. Physics*, Vol. 12 (Lawrence, J. & Gofman, J. W., eds), 1. New York: Academic Press.

KOEHLER, J. K. (1972). The freeze-etching technique. In *Principles and Techniques of Electron Microscopy* (Hayat, M. A., ed.), 53. New York: Van Nostrand Reinhold Co.

KREUTZIGER, G. O. (1970). The cryopump freeze-etch device. *J. Cell Biol.*, **47,** 111A.

LEVITT, J. & DEAR, J. (1970). The role of membrane proteins in freezing injury and resistance. In *The Frozen Cell.* A CIBA Foundation Symposium, (Wolstenholme, G. E. & O'Connor, M., eds), 149. London: J. & A. Churchill.

MAZUR, P. (1966). Physical and chemical basis of injury in single-celled microorganisms subjected to freezing and thawing. In *Cryobiology* (Meryman, H. T., ed.), 214. New York: Academic Press.

MAZUR, P. (1970). Cryobiology—the freezing of biological systems. *Science*, **168,** 939.

MOOR, H., MÜHLETHALER, K., WALDNER, H. & FREY-WYSSLING, A. (1961). A new freezing ultramicrotome. *J. Biophys. Biochem. Cytol.*, **10,** 1.

MOOR, H. (1966). Use of freeze etching in the study of biological ultrastructure. In *Inter. Rev. of Exp. Path.*, 179. New York: Academic Press.

MÜHLETHALER, K., WEHRLI, E. & MOOR, H. (1970). Double fracturing methods for freeze etching. *Proc. 7th Int. Cong. Electron Micros.*, (Farard, P. ed.), Societe Francaise de Microscopie Electronique Paris, France, **1,** 449.

NASH, T. (1966). Chemical constitution and physical properties of compounds able to protect living cells against damage due to freezing and thawing. In *Cryobiology* (Meryman, H. T., ed.), 116. New York: Academic Press.

PEASE, D. C. (1967). Eutectic ethylene glycol and pure propylene glycol as substituting media for the dehydration of frozen tissue. *J. Ultrastruct. Res.*, **21,** 75.

PINTO DE SILVA, P. & BRANTON, D. (1970). Membrane splitting in freeze etching. Covalently bound ferritin as a membrane marker. *J. Cell Biol.*, **45,** 598.

REBHUN, L. I. & SANDER, G. (1971). Electron microscope studies of frozen substituted marine eggs. I. Conditions for avoidance of intracellular ice crystallisation. *Am. J. Anat.*, **130,** 1.

REBHUN, L. I. (1972). Freeze-substitution and freeze-drying. In *Principles and Techniques of electron microscopy* (Hayat, M. A., ed.), 3. New York: Van Nostrand Reinhold Co.

SJÖSTRAND, F. S. (1967). *Electron Microscopy of cells and tissues*, Vol. I. New York: Academic Press.

STEERE, R. L. (1969). Freeze-etching and direct observation of freezing damage. *Cryobiology*, **6,** 137.

WEHRLI, E., MÜHLETHALER, K. & MOOR, H. (1970). Membrane structure as seen with a double replica method for freeze-fracturing. *Exp. Cell Res.*, **59,** 336.

ZINGSHEIM, H. P., ABERMANN, R. & BACHMANN, L. (1970). Apparatus for ultrashadowing of freeze-etched electron microscopic specimens. *J. Phys. Scient. Inst.*, **3,** 39.

Ultrathin Sectioning

Probably the most popular and practised of all the preparative techniques, more observations on the ultrastructure of biological specimens have been made using ultrathin sections than any other technique. It is not surprising, therefore, that a great deal of pertinent information exists regarding the principles and methods involved. A good knowledge of these is an essential prerequisite if the parameters of other techniques such as histochemistry, immunocytochemistry, etc. are to be explored.

The reader may also find that some basic knowledge of preparative methods and chemical agents employed in histological routines for light microscopy helpful as some of the principles are directly related to electron microscopy.

Ultrathin sectioning (Figs 8 and 9) requires fixation (which includes freeze drying and freeze substitution as well as chemical fixation) possibly dehydration, embedding in a suitable medium, polymerization of that medium, thin sectioning and possibly staining of these sections by a particular chosen method.

For convenience the references given have been split into some of these various categories.

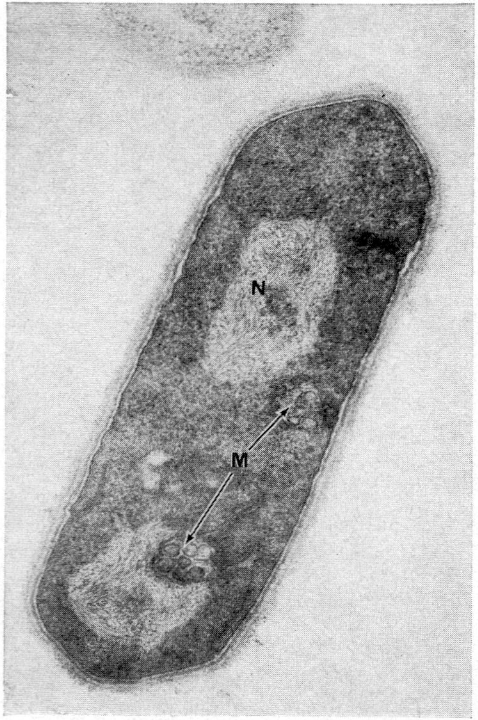

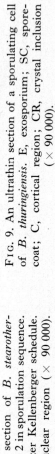

FIG. 8. An ultrathin section of *B. stearothermophilus* showing Stage 2 in sporulation sequence. Fixed according to Ryter Kellenberger schedule. M, mesosome; N, nuclear region (× 90 000).

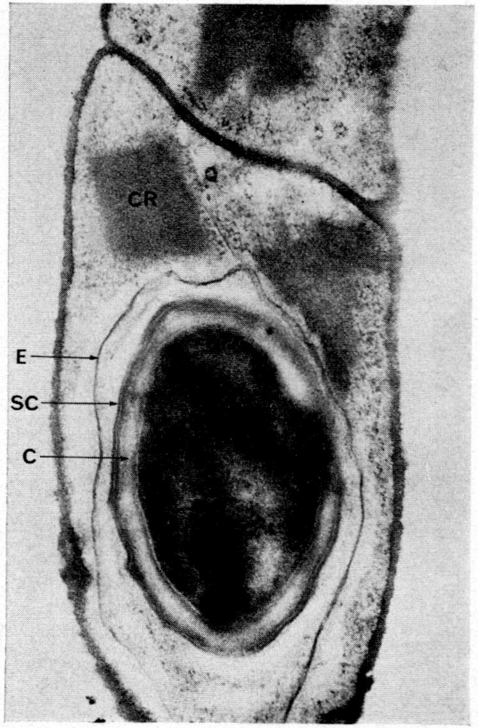

FIG. 9. An ultrathin section of a sporulating cell of *B. thuringiensis*. E, exosporium; SC, spore-coat; C, cortical region; CR, crystal inclusion (× 90 000).

References for Thin Sectioning Fixation and Dehydration

BAHR, G. F., BLOOM, G. & FRIBERG, U. (1957). Volume changes of tissues in physiological fluids during fixation in osmium tetroxide or formaldehyde and during subsequent treatment. *Exptl. Cell Res.*, **12,** 342.

BAHR, G. F., BLOOM, G. & JOHANNISSON, E. (1958). Further studies on fixation with osmium tetroxide. *Histochemie*, **1,** 113.

BARNETT, R. J., PERNEY, D. P. & HAGSTRÖM, P. E. (1964). Additional new aldehyde fixatives for histochemistry and electron microscopy. *J. Histochem. Cytochem.*, **12,** 36.

FEDER, N. (1960). Some modifications in convenitonal techniques of tissue preparation. *J. Histochem. Cytochem.*, **8,** 309.

FRANKE, W. W., KREIN, S. & BROWN, JNR., R. M. (1969). Simultaneous glutaraldehyde-osmium fixation with postosmication. *Histochemie*, **19,** 162.

GAYLARDE, P. & SARKANY, I. (1968). Reuthenium tetroxide for fixing and staining cytoplasmic membranes. *Science*, **161,** 1157.

GLAUERT, A. M. (1965). The fixation and embedding of biological specimens. In *Techniques for electron microscopy* (Kay, D. M., ed.), 166. Oxford: Blackwell Scientific Publications.

HAYAT, M. A. (1970). *Principles and techniques of electron microscopy*, Vol. I, 5. New York: Van Nostrand Reinhold Co.

HIRSCH, J. G. & FEDORKO, M. E. (1968). Ultrastructure of human leucocytes after simultaneous fixation with glutaraldehyde and osmium tetroxide and "post fixation" in uranyl acetate. *J. Cell Biol.*, **38,** 615.

HOPWOOD, D. (1968). Some aspects of fixation by glutaraldehyde and formaldehyde. *J. Anat.*, **103,** 581.

KUSHIDA, H. (1965*a*). Durcupan as a dehydrating agent for embedding with polyester, styrene and methacrylate resins. *J. Electronmicroscopy*, **14,** 52.

KUSHIDA, H. (1965*b*). Dehydration and embedding for electron microscopy. I. Dehydration. *J. Electronmicroscopy*, **14,** 167.

KELLENBERGER, E., RYTER, A. & SECHAUD, J. (1958). Electron microscope study of DNA—containing plasms. II. Vegetative and mature phage DNA as compared with normal bacterial nucleoids in different physiological states. *J. Biophys. Biochem. Cytol.*, **4,** 671.

LUFT, J. H. (1956). Permanganate—a new fixative for electron microscopy. *J. Biophys. Biochem. Cytol.*, **2,** 799.

LUFT, J. H. (1959). The use of acrolein as a fixative for light and electron microscopy. *Anat. Rec.*, **113,** 305.

LUFT, J. H. & WOOD, R. L. (1963). The extraction of tissue protein during and after fixation with osmium tetroxide in various buffer systems. *J. Cell Biol.*, **19,** 64A.

MILLONIG, G. & MARINOZZI, V. (1968). Fixation and embedding in Electron Microscopy. In *Advances in Optical and Electron Microscopy*, Vol. II (Barer, R. and Cosslett, V. E., eds), 251. London: Academic Press.

PALADE, G. E. (1952). A study of fixation for electron microscopy. *J. Exp. Med.*, **95,** 285.

PEASE, D. C. (1964). *Histological techniques for electron microscopy*, 31. London: Academic Press.

ROSENBLUTH, J. (1963). Contrast between osmium fixed and permanganate fixed toad spinal ganglia. *J. Cell Biol.*, **16,** 143.

SABATINI, D. D., BENSCH, K. & BARNETT, R. J. (1963). Cytochemistry and electron microscopy. The preservation of cellular structure and enzymatic activity by aldehyde fixation. *J. Cell Biol.*, **17**, 19.

SABATINI, D. D., MILLER, F. & BARRNETT, R. J. (1964). Aldehyde fixation for morphological and enzyme histochemical studies with the electron microscope. *J. Histochem. Cytochem.*, **12**, 57.

TORMLEY, J. M. (1964). Differences in membrane configuration between osmium tetroxide fixed and glutaraldehyde fixed ciliary epithelium. *J. Cell Biol.*, **23**, 658.

VAN HARREVELD, A. & KHATTAB, F. I. (1969). Changes in the extracellular space of the mouse cerebral cortex during hydroxyadipaldehyde fixation and osmium tetroxide post fixation. *J. Cell Sci.*, **4**, 437.

References for Embedding

COLE, M. B. (1968). A simple apparatus for ultraviolet polymerisation of soluble embedding media employed in electron microscopy. *J. Microscopie*, **7**, 441.

COULTER, H. D. (1967). Rapid and improved methods for embedding biological tissues in Epon 812 and Araldite 502. *J. Ultrastruct. Res.*, **20**, 346.

ESTES, L. W. & APICELLA, J. V. (1969). A rapid embedding technique for electron microscopy. *Lab. Invest.*, **20**, 159.

FARRANT, J. L. & MCLEAN, J. D. (1969). Albumins as embedding media for electron microscopy. *Proc. 27th Ann. Meet. Elect. Micros. Soc. Amer.*, 422.

FREEMAN, J. A. & SPURLOCK, B. O. (1962). A new epoxy embedment for electron microscopy. *J. Cell Biol.*, **13**, 437.

GIBBONS, I. R. (1959). An embedding resin miscible with water for electron microscopy. *Nature*, **184**, 375.

GILËV, V. P. (1958). The use of gelatin for embedding biological specimens in preparation of ultrathin sections for electron microscopy. *J. Ultrastruct. Res.*, **1**, 349.

GLAUERT, A. M., ROGERS, G. E. & GLAUERT, R. H. (1956). A new embedding medium for electron microscopy. *Nature*, **178**, 803.

KUSHIDA, H. (1960). A new polyester embedding method for ultrathin sectioning. *J. Electron Microscopy*, **9**, 113.

KUSHIDA, H. (1965). Dehydration and embedding for electron microscopy. II. Embedding. *J. Electron Microscopy*, **14**, 251.

LEDUC, E. H. & HOLT, S. J. (1965). Hydroxypropyl methacrylate a new miscible embedding medium for electron microscopy. *J. Cell Biol.*, **26**, 137.

LEDUC, E. H. & BERNHARD, W. (1967). Recent modifications of the glycol methacrylate embedding procedure. *J. Ultrastruct. Res.*, **19**, 196.

LUFT, J. H. (1961). Improvements in epoxy resin embedding methods. *J. Biophys. Biochem. Cytol.*, **9**, 409.

MCLEAN, J. D. & SINGER, S. J. (1964). Crosslinked polyampholytes. New water soluble embedding media for electron microscopy. *J. cell. Biol.*, **20**, 518.

RYTER, A. & KELLENBERGER, E. (1958). L'inclusion au polyester pour ultramicrotomie. *J. Ultrastruct. Res.*, **2**, 200.

SANDSTRÖM, B. & WESTMAN, J. (1969). Non-freezing light and electron microscopic emzyme histochemistry by means of polyethylene glycol embedding. *Histochemie*, **19**, 181.

STÄUBLI, W . (1963). A new embedding technique for electron microscopy combining a water soluble epoxy resin (Durcupan) with water insoluble Araldite. *J. Cell Biol.*, **16**, 197.

WARD, R. T. (1958). Prevention of polymerisation damage in methacrylate embedding media. *J. Histochem. Cytochem.*, **6,** 398.

WINBORN, W. B. (1965). Dow epoxy resin with Triallyl cyanurate and similarly modified. Araldite and Maraglas mixtures as embedding media for electron microscopy. *Stain Technol.*, **40,** 227.

References for Thin Sectioning and Staining

FRASCA, J. M. & PARKS, V. R. (1965). A routine technique for double staining ultrathin sections with uranyl and lead salts. *J. Cell Biol.*, **25,** 157.

FULLAGAR, K. (1966). The role of the L.K.B. Knifemaker. *Ultramicrotomy Sci. Tools*, **13,** 39.

GLAUERT, A. M. & PHILLIPS, R. (1965). The preparation of thin sections. In *Techniques for electron microscopy* (Kay, D. H., ed.), 213. Oxford: Scientific Publications.

HAYAT, M. A. (1970). Sectioning and Staining. In *Principles and techniques of electron microscopy*, 183 and 241. New York: Van Nostrand Reinhold Co.

HELANDER, H. F. (1969). Surface topography of ultra-microtome sections. *J. Ultrastruct. Res.*, **29,** 373.

HUXLEY, H. E. & ZUBAY, G. (1961). Preferential staining of nucleic acid containing structures for electron microscopy. *J. Biophys. Biochem. Cytol.*, **11,** 273.

KARNOVSKY, M. J. (1961). Simple methods for "staining with lead" at high pH in electron microscopy. *J. Biophys. Biochem. Cytol.*, **11,** 729.

KUSHIDA, H. (1966). Staining of thin sections with lead acetate. *J. Electron Microscopy*, **15,** 93.

MARINOZZI, V. (1967). Reaction de l'acide phosphotungstic avec la mucine et les glycoproteins des plasma membranes. *J. Microscopie*, **6,** 68A.

MILLONIG, G. (1961). A modified procedure for lead staining of thin sections. *J. Biophys. Biochem. Cytol.*, **11,** 736.

REYNOLDS, E. S. (1963). The use of lead citrate at high pH as an electron opaque stain in electron microscopy. *J. Cell Biol.*, **17,** 208.

SATO, T. (1967). A modified method for lead staining of thin sections. *J. Electron Microscopy*, **16,** 133.

SILVERMAN, L., SCHREINER, B. & GLICK, D. (1969). Measurement of thickness within sections by quantative electron microscopy. *J. Cell Biol.*, **40,** 768.

SUTTON, J. S. (1968). Potassium permanganate staining of ultrathin sections for electron microscopy. *J. Ultrastruct. Res.*, **21,** 424.

VENABLE, J. H. & COGGESHALL, R. (1965). A simplified lead citrate stain for use in electron microscopy. *J. Cell Biol.*, **25,** 407.

WATSON, M. L. (1958). Staining of tissue sections for electron microscopy with heavy metals. II. Application of solutions containing lead and barium. *J. Biophys. Biochem. Cytol.*, **4,** 727.

Immunocytochemistry

The discipline enables the cellular and sub-cellular structure of biological systems to be studied by immunological methods. This entails the labelling of antibodies or antibody fragments with specific markers which may be clearly demonstrated or recognized in both the light and electron microscope (Figs 10 and 11).

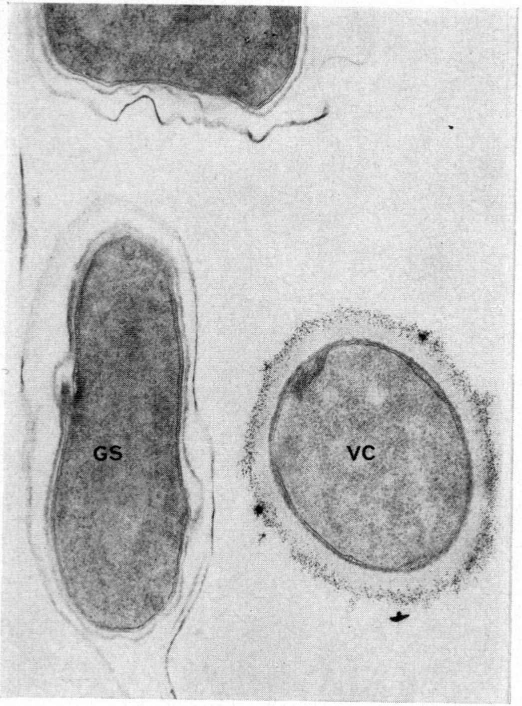

FIG. 10. An ultrathin section of a *B. cereus* culture stained with ferritin labelled vegetative antibody prior to embedding. VC, vegetative cell; GS, germinating spore prior to outgrowth ($\times$ 27 000).

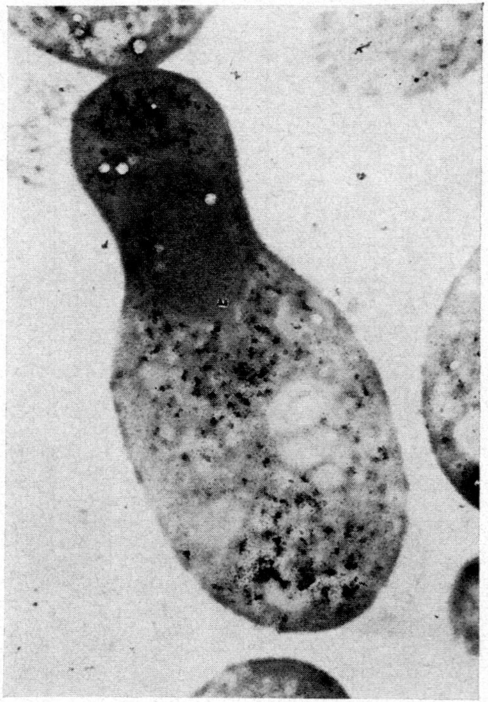

FIG. 11. An ultrathin section of a sporulating cell of *B. cereus*. Fixed 2·5% glutaraldehyde embedded in glycol methacrylate stained with peroxidase anti-peroxidase complex (PAP) to show sites of spore antigen ($\times$ 27 000).

Of the reagents used probably the best known are the fluorescent labelled globulins employed in light microscopy. It was not until the advent of methods that enabled the labelling of similar globulins with ferritin that comparable studies were possible using the electron microscope. From that time various research workers have sought to improve the sensitivity, specificity and methodology of the technique until to-day there is an array of methods available. These include the use of ferritin- and enzyme-conjugated antibodies, the use of hybrid antibody, cytochrome labelling and enzyme-antibody complexes.

References for Immunocytochemistry

AVRAMEAS, S. & TERNYCK, T. (1971). Peroxidase labelled antibody and FAb conjugates with enhanced intracellular penetration. *Immunocytochemistry*, **8,** 1175.

BIRNBAUM, U., VOGT, A. & MARINIS, S. (1970). Isolation and characterization of immunoferritin conjugates. II. Antibody binding capacity *in vitro*. *Immunology*, **18,** 443.

DE GRANDI, P. B., KRAEHENBUHL, J. P. & CAMPICHE, M. A. (1971). Ultrastructural localization of calcitonin in the parafolicular cells of pig thyroid gland with cytochrome-C labelled antibody fragments. *J. Cell Biol.*, **50,** 446.

HÄMMERLING, U., AOKI, T., DE HARVEN, E., BOYSE, E. A. & OLD, L. J. (1968). Use of hybrid antibody with anti-IgG and anti-ferritin specificities in locating cell surface antigens by electron microscopy. *J. Exp. Med.*, **128,** 1461.

KNÜSEL, A., BÄCHLI, T. L., GITZELMANN, R. & LINDENMANN, J. (1971). Electron microscopic recognition of surface antigen by direct reaction and ferritin capture with guinea pig hybrid antibody. *J. Immunology*, **106,** 583.

MASON, T. E., PHIFER, R. F., SPICER, S. S., SWALLOW, R. A. & DRESKIN, R. B. (1969). An immunoglobulin-enzyme bridge method for localizing tissue antigens. *J. Histochem. Cytochem.*, **17,** 563.

MORIARTY, G. C., MORIARTY, M. & STERNBERGER, L. A. (1973). Ultrastructural immunocytochemistry with unlabelled antibodies and the peroxidase anti-peroxidase complex. A technique more sensitive than radioimmunoassay. *J. Histochem. Cytochem.*, **21,** 825.

NAKANE, P. K. & PIERCE, JNR., G. B. (1966). Enzyme labelled antibodies—preparation and application for the localization of antigens. *J. Histochem. Cytochem.*, **14,** 929.

NAKANE, P. K. & PIERCE, JNR., G. B. (1967). Enzyme labelled antibodies for the light and electron microscopic localization of tissue antigens. *J. Cell Biol.*, **33,** 307.

NOVIKOFF, A. B., NOVIKOFF, P., QUINTANA, N. & DAVIS, C. (1972). Diffusion artefacts in 3–3′-diaminobenzidine cytochemistry. *J. Histochem. Cytochem.*, **20,** 745.

PIERCE, T. B., RAM, J. S. & MIDGLEY, A. R. (1964). Labelled antibodies in electron microscopy. *Int. Rev. Exp. Path.*, **3,** 1.

SINGER, S. J. & SCHICK, A. F. (1961). The properties of specific stains for electron microscopy prepared by conjugation of antibody molecules with ferritin. *J. Biophys. Biochem. Cytol.*, **9,** 519.

STERNBERGER, L. A. (1967). Electron microscopic immunocytochemistry. *J. Histochem. Cytochem.*, **15,** 139.

STERNBERGER, L. A., HARDY, JNR., P. H., CUCULIS, J. J. & MEYER, H. G. (1970). The unlabelled antibody enzyme method of immunocytochemistry. Preparation and properties of soluble antigen-antibody complex (horseradish peroxidase-antihorseradish peroxidase), and its use in the identification of spirochaetes. *J. Histochem. Cytochem.*, **18,** 315.

STERNBERGER, L. A. (1974). *Immunocytochemistry*. Englewood Cliffs, New Jersey: Prentice-Hall Inc.

TANAKA, H. (1968). The ferritin labelled antibody method—its advantages and disadvantages. A methodological review. *Acta Haematol. Japonica.*, **31,** 125.

Histochemistry and Cytochemistry

Much progress has been made in the development of techniques for the localization of enzymes and chemical components in tissues by specific staining techniques.

These can be divided into enzyme digestion studies, specific staining of cellular components with heavy metal ions, e.g. silver, iron and thorium, and classical histochemical reactions which rely on one of the soluble end products of an enzyme reaction being captured by a metallic ion to form an insoluble electron opaque deposit (Figs 12 and 13).

It is beyond the scope of this text to detail all the advances made and the references given are, therefore, general in nature with the exception of those techniques which have been applied to microbiological problems.

References for Histochemistry and Cytochemistry

ANDERSON, W. A. & ANDRE, J. (1968). The extraction of some cell components with pronase and pepsin from thin sections of tissue embedded in an Epon-Araldite mixture. *J. Microscopie.*, **7,** 343.

BAILLIE, A., THOMSON, R. O., BATTY, I. & WALKER, P. D. (1967). Some preliminary observations on the location of esterases in *Bacillus cereus*. *J. appl. Bact.*, **30,** 312.

GRANBOULAN, P. & LEDUC, E. H. (1967). Ultrastructural cytochemistry of *Bacillus subtilis*. *J. Ultrastruct. Res.*, **20,** 111.

HANKER, J. S., SEAMAN, A. R., WEISS, L. P., UENO, H., BERGMANN, R. A. & SELIGMAN, A. M. (1964). Osmiophilic reagents—new cytochemical principles for light and electron microscopy. *Science*, **146,** 1039.

HAYAT, M. A. (1970). In *Principles and Techniques for electron microscopy*, Vol. 1, 241. New York: Van Nostrand Reinhold Co.

HOLT, S. J. & HICKS, R. M. (1962). Specific staining methods for enzyme location at the subcellular level. *Brit. Med. Bull.*, **18,** 214.

VAN ITERSEN, W. (1965). Symposium on the fine structure and replication of bacteria and their parts. II Cytoplasm. *Bact. Rev.*, **28,** 299.

MARINOZZI, V. (1961). Silver impregnation of ultrathin sections for electron microscopy. *J. Biophys. Biochem. Cytol.*, **9,** 121.

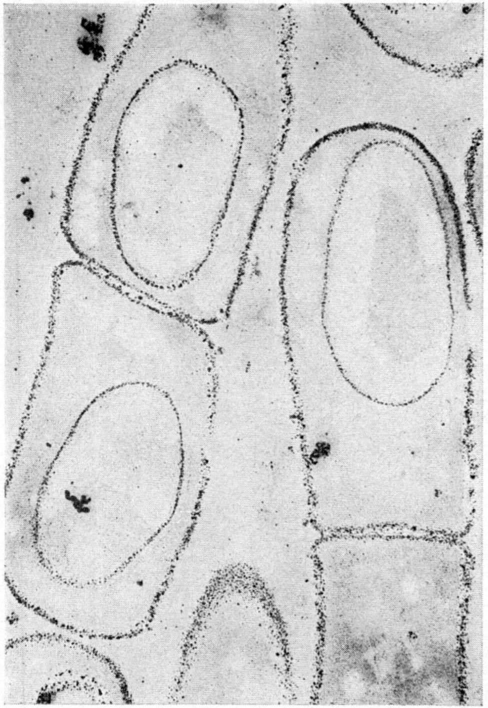

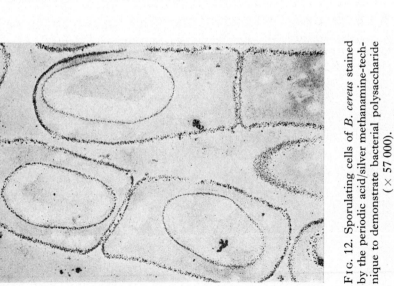

Fig. 12. Sporulating cells of *B. cereus* stained by the periodic acid/silver methanamine-technique to demonstrate bacterial polysaccharide (× 57 000).

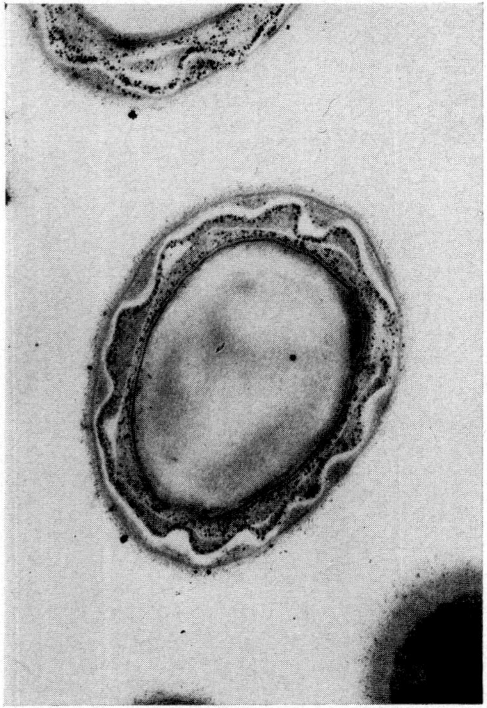

Fig. 13. Developing spore of *B. cereus* stained to show the location of esterase activity (× 122 000).

de Martino, C. & Zamboni, L. (1967). Silver methanamine stain for electron microscopy. *J. Ultrastruct. Res.*, **19,** 273.

Pearse, A. G. (1963). Some aspects of the localization of enzyme activity with the electron microscope. *J. Roy. Microscop. Soc.*, **81,** 107.

Pearse, A. G. (1968). *Histochemistry Theoretical and Applied.* London: J. & A. Churchill Ltd.

Scarpelli, D. G. & Kanezak, N. M. (1965). Ultrastructural cytochemistry: principles, limitations and applications. *Inter. Rev. Exptl. Path.*, **4,** 55.

Sedar, A. W. & Burde, R. M. (1965). The demonstration of the succinic dehydrogenase system in *Bacillus subtilis* using tetranitro-blue tetrazolium combined with techniques of electron microscopy. *J. Cell Biol.*, **27,** 53.

Walker, P. D. (1969). The location of chemical components on ultrathin sections of *Bacillus cereus* embedded in glycol methacrylate. *J. appl. Bact,*, **32,** 463.

Walker, P. D. & Short, J. A. (1969). Location of bacterial polysaccharide during various phases of growth. *J. Bact.*, **98,** 1342.

Wischnitzer, S. (1967). Current techniques in biomedical electron microscopy. *Inter. Rev. Cytol.*, **22,** 1.

Scanning Electron Microscopy of Microbial Colonies

Susan M. Passmore and Barbara Bole

Long Ashton Research Station, University of Bristol, England

The gross colony morphology of micro-organisms is an integral part of present schemes for identification and classification, yet little is known about the basic relationships between individual cells and their influence on these semi-artificial formations.

Observations of intact colonies with the light microscope are limited by the shallow depth of field at high magnifications, unless a technique such as the impression preparations of Bisset (1938) is used and this is restricted to very small colonies. The transmission electron microscope has been used to study colony structure by growth of micro-colonies on collodion film (Bisset and Pease, 1957) and by observation of sections through embedded colonies (Swanson and McCarty, 1969; Knudson and Macleod, 1970).

The three-dimensional structure of colonies and inter-relationships of the constituent organisms can be ideally studied using the scanning electron microscope (SEM). The principles and uses, including some biological applications, of this instrument have been reviewed by Nixon (1969). The SEM not only has the advantage of continuously variable useful magnifications between 20 × and 20 000 ×, but also has a considerably greater depth of field than the optical microscope, so enabling production of apparently three-dimensional images. The original applications of the instrument were to studies of rigid, non-biological structures, but in the last few years techniques have been evolved which allow observations of even soft biological materials (Marszalek and Small, 1969). The surface morphology of individual micro-organisms has been studied in the SEM by a variety of techniques including unfixed preparations, chemical fixation and freeze drying, and these methods are reviewed by Kormendy and Wayman (1972). The study of colony morphology has been restricted to a few workers who have developed techniques to prevent distortion of the delicate colony structures. Whittaker and Drucker (1970) examined several available methods and concluded that freeze drying without prior chemical fixation best preserved the structure of the colonies they examined. They have continued with this technique for colonies of other organisms (Drucker and Whittaker, 1971) and in a study of modifications of colonial morphology during growth (Whittaker and Drucker, 1972). Springer and Roth (1972) published detailed micrographs of bacterial colonies treated with buffered glutaraldehyde to fix the structures. These workers commented on the distortion of the agar medium and cavities in the colonies shown by Whittaker and Drucker (1970) and stated that this was "indicative of incomplete fixation and (or) drying artifacts".

The present chapter is based on a survey of methods for observing surface and colony morphology of the acetic acid bacteria (Passmore, 1973*b*). These organisms are notoriously variable in colony morphology, both with age and culture medium, and part of the survey was an attempt to find an appropriate technique to show up any differences in the basic structures of colony variants. This involved assessment of available methods on a range of micro-organisms and some of the findings are detailed below and elsewhere (Passmore, 1973*a*, 1973*b*; Passmore and Haggett, 1973).

Materials and Methods

Organisms

The surface morphology of a wide range of discrete micro-organisms has been observed in the SEM, but studies of colony morphology have largely been confined to bacterial species and a few yeasts (Whittaker and Drucker, 1970; Passmore, 1973*a*). We have successfully observed colonies of bacteria, yeasts and moulds with some experimental variation of the methods where appropriate.

Methods

The micro-organisms to be studied are inoculated on to dry agar plates to obtain discrete colonies. In general, the agar normally used for cultivation of the specific organisms is suitable for SEM observations and micrographs have been produced of colonies on a variety of media. After normal incubation, suitable small colonies (preferably < 3 mm) are cut from the plates using a fine scalpel. The cut should not be made so close to the colony that disruption occurs when the agar is moved, but too much frill (> 1 mm) of agar is detrimental to observations. For rapid drying the pieces of agar bearing the colonies should be 3–4 mm cubes, but obviously this is dictated by the size of the colony.

The excised colonies can be prepared for the SEM in three ways:

Untreated—The agar blocks bearing the colonies are mounted directly onto aluminium SEM stubs (Alan W. Agar, 127a Rye Street, Bishop's

Stortford, Herts.) with a suitable adhesive (e.g. "Durofix" or "Silver Dag" from Polaron Equipment Ltd, 60/62 Greenhill Crescent, Holywell Industrial Estate, Watford, Herts.). The block should be embedded in a drop of the adhesive up to the agar surface, so that the top is held flat and cannot curl during drying. A layer of metal (approximately 20nm thick) is then evaporated on to the mounted colony in a vacuum coating unit (e.g. Edwards Model 12E6), modified to allow rotation of the specimen at 50–100 revs/min through varying angles in a vacuum of 8×10^{-5} Torr. The material most often used and recommended for coating is gold–palladium (60:40) alloy (Johnson Matthey Metals Ltd, 81 Hatton Garden, London, E.C.1.) although aluminium has been used by some workers (Rousseau *et al.*, 1972). Specimens prepared in this way can then be examined in the SEM.

Chemical fixation—Microbial colonies which are susceptible to distortion during vacuum drying can be preserved using a variety of chemical fixatives. Osmium tetroxide has been used either alone (Williams, 1970) or in combination with glutaraldehyde (Marszalek and Small, 1969; Kormendy and Wayman, 1972). This substance was used as an unbuffered 2% solution (Glauert, 1965) in some of our preliminary investigations (Passmore, 1973*b*), but was found to give no significant advantage over the easier to handle glutaraldehyde fixatives.

Glutaraldehyde has been used to fix micro-organisms for SEM observations in a variety of concentrations and buffers and for different times (Greenwood and O'Grady, 1969; Whittaker and Drucker, 1970; Kormendy and Wayman, 1972; Springer and Roth, 1972). We have found that a 2·5% solution prepared from 70% glutaraldehyde and buffered to pH 6·5 with 0·2M cacodylate buffer gives satisfactory results. The colonies on agar blocks are fixed in the solution for 12–18 h at room temperature in covered disposable microtiter plates (Sterilin Ltd, 43–45 Broad Street, Teddington, Middlesex, TW11 8QZ). These containers have the advantages of using minimal quantities of fixative and allowing separation and labelling of the colonies. After fixing the blocks are rinsed in buffer and dried on filter paper before mounting and coating.

Freeze drying—Freeze drying has been recommended as the best method for preserving surface structure of discrete cells and many methods have been used (Kormendy and Wayman, 1972). Whittaker and Drucker (1970) concluded that quenching in isopentane in liquid nitrogen without any prior chemical fixation and then freeze drying was a satisfactory method of preparation of colonies for scanning electron microscopy. In our survey the colonies were quenched in liquid nitrogen and dried at

$-60°$ at 10^{-2} Torr for a maximum of 18 h in an Edwards-Pearse Tissue Dryer. When fully dried the specimens were mounted and coated for observation.

Examination of the colonies

The colonies were examined in a Cambridge "Stereoscan" Scanning Electron Microscope type 96113, Mark IIA, at an accelerating voltage of 10 kV. Voltages of 20–40 kV have been used by most workers studying microbial colonies, but we have found that the lower voltage gives less beam penetration and charging of the specimen. This is especially necessary for examination of freeze-dried specimens as these rapidly accumulate surface charge which obliterates surface detail when the cells begin to "glow" with adsorbed electrons (see for example Figs 6 and 18 and Drucker and Whittaker, 1970).

All colonies can be examined at any angle in the "Stereoscan", but for high magnifications a beam-to-specimen angle of 45° is recommended (Passmore and Haggett, 1973).

Discussion

Results obtained with bacteria and yeasts using the three basic methods given above are shown in Figs 1–18. Many colonies can be studied by these techniques (Passmore, 1973*a*, 1973*b*; Passmore and Haggett, 1973) but some treatments are more suitable for certain organisms. Figures 1–6 illustrate colonies of a *Lactobacillus* sp. prepared by the three methods. Figures 1 and 2 are micrographs of an untreated colony exhibiting some collapse under vacuum, but with the cells arranged in small palisades under a covering of dense slime. When treated with glutaraldehyde (Figs 3 and 4) the colony shape is retained and some of the slime covering is washed off revealing the arrangement of the cells more clearly. Freeze drying (Figs 5 and 6) also preserves the colony shape but the agar is badly distorted, as criticized by Springer and Roth (1972). "Charging" is a major problem in resolving the cell surface, due to poor surface coating in the cavities between the cells where the slime has dried down and consequent lack of conductivity. The Diode Sputtering System (Polaron Ltd) for specimen coating helps to cure this fault, because this apparatus gives a more uniform coating even of optically hidden surfaces and therefore reduces charging considerably.

Yeast colony structure can be observed adequately with untreated specimens as shown in Figs 7, 9 and 11, where the bud and birth scars of *Saccharomyces cerevisiae* AWY 350R are well illustrated at a magnification

FIGS 1–6. A *Lactobacillus* sp. observed by the three techniques.

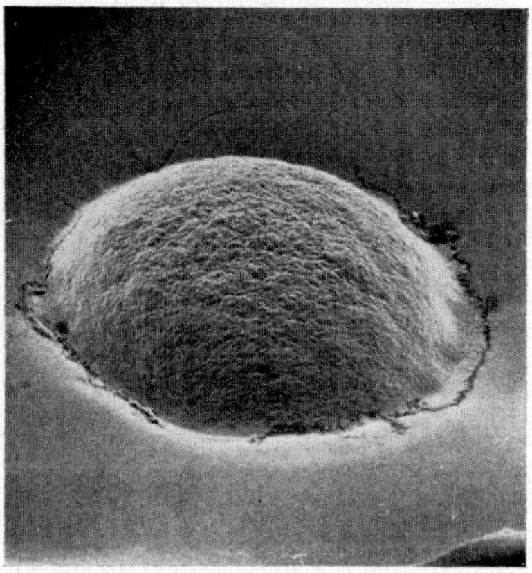

1. Untreated (× 54)

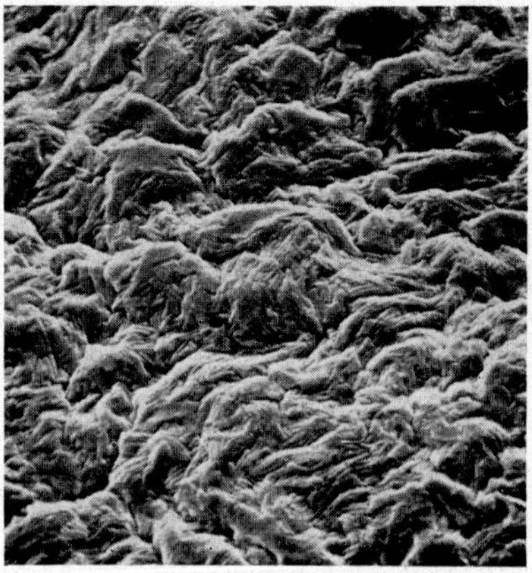

2. Untreated (× 640)

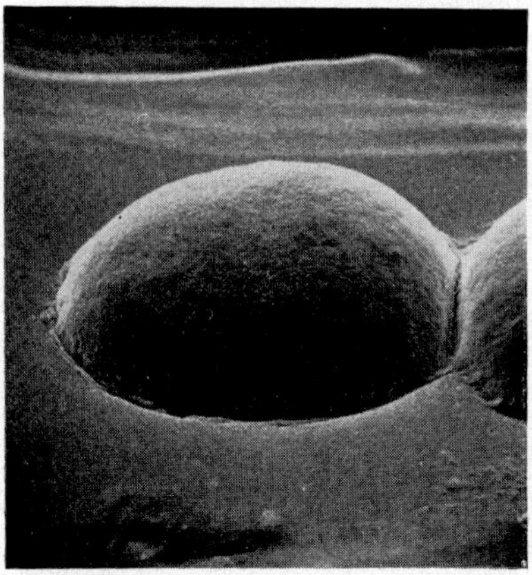

3. Glutaraldehyde fixed (× 60)

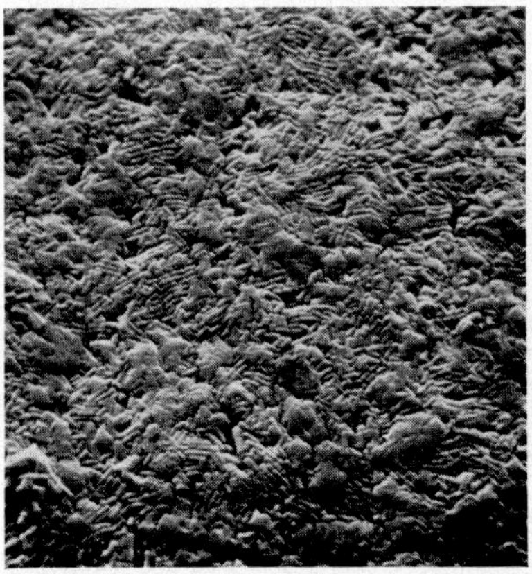

4. Glutaraldehyde fixed (× 800)

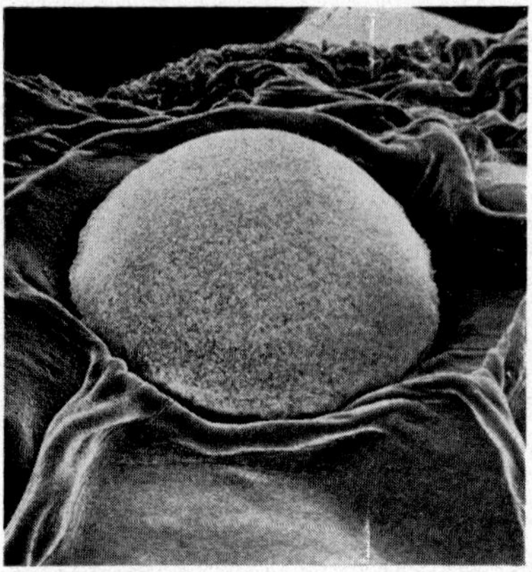

5. Freeze-dried (× 63)

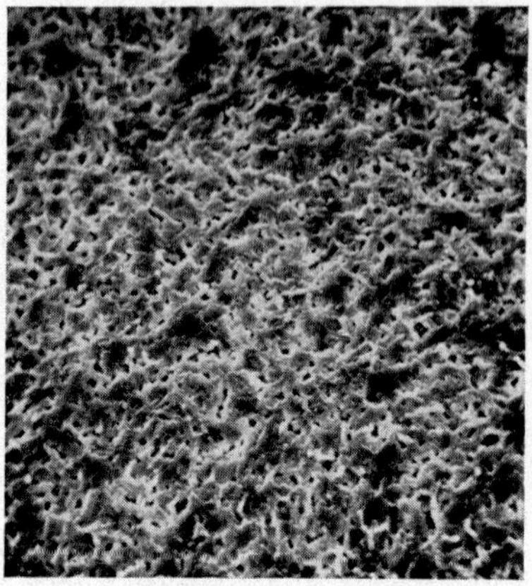

6. Freeze-dried (× 666)

FIGS 7–12. *Sacch. cerevisiae* AWY 350 R.

(Figs 7, 9 and 11 are reproduced by permission of the Journal of the Institute of Brewing.)

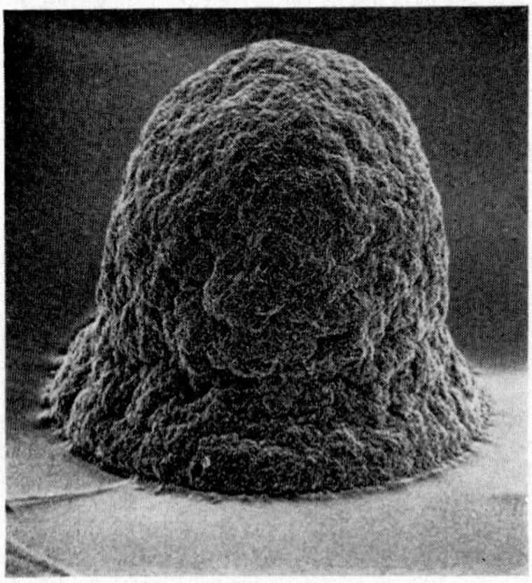

7. Untreated (× 113)

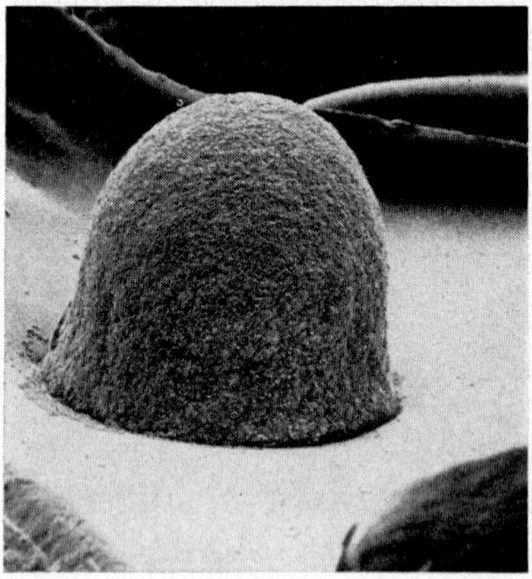

8. Glutaraldehyde fixed (× 73)

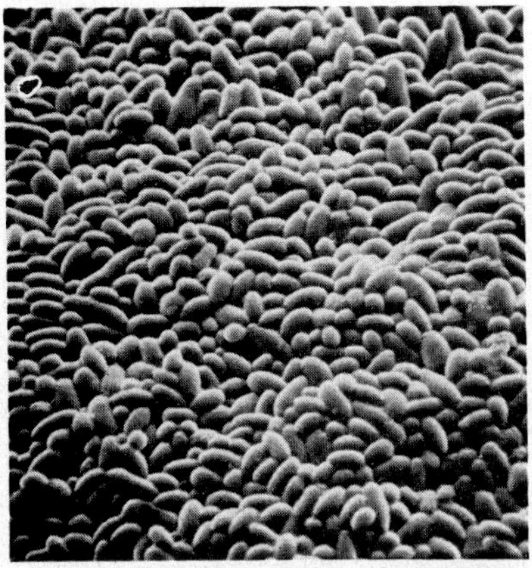

9. Untreated (× 747)

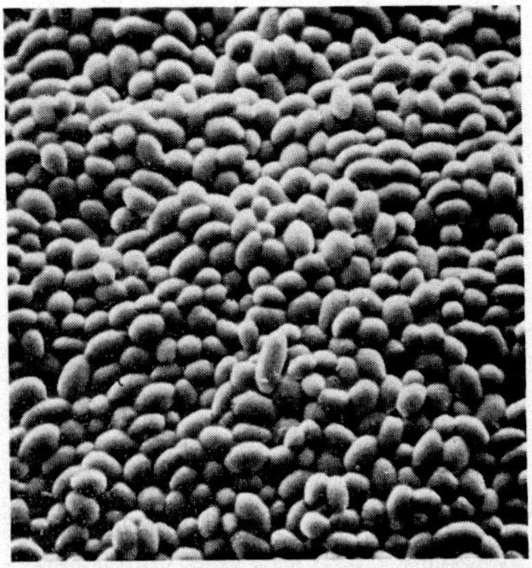

10. Glutaraldehyde fixed (× 800)

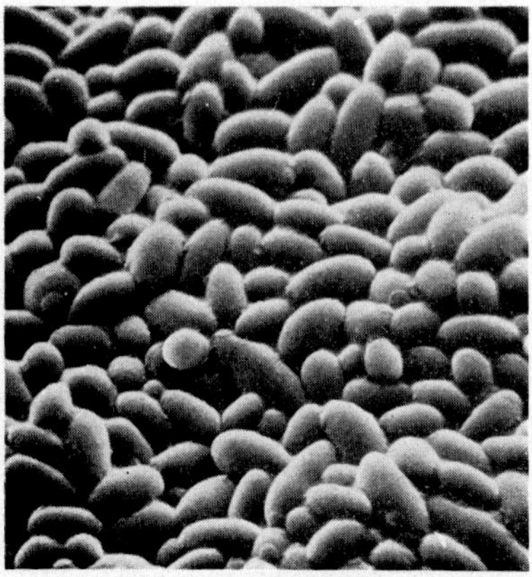

11. Untreated (× 1533)

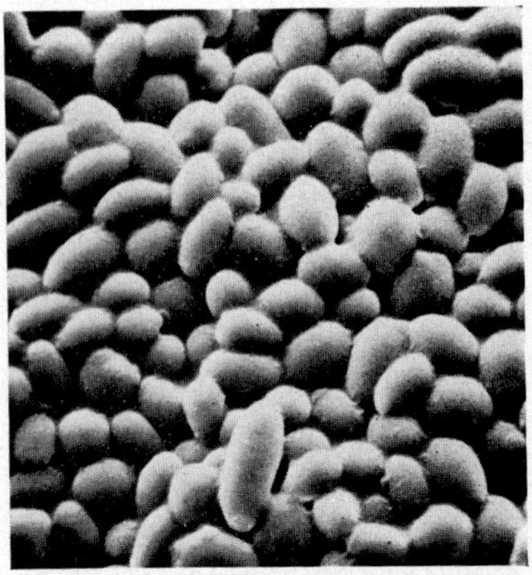

12. Glutaraldehyde fixed (× 1400)

Figs 13–18. A strain of *Acetobacter rancens* observed by the three techniques.

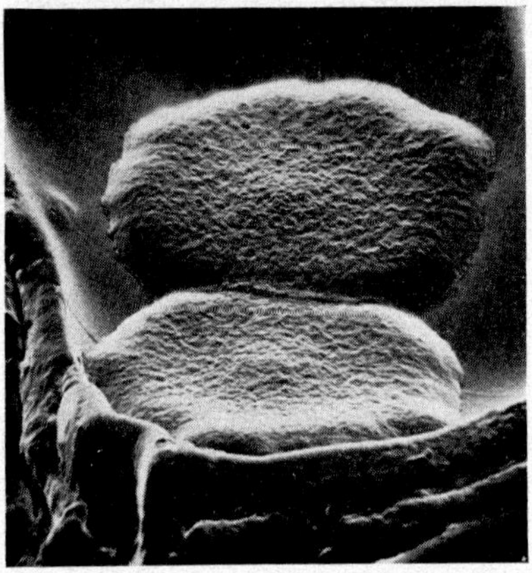

13. Untreated (× 70)

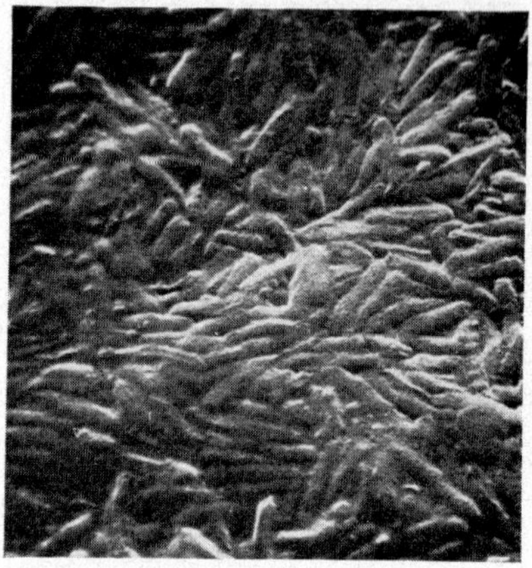

14. Untreated (× 3600)

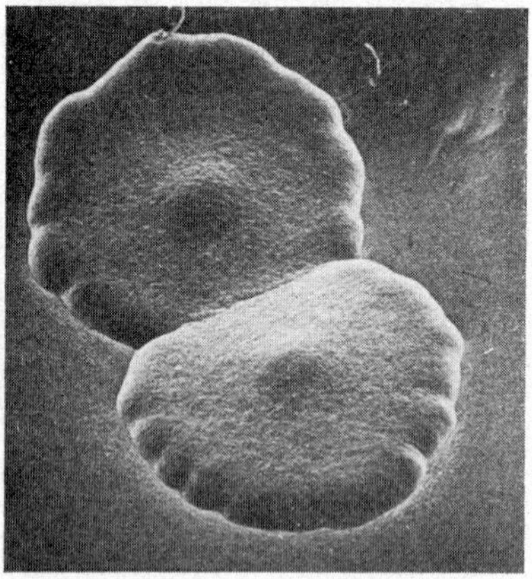

15. Glutaraldehyde fixed (× 40)

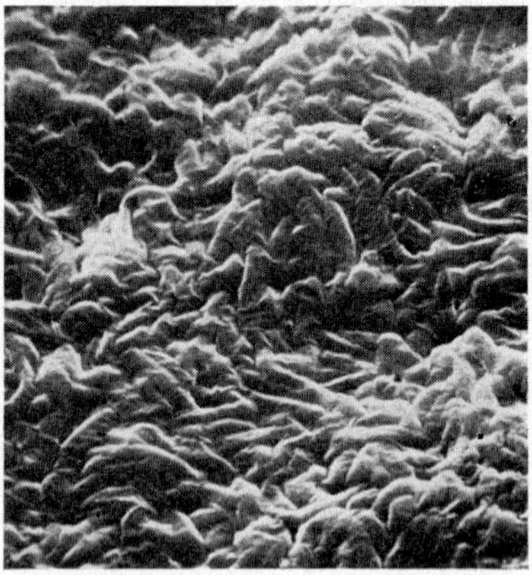

16. Glutaraldehyde fixed (× 3200)

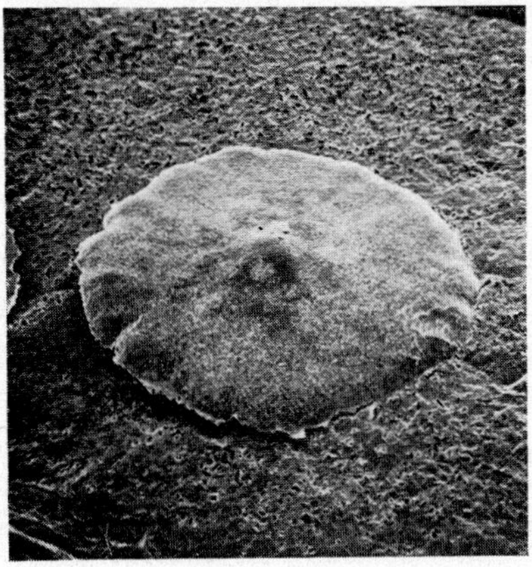

17. Freeze-dried (× 34)

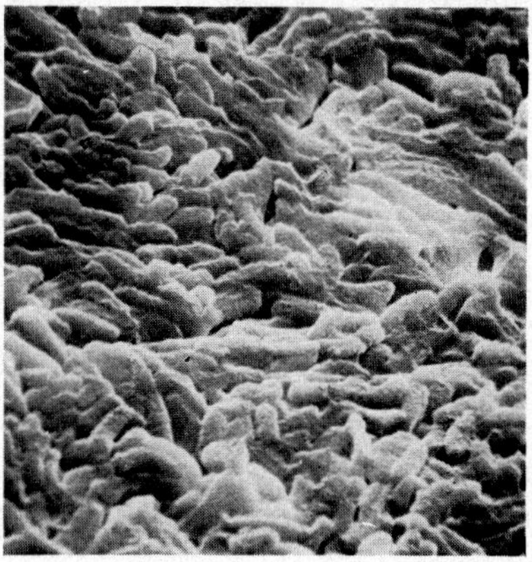

18. Freeze-dried (× 3133)

of almost 2000×. However, some collapse has occurred and the colony shape is better preserved with glutaraldehyde treatment (Figs 8, 10 and 12), although immersion in the fixative does tend to wash some of the cells from the colony surface and deposit them on the agar. The higher magnification images are also somewhat clearer after prior fixation.

Colonies of the acetic acid bacteria were difficult to observe in the SEM, due to their copious production of extracellular polysaccharides which often left the cells engulfed in a watery matrix. One of the easier species to examine, *Acetobacter rancens*, is illustrated in Figs 13–18. Even with this organism the cells are firmly embedded in dried down slime in the untreated colony (Figs 13 and 14) and the characteristic umbonate colony form is almost unrecognizable. Glutaraldehyde treatment again retains the colony shape well (Fig. 15) and makes the cells (Fig. 16) slightly more visible under the slime covering. Freeze drying is probably the most suitable method for these organisms and others which produce even greater amounts of slime, as the colony is reasonably well-preserved (Fig. 17) and individual organisms can be discerned at fairly high magnifications (Fig. 18).

In general it should be possible to observe untreated colonies of those organisms which produce densely packed structures, and therefore give minimal collapse and distortion. We have rarely observed "lysis" of cells in untreated colonies as Kormendy and Wayman (1972) noted with discrete cell preparations. Organisms which produce small amounts of intracellular material are best observed after chemical fixation, and glutaraldehyde is perhaps the simplest effective treatment. Freeze drying can give good results with single cells and colonies, but the "charging" effect rather limits resolution. Also it is a complicated process requiring special equipment and we have routinely used it only for organisms which could not be observed by any other method. Recent work (Parsons *et al.*, 1974) has indicated that the critical point drying method (Anderson, 1951; Nemanic, 1972) may be useful in preserving microbial colonies for scanning electron microscopy and preliminary investigations have been carried out, but this method also requires specialized equipment.

References

ANDERSON, T. F. (1951). Techniques for the preservation of three-dimensional structure in preparing specimens for the electron microscope. *Trans. N. Y. Acad. Sci.*, Ser. II, **13,** 130.

BISSET, K. A. (1938). The structure of "rough" and "smooth" colonies. *J. Path. Bact.*, **47,** 223.

BISSET, K. A. & PEASE, P. (1957). The distribution of flagella in dividing bacteria. *J. gen. Microbiol.*, **16,** 382.

DRUCKER, D. B. & WHITTAKER, D. K. (1971). Microstructure of colonies of rod-shaped bacteria. *J. Bact.*, **108**, 515.

GLAUERT, A. M. (1965). Section staining, cytology, autoradiography and immunochemistry for biological specimens. In *Techniques for electron microscopy.* (Kay, D. H., ed.), 254. Oxford: Blackwell Scientific Publications.

GREENWOOD, D. & O'GRADY, F. (1969). Antibiotic-induced surface changes in micro-organisms demonstrated by scanning electron microscopy. *Science, New York*, **163,** 1076.

KNUDSON, D. L. & MACLEOD, R. (1970). *Mycoplasma pneumoniae* and *Mycoplasma salivarium*: electron microscopy of colony growth in agar. *J. Bact.*, **101,** 609.

KORMENDY, A. C. & WAYMAN, M. (1972). Scanning electron microscopy of micro-organisms. *Micron*, **3,** 33.

MARSZALEK, D. S. & SMALL, E. B. (1969). Preparation of soft biological materials for scanning electron microscopy. In *Proc. 2nd Ann. Symp. Scanning Electron Microscopy*. (Johari, O., ed.), 231. Chicago: Illinois Institute of Technology Press.

NEMANIC, M. K. (1972). Critical point drying, cryofracture and serial sectioning. In *Proceedings of the Workshop on Biological Specimen Preparation Techniques for Scanning Electron Microscopy*. (Johari, O. and Korvin, Irene, eds), 297. Chicago: Illinois Institute of Technology Press.

NIXON, W. C. (1969). Scanning electron microscopy. *Contemp. Phys.*, **10,** 71.

PARSONS, E., BOLE, B., HALL, D. J. & THOMAS, W. D. E. (1974). A comparative study of techniques for preparing plant surfaces for the scanning electron microscope. *J. Microsc.*, **101,** 59.

PASSMORE, S. M. (1973*a*). Scanning electron microscopy of yeast colonies. *J. Inst. Brew.*, **79,** 237.

PASSMORE, S. M. (1973*b*). The acetic acid bacteria: ecology, taxonomy and morphology. Ph.D. Thesis, University of Bristol.

PASSMORE, S. M. & HAGGETT, B. (1973). The use of scanning electron microscopy to show confluent growth of a *Saccharomyces* and a *Leuconostoc* species. *J. appl. Bact.*, **36**, 89.

ROUSSEAU, P., HALVORSON, H. O., BULLA, L. A. & ST. JULIAN, G. (1972). Germination and outgrowth of single spores of *Saccharomyces cerevisiae* viewed by scanning electron and phase-contrast microscopy. *J. Bact.*, **109,** 1232.

SPRINGER, E. L. & ROTH, I. L. (1972). Scanning electron microscopy of bacterial colonies. Part I. *Diplococcus pneumoniae* and *Streptococcus pyogenes*. *Can. J. Microbiol.*, **18,** 219.

SWANSON, J. & MCCARTY, M. (1969). Electron microscopic studies on opaque colony variants of Group A streptococci. *J. Bact.*, **100,** 505.

WHITTAKER, D. K. & DRUCKER, D. B. (1970). Scanning electron microscopy of intact colonies of micro-organisms. *J. Bact.*, **104**, 902.

WHITTAKER, D. K. & DRUCKER, D. B. (1972). Modifications of colonial morphology during growth: a study of *Streptococcus mutans* by scanning electron microscopy. *Micron*, **3,** 150.

WILLIAMS, S. T. (1970). Further investigations of actinomycetes by scanning electron microscopy. *J. gen. Microbiol.*, **62,** 67.

The Surface Structure of Bacteria

AUDREY M. GLAUERT*, MARGARET J. THORNLEY††,
KAREEN J. I. THORNE* AND U. B. SLEYTR*

* *Strangeways Research Laboratory, Cambridge*
†† *Immunology Division, Department of Pathology, University of Cambridge, Cambridge, England*

Introduction

This contribution illustrates the results which may be obtained when studying the surface layers of bacteria by electron microscopy. Three techniques have most commonly been used for the preparation of bacteria in order to provide information about their surface structure.

1. To make thin sections, intact bacteria or isolated cell walls are fixed, embedded in epoxy resin and then cut into sections 50 to 100 nm thick.
2. For examination by the freeze-etching method, a pellet of intact bacteria is frozen rapidly and then fractured. After a period of etching, during which some of the ice surrounding the bacteria sublimes away, a platinum-carbon replica of the exposed surface is made. This surface includes areas of ice, and, where the bacteria are embedded in it, some of their internal structure is revealed by the initial fracture, and parts of their outer surfaces are exposed by etching. The replica is removed, cleaned and examined in the electron microscope.
3. The technique of negative staining is most useful when applied to thin specimens such as cell walls or fragments of cell walls. An electron microscope grid coated with a collodion-carbon film is floated face-downwards on the surface of a suspension of cell walls and then on a negative-staining solution (e.g. 1% ammonium molybdate at pH 7). After drying, the preparation is ready for examination in the electron microscope.

The bacteria chosen for illustration in this paper all carry external layers of regularly arranged sub-units which are found in many genera of both Gram-negative and Gram-positive bacteria (Glauert and Thornley, 1969; Holt and Leadbetter, 1969; Remsen and Watson, 1972; Thornley, Glauert and Sleytr, 1974). The Gram-negative bacteria illustrated are two strains of *Acinetobacter* originally isolated from chicken carcasses (Thornley, Ingram and Barnes, 1960). One of these strains, MJT/F5/5, has a hexagonally arranged surface layer which proved useful as a morphological marker in a study of the relationships between the layers of the cell envelope as revealed by freeze etching and those seen in thin sections (Sleytr, Thornley and Glauert, 1974). The other, *Acinetobacter* strain MJT/F5/199A, was used to study the separation of and chemical composition of the different layers of the cell envelope, including the outer membrane (Thornley, Glauert and Sleytr, 1973; Thorne, Thornley and Glauert, 1973). This *Acinetobacter* has an outer layer of tetragonally arranged protein subunits and when these sub-units are detached they are able to assemble into regular arrays with the same dimensions as the patterns on intact bacteria. The requirements for detachment and self-assembly of the sub-units, and the mode of their attachment to the underlying surface, have been investigated (Thornley, Thorne and Glauert, 1974; Thorne *et al.*, 1975). Other strains of *Acinetobacter* which lack the regularly arranged surface sub-units are known, and so this property is not characteristic of the genus as a whole.

The structure of the envelope of a Gram-negative bacterium with regularly arranged surface sub-units, *Acinetobacter* strain MJT/F5/5

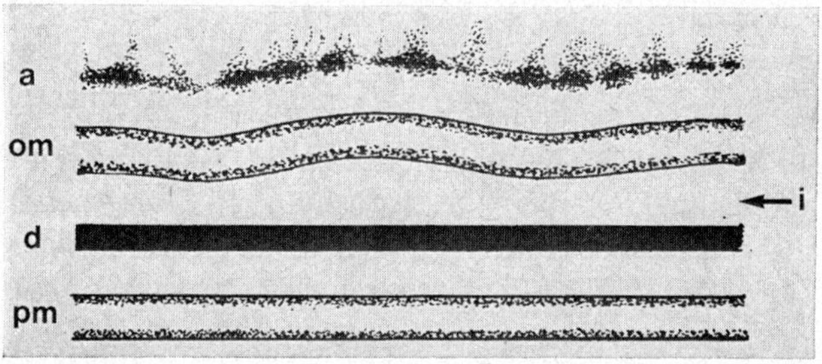

FIG. 1. Diagram of the surface layers of *Acinetobacter* strain MJT/F5/5 as seen in thin sections. The structure shown is that typical of Gram-negative bacteria, apart from the additional layer (a). (pm) is the plasma membrane, which is adjacent to the cytoplasm. The cell wall consists of: (d) the dense layer, in which the peptidoglycan is located; (i) the intermediate region; (om) the outer membrane, which contains lipopolysaccharide, protein and lipid; and (a) the additional layer composed of regularly arranged sub-units. (From Glauert and Thornley, 1971.)

FIGS 2, 3 AND 4 (*opposite*). Electron micrographs of *Acinetobacter* strain MJT/F5/5.

FIG. 2. The plasma membrane (pm), dense layer (d), outer membrane (om) and additional layer (a) are visible in a thin section of the surface layers of an intact cell (see Fig 1). (× 252 000). (From Sleytr, Thornley and Glauert, 1974.)

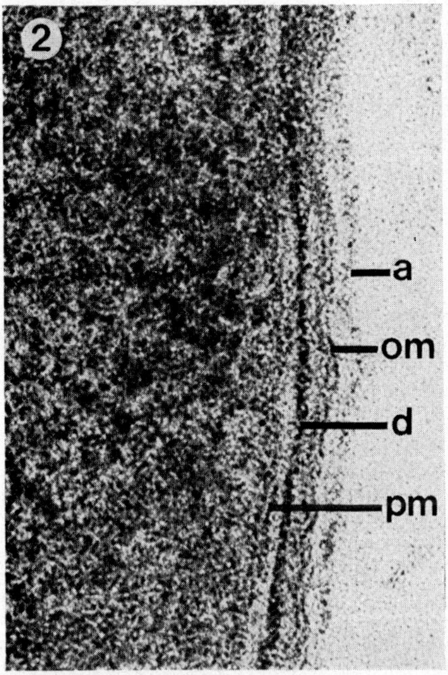

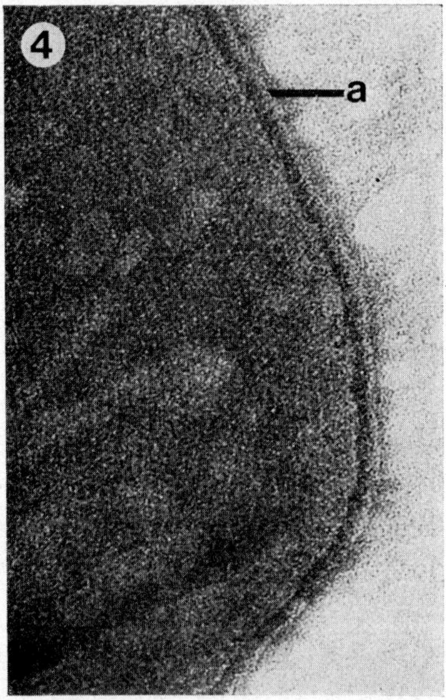

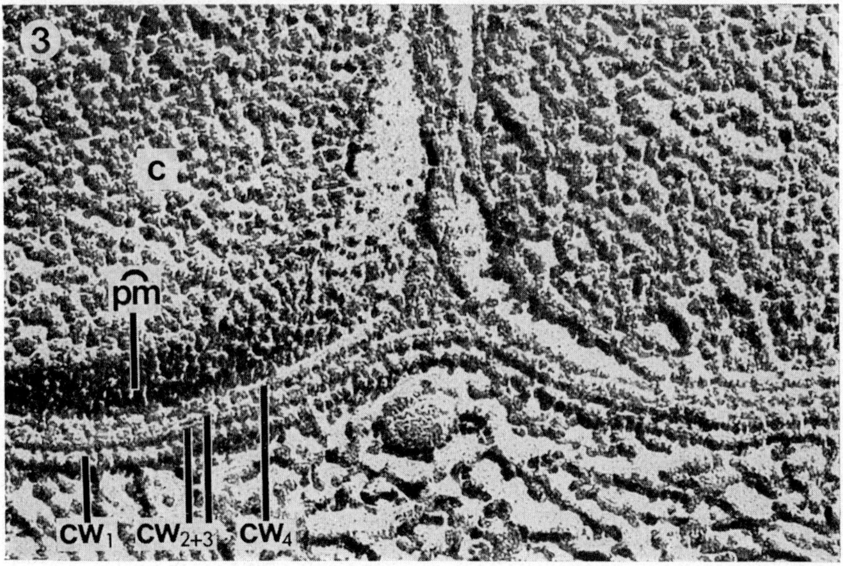

FIG. 3. The envelope of an intact cell, cross-fractured in a preparation freeze-etched in the presence of glycerol. The cell wall shows three main ridges, cw_1, cw_{2+3}, whose double structure is visible in places, and cw_4. The relationships between these ridges and the layers seen in thin sections (Figs 1 and 2) are shown in the diagram in Fig. 5. The fracture has also revealed a small region of the convex fracture face of the plasma membrane ($\widehat{pm}$). c, cytoplasm (× 135 000). (From Sleytr, Thornley and Glauert, 1974.)

FIG. 4. The additional layer (a) is visible at the folded edge of the cell wall of a heat-treated cell in a negatively-stained preparation. The cytoplasm has retracted and the regular array of surface sub-units is visible (× 135 000). (From Sleytr, Thornley and Glauert, 1974.)

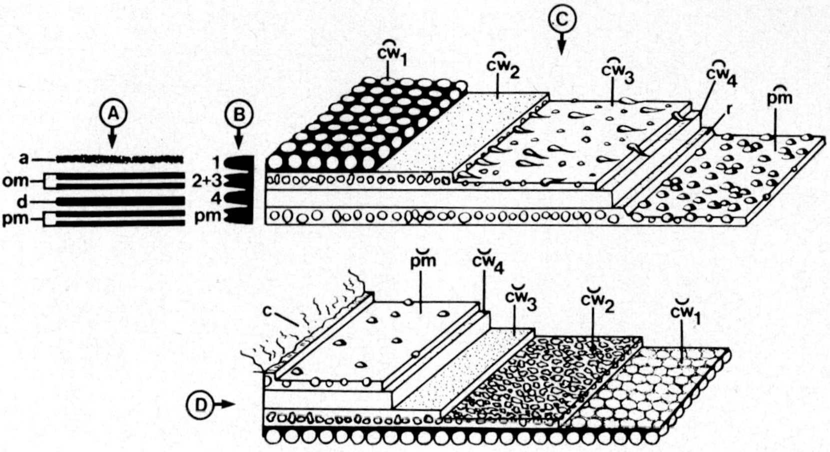

FIG. 5. Diagram showing the relationships between the layers of the cell envelope of *Acinetobacter* strain MJT/F5/5 revealed in thin sections and in freeze-etched preparations. (A) Cell envelope as seen in thin sections. (pm) plasma membrane; (d) dense layer; (om) outer membrane; (a) additional layer. (see Figs 1 and 2). (B) Cell envelope as it appears in a freeze-etched preparation after cross-fracture (see Fig. 3). The four main ridges represent the plasma membrane (pm) and the layers of the cell wall: cw_1, which represents the additional layer; cw_{2+3}, which represents the outer membrane; and cw_4 which corresponds to the dense layer and the intermediate region between the dense layer and the outer membrane. (C) An obliquely fractured cell envelope seen from the convex side. The etched outer surface ($\widehat{cw}_1$) consists of the hexagonally arranged sub-units of the additional layer. The underlying layers of the envelope are visible as alternating ridges and fracture faces. The fracture face $\widehat{cw}_2$ is only revealed occasionally. The main convex fracture face in the cell wall, $\widehat{cw}_3$, appears fibrillar, and it is separated by ridges (cw_4 and r, the outer portion of the plasma membrane) from the internal fracture face of the plasma membrane ($\widehat{pm}$), which carries irregularly arranged particles. (D) An obliquely fractured cell envelope seen from the concave side shows the cytoplasm (c), the concave fracture face of the plasma membrane ($\breve{pm}$), and the main concave fracture face within the cell wall ($\breve{cw}_2$), which consists of small densely packed particles. The fracture faces $\breve{cw}_3$ and $\breve{cw}_1$ are only revealed occasionally. (From Sleytr, Thornley and Glauert, 1974.)

The fracture faces most often revealed in the envelopes of both Gram-negative and Gram-positive bacteria are the internal fracture faces of the plasma membrane, which show irregularly arranged intra-membraneous particles (Figs 5, 6, 7, 8, 18 and 19). In addition, Gram-negative bacteria very often show another main fracture plane within the cell wall and in a wide variety of Gram-negative bacteria the concave fracture face has the characteristic particulate appearance shown here (Figs 5 and 8, $\breve{cw}_2$). This suggests that the fracture is formed in the same part of the wall in all cases, but its location in relation to the layers seen in thin sections has been a matter of some doubt (Remsen and Watson, 1972). Experiments with *Acinetobacter* strain MJT/F5/5 established that the convex face $\widehat{cw}_3$ and the con-

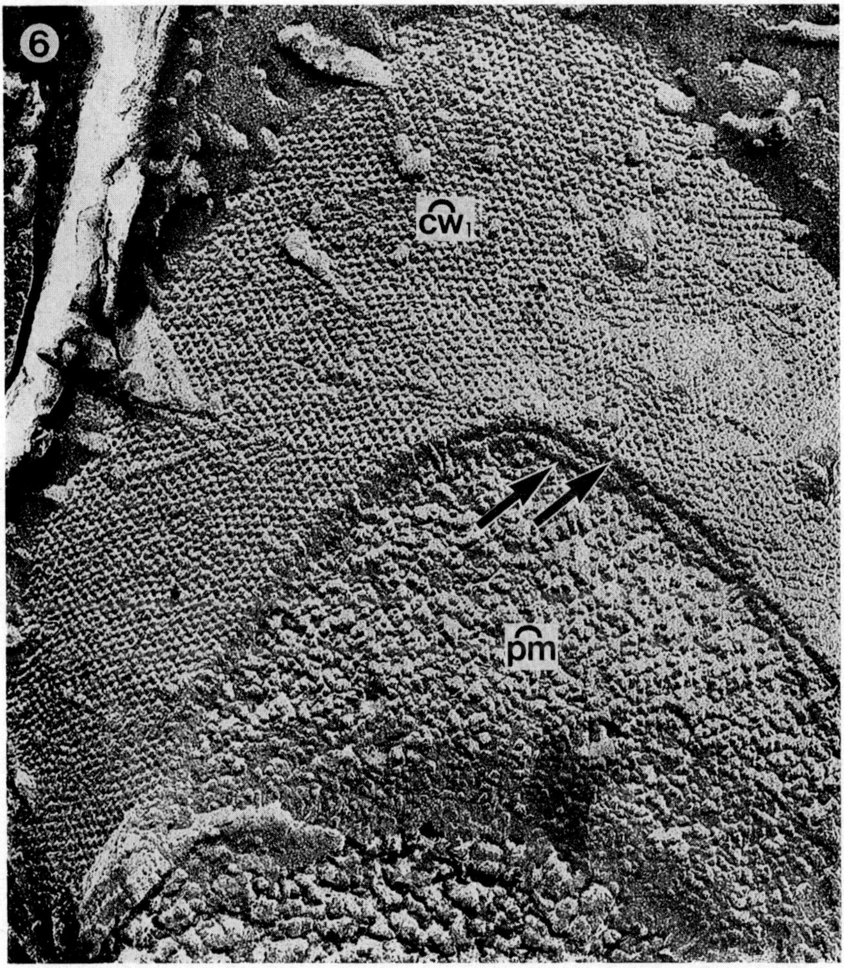

cave face $\overset{\smile}{\mathrm{cw}}_2$ arose by internal fracture of the outer membrane, as indicated in Fig. 5 (Sleytr, Thornley and Glauert, 1974). Subsequent experiments with *Pseudomonas fluorescens* NCTC 10038, an organism lacking surface sub-units, gave similar results regarding the location and surface structure of both fracture faces (Thornley and Sleytr, 1974). It seems very probable that the internal fracture of the outer membrane is usual in Gram-negative bacteria.

FIG. 6. Electron micrograph of a freeze-etched preparation of *Acinetobacter* strain MJT/F5/5. The outer surface, $\overset{\frown}{\mathrm{cw}}_1$, has been revealed by etching and is seen to consist of hexagonally arranged sub-units (see Fig. 5). The fractured edge of the cell wall shows two ridges (arrows). $\overset{\frown}{\mathrm{pm}}$, internal fracture face of the plasma membrane (× 112 500). (From Sleytr, Thornley and Glauert, 1974.)

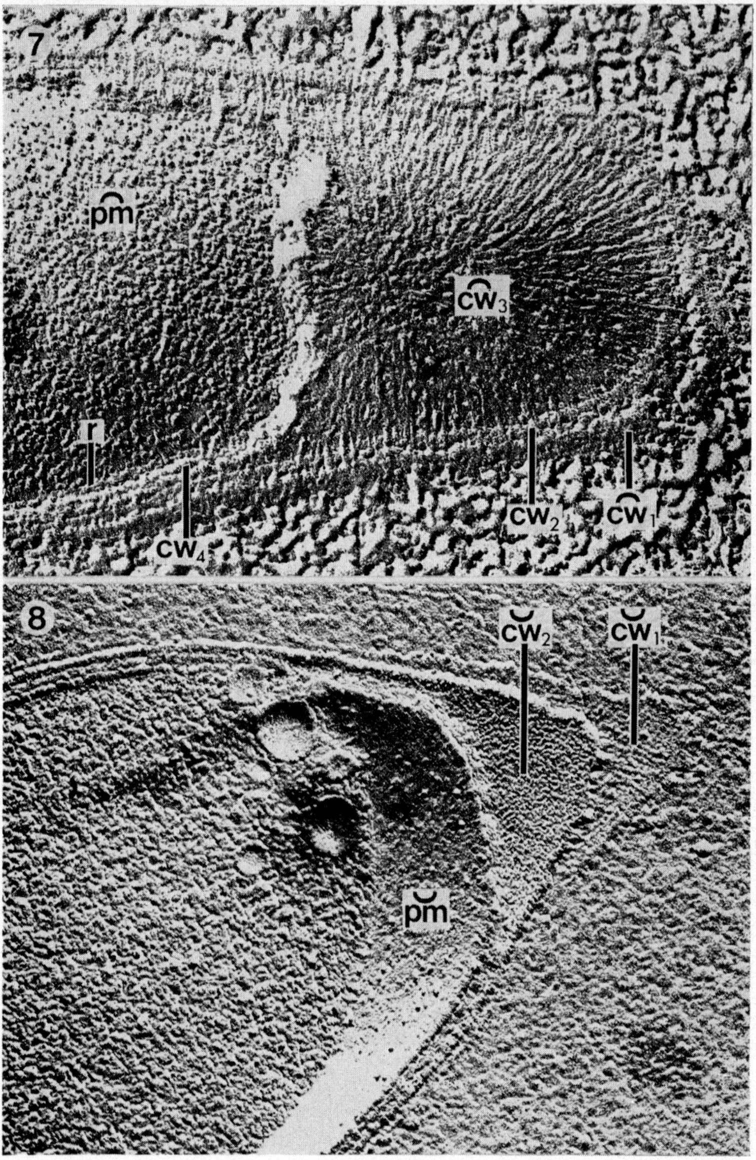

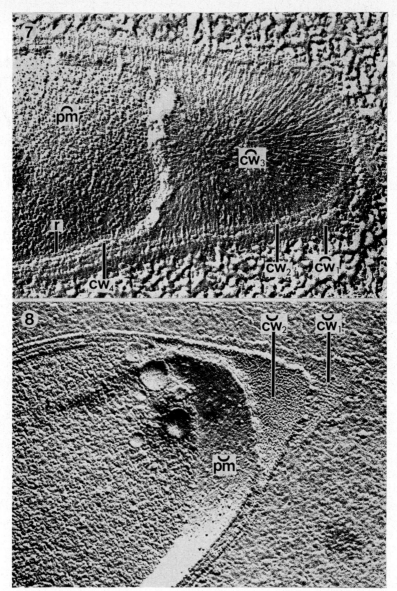

FIGS 7 AND 8. Electron micrographs of freeze-etched preparations of *Acinetobacter* strain MJT/F5/5.

FIG. 7. Convex view of an obliquely fractured cell envelope in a preparation freeze-etched in the presence of glycerol (see Fig. 5C). Due to the glycerol, only a small area of the outer surface ($\overset{\frown}{cw}_1$) has been exposed by etching; it shows the regular array of sub-units. The edge of cw_2 adjoins the main convex fracture face in the cell wall, $\overset{\frown}{cw}_3$, whose surface shows radially arranged fibrils and a few granules and depressions. The edges of cw_4 and of the outer portion of the plasma membrane (r) lie next to the internal fracture face of the plasma membrane ($\overset{\frown}{pm}$) ($\times$ 80 000). (From Sleytr, Thornley and Glauert, 1974.)

FIG. 8. Concave view of an oblique fracture through the envelope of a cell treated briefly with lysozyme and freeze-etched in the presence of glycerol. The concave fracture faces, $\overset{\smile}{pm}$, $\overset{\smile}{cw}_2$ and $\overset{\smile}{cw}_1$ are visible, and cw_2 appears to consist of densely packed particles (see Fig. 5D) ($\times$ 80 000). (From Sleytr, Thornley and Glauert, 1974)

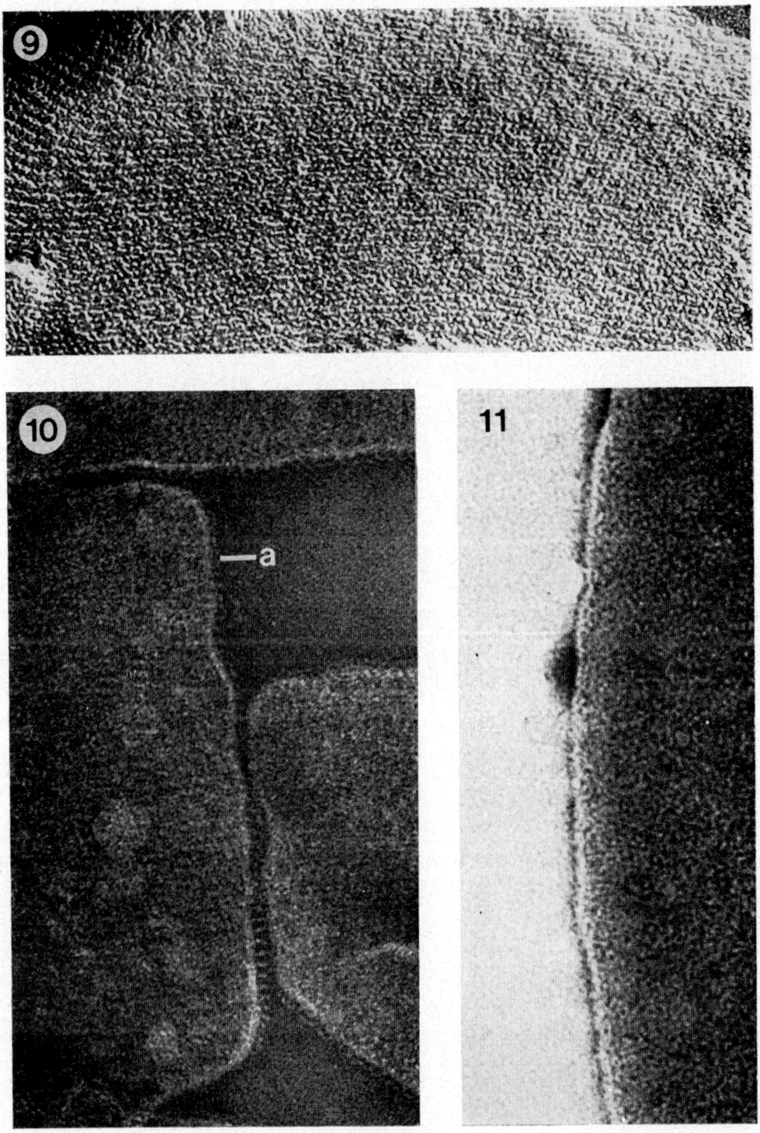

FIGS 9, 10 AND 11. Electron micrographs of *Acinetobacter* strain MJT/F5/199A.

FIG. 9. The tetragonal array of surface sub-units is visible in a freeze-etched preparation. The sub-units are arranged in rows at an angle of about 80° to each other (× 160 000). (From Sleytr and Thornley, 1973.)

FIG. 10. A negatively-stained preparation of isolated cell walls. The walls are covered with a regular array of surface sub-units which are particularly clearly seen at folded edges of fragments of cell wall (a) (× 160 000).

FIG. 11. Sub-units are no longer visible on the surface of a cell wall treated with 1 M urea for 2 h at 37° and then washed with distilled water (× 160 000).

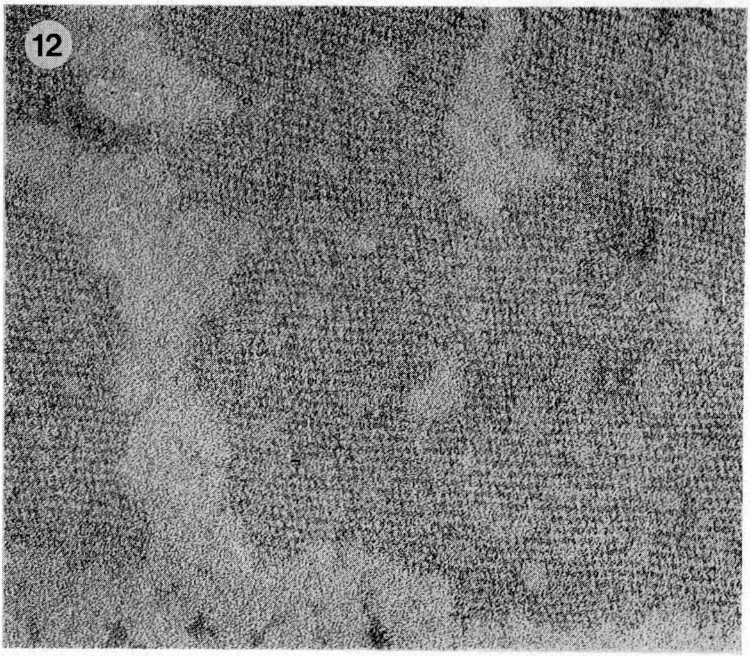

FIG. 12. Arrays of sub-units in a negatively-stained preparation of the material extracted from cell walls of *Acinetobacter* strain MJT/F5/199A by treatment with urea. The extract was dialysed and then allowed to stand at room temperature for 30 min in the presence of 20 mM $MgCl_2$. The isolated surface protein has assembled spontaneously into arrays with the same dimensions as those seen on the surface of the cell wall (compare with Fig. 10) ($\times$ 160 000).

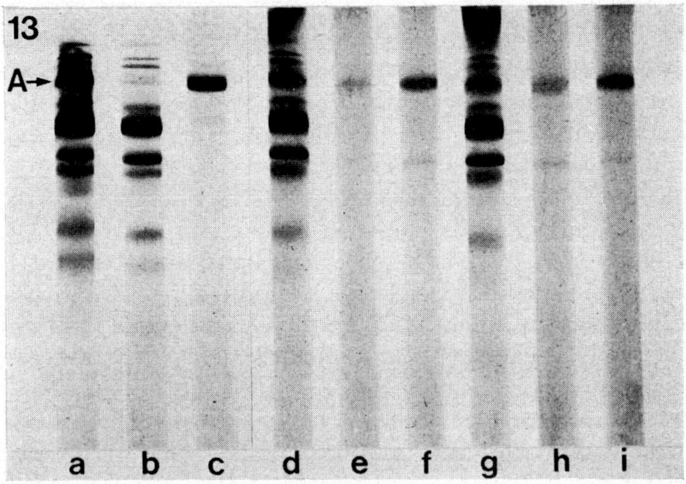

FIG. 13. SDS acrylamide gel electrophoresis of chemically treated cell walls. (a) Inact cell wall; (b) wall treated with 2 M urea; (c) 2 M urea extract; (d) wall treated with 10 mM EDTA and washed with distilled water; (e) 10 mM EDTA extract; (f) water wash after EDTA treatment; (g) wall treated with 10 mM EGTA and washed with distilled water; (h) 10 mM EGTA extract; (i) water wash after EGTA treatment.

The band (A) is solubilized by urea, EDTA and EGTA and represents the surface sub-units. The sub-units are an acidic protein of molecular weight 65 000. (From Thornley, Thorne and Glauert, 1974.)

The Gram-positive (or Gram-variable) bacteria shown here belong to the genera *Bacillus* and *Clostridium* which both contain many species with regularly arranged sub-units on the surfaces of the vegetative cells (Holt and Leadbetter, 1969; Sleytr and Glauert, 1975), as well as some which lack sub-units. Brinton, McNary and Carnahan (1969) described a strain of *B. brevis* with tetragonally arranged sub-units which, when removed from the bacterium, had the property of self-assembly into curved sheets with the same pattern as that found on the bacterial surface. Details of the pattern, as revealed by optical diffraction and filtering techniques, were described by Aebi *et al.* (1973), and the same strain is illustrated here. Chemical studies by Brinton, McNary and Carnahan (1969) and by Howard and Tipper (1973) (who classified Brinton's strain as *B. sphaericus*) showed that the sub-units were composed of a protein with a molecular weight of 150 000 which formed 16% of the total protein of the cell. The two species of *Clostridium* illustrated here are thermophiles which were isolated from an extraction plant for beet sugar where they were growing at a temperature of 68°. In *C. thermohydrosulfuricum* the surface sub-units are arranged hexagonally, while in *C. thermosaccharolyticum* the pattern is tetragonal. It was suggested by Hollaus and Sleytr (1972) that this difference in arrangement of the surface sub-units should be regarded as an additional taxonomic character for the differentiation of these two closely related species.

The structure of the envelopes of Gram-positive bacteria with regularly arranged surface sub-units

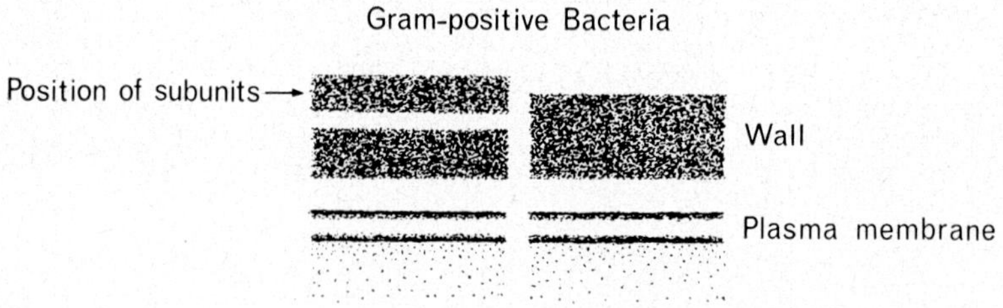

FIG. 14. Diagrams illustrating the structure of the envelopes of Gram-positive bacteria as seen in thin sections. The cell wall may appear as two densely staining layers, separated by a less dense layer, or as a single layer of uniform density. The regular arrays of sub-units are located in the outer regions of the dense layers at the surfaces of the bacteria. (From Thornley, Glauert and Sleytr, 1974.)

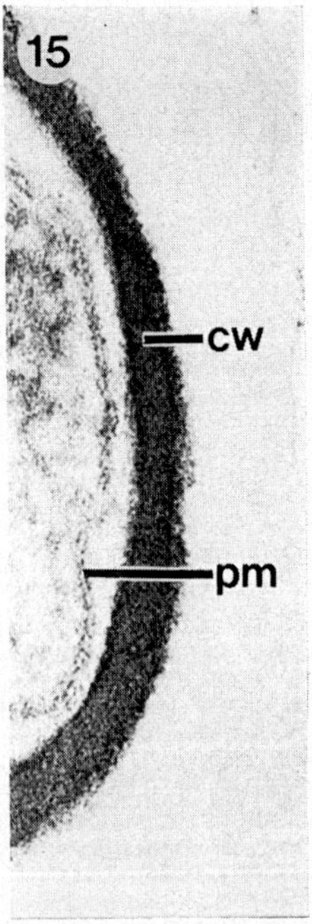

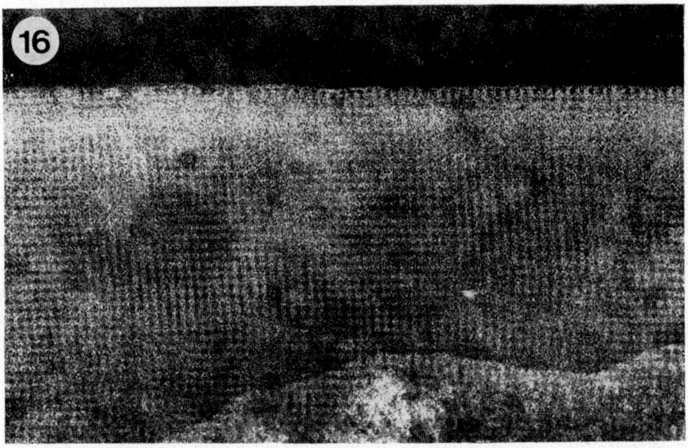

FIGS 15, 16 AND 17. Electron micrographs of *B. brevis*.

FIG. 15. The plasma membrane (pm) and cell wall (cw) are visible in a thin section of the surface layers of a lysed cell. The cell wall appears as two thin dense layers, separated by a wide layer of lower density (× 160 000).

FIG. 16. A negatively-stained preparation of a lysed cell. The cytoplasm has retracted and the regular array of sub-units is visible on the surface of the cell wall (× 80 000).

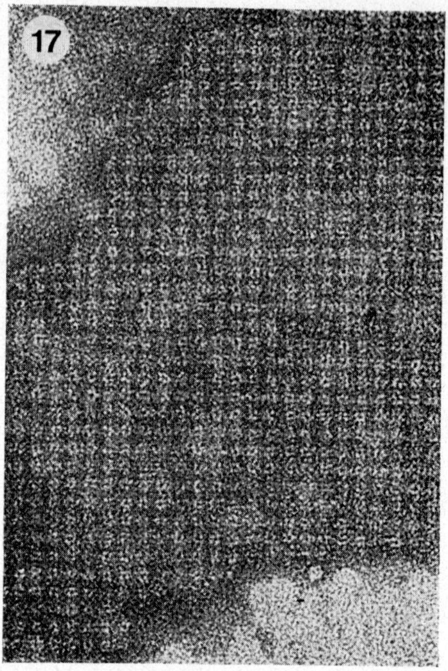

FIG. 17. A negatively-stained preparation of an array of surface sub-units which has become detached from a cell ($\times$ 160 000).

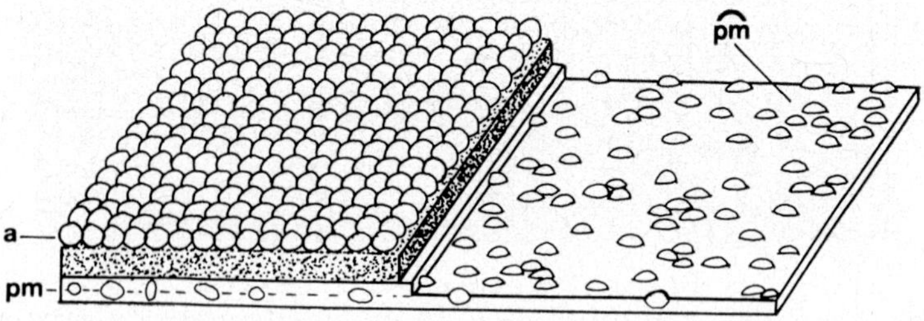

FIG. 18. Diagram of the envelope of a Gram-positive (or Gram-variable) bacterium with a regular array of surface sub-units, as seen in a freeze-etched preparation. The pattern of sub-units (a) is seen on the etched outer surface of the bacterium. p͡m, internal fracture face of the plasma membrane.

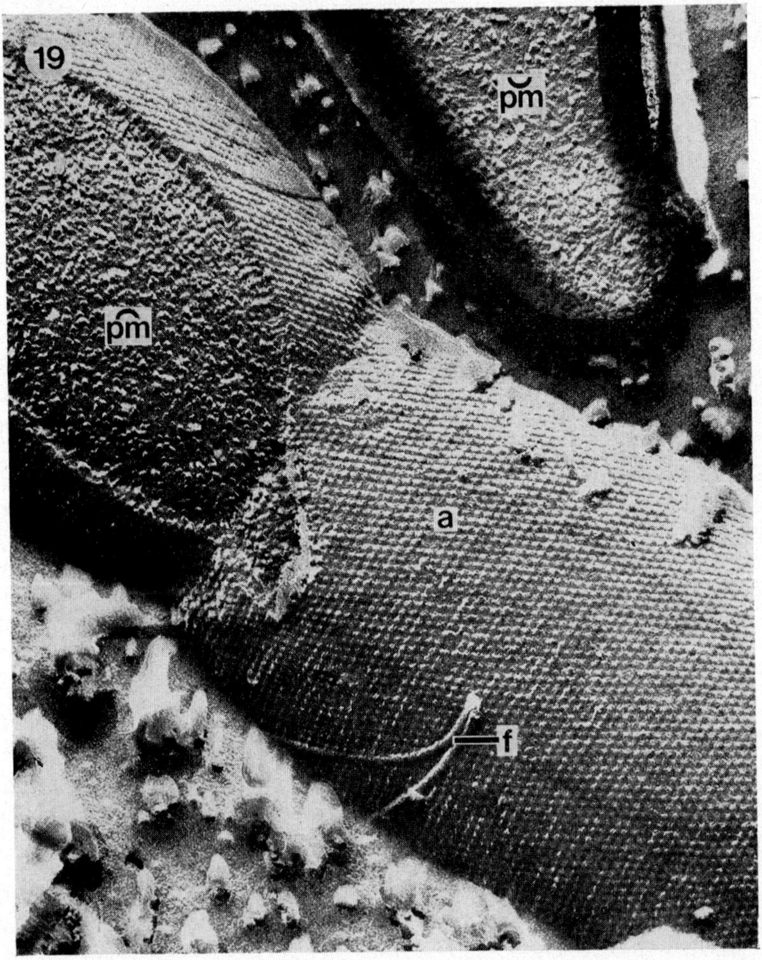

FIG. 19. Electron micrograph of a freeze etched preparation of the Gram-variable bacterium *C. thermohydrosulfuricum*. An array of hexagonally arranged sub-units is visible on the etched outer surface (a) of the cell. $\widehat{pm}$, convex fracture face of the plasma membrane (see Fig. 18). The concave fracture face of the plasma membrane ($\smile{pm}$) of an adjacent bacterium is also visible. f, flagella ($\times$80 000).

FIGS 20, 21, 22 AND 23 (*opposite*). Electron micrographs of freeze-etched preparations of two species of *Clostridium*.

FIG. 20. At a site of division the cell surface of *C. thermohydrosulfuricum* is composed of a mosaic of small crystallites in which the regular hexagonal pattern of sub-units is in different orientations. f, flagellum (× 96 000). (From Sleytr and Glauert, 1975.)

FIG. 21. The sub-units on the surface of *C. thermosaccharolyticum* are arranged in a tetragonal pattern. At a site of division the surface pattern is composed of small crystallites. f, flagellum (× 96 000). (From Sleytr and Glauert, 1975.)

FIG. 22. Site of insertion of a flagellum in the surface of *C. thermohydrosulfuricum*. The hook region, which is just outside the bacterial surface, has a characteristic banded structure. The regular hexagonal array of surface sub-units is slightly disturbed by the flagellum (× 160 000). (From Sleytr and Glauert, 1975.)

FIG. 23. Site of insertion of a flagellum in the surface of *C. thermosaccharolyticum*. The tetragonal array of surface sub-units is distorted by the flagellum and some of the rows of sub-units are curved (× 160 000). (From Sleytr and Glauert, 1975.)

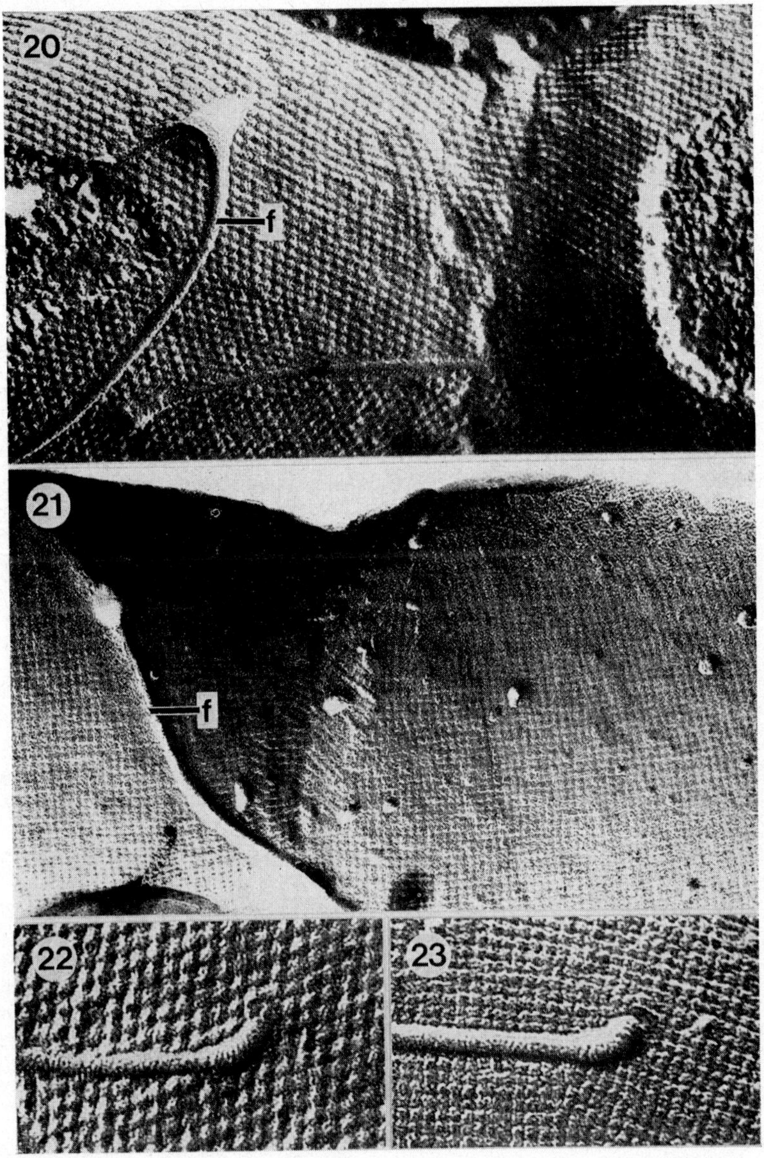
20
f
21
f
22
23

Acknowledgement

The skilled technical assistance of Mr. R. A. Parker in the preparation of the illustrations is gratefully acknowledged.

References

AEBI, U., SMITH, P. R., DUBOCHET, J., HENRY, C. & KELLENBERGER, E. (1973). A study of the structure of the T-layer of *Bacillus brevis*. *J. supramol. Struct.*, **1,** 498.

BRINTON, C. C., MCNARY, J. C. & CARNAHAN, J. (1969). Purification and *in vitro* assembly of a curved network of identical protein subunits from the outer surface of a *Bacillus*. *Bact. Proc.*, 48.

GLAUERT, A. M. & THORNLEY, M. J. (1969). The topography of the bacterial cell wall. *Ann. Rev. Microbiol.*, **23,** 159.

GLAUERT, A. M. & THORNLEY, M. J. (1971). Fine structure and radiation resistance in *Acinetobacter*: a comparison of a range of strains. *J. Cell Sci.*, **8,** 19.

HOLLAUS, F. & SLEYTR, U. (1972). On the taxonomy and fine structure of some hyperthermophilic saccharolytic clostridia. *Arch. Mikrobiol.*, **86,** 129.

HOLT, S. C. & LEADBETTER, E. R. (1969). Comparative ultrastructure of selected aerobic spore-forming bacteria; a freeze-etching study. *Bact. Rev.*, **33,** 346.

HOWARD, L. & TIPPER, D. J. (1973). A polypeptide bacteriophage receptor: modified cell wall protein subunits in bacteriophage-resistant mutants of *Bacillus sphaericus* strain P-1. *J. Bact.*, **113,** 1491.

REMSEN, C. C. & WATSON, S. W. (1972). Freeze-etching of bacteria. *Int. Rev. Cytol.*, **33,** 253.

SLEYTR, U. B. & GLAUERT, A. M. (1975). Analysis of regular arrays of sub-units on bacterial surfaces; evidence for a dynamic process of assembly. *J. Ultrastruct. Res.*, **50,** 103.

SLEYTR, U. B. & THORNLEY, M. J. (1973). Freeze-etching of the cell envelope of an *Acinetobacter* species which carries a regular array of surface subunits. *J. Bact.*, **116,** 1383.

SLEYTR, U. B., THORNLEY, M. J. & GLAUERT, A. M. (1974). Location of the fracture faces within the cell envelope of *Acinetobacter* species strain MJT/F5/5. *J. Bact.*, **118,** 693.

THORNE, K. J. I., THORNLEY, M. J. & GLAUERT, A. M. (1973). Chemical analysis of the outer membrane and other layers of the cell envelope of *Acinetobacter* sp. *J. Bact.*, **116,** 410.

THORNE, K. J. I., THORNLEY, M. J., NAISBITT, P. & GLAUERT, A. M. (1975). The nature of the attachment of a regularly arranged surface protein to the outer membrane of an *Acinetobacter* sp. *Biochem. Biophys. Acta*, **389,** 97.

THORNLEY, M. J., GLAUERT, A. M. & SLEYTR, U. B. (1973). Isolation of outer membranes with an ordered array of surface subunits from *Acinetobacter*. *J. Bact.*, **114,** 1294.

THORNLEY, M. J., GLAUERT, A. M. & SLEYTR, U. B. (1974). Structure and assembly of bacterial surface layers composed of regular arrays of subunits. *Phil. Trans. R. Soc. Lond.* B, **268,** 147.

THORNLEY, M. J., INGRAM, M. & BARNES, E. M. (1960). The effects of antibiotics and irradiation on the *Pseudomonas-Achromobacter* flora of chilled poultry. *J. appl. Bact.*, **23,** 487.

THORNLEY, M. J. & SLEYTR, U. B. (1974). Freeze-etching of the outer membranes of *Pseudomonas* and *Acinetobacter*. *Arch. Microbiol.*, **100,** 409.

THORNLEY, M. J., THORNE, K. J. I. & GLAUERT, A. M. (1974). Detachment and chemical characterization of the regularly arranged subunits from the surface of an *Acinetobacter*. *J. Bact.*, **118,** 654.

Bacterial Surface Structures

W. Hodgkiss
Torry Research Station, Aberdeen, Scotland
AND
J. A. Short and P. D. Walker
Wellcome Research Laboratories, Beckenham, Kent, England

The presence of flagella on some bacteria has been noted since the early part of the 19th century. That these organelles were associated with motility was not, however, elucidated until much later with the advent of specific flagella stains and the introduction of more sophisticated microscopy techniques such as dark ground illumination.

Using the light microscope, recognition of both motility and the form of flagellation became criteria of taxonomic importance in the early systems of bacterial classification. Numerous studies were carried out, some of the more comprehensive of which were those of Leifson (1960), Lautrop and Jessen (1964) and Rhodes (1965).

The view that flagella are the organelles of bacterial motility did not, however, go unchallenged (Pijper, 1946) and it was only natural that such controversial views and divergent opinions should stimulate further interest in bacterial flagellation. With the development of the electron microscope as a research tool for the examination of biological specimens it was quickly realized that here was an instrument which, having a greater magnification and resolving power than the light microscope, might be able to shed some light on bacterial flagella. Kingma Boltjes (1948) and Rinker and Koffler (1951) were amongst the first to publish electron micrographs specifically on bacterial flagella. The most important contribution around this time, however, came from Houwink and van Iterson (1950). Not only did these authors demonstrate flagella, flagella hooks, basal bodies and flagella bundles (fasicles, Riesenzöpfe) but they were the first to describe the much finer structures later to be termed fimbriae (Duguid, 1955) or pili (Brinton, 1959). They also put forward the hypothesis that these structures could be organelles of adhesion.

Flagella

Research on flagella may now be divided into two main disciplines. The first of these is an investigation of the ultrastructure of the organelle itself whilst the second is the study of the flagellation of bacterial cells in relation to identification and taxonomy.

Flagella ultrastructure

In the electron microscope flagella may be demonstrated by the technique of metal shadowing or negative staining and can, with care, be detected in ultrathin sections. They appear as sinuous filaments usually 12–20 nm in diameter and about 6–8 μm in length. Whilst the diameter of individual flagella in a culture is usually constant, the length often varies markedly. The bacterial flagellum is composed of three distinct parts, the filament, the proximal hook and the basal granule.

The filament is a simple fibre composed of a protein known as flagellin. Many elegant experiments have been reported using flagellin to determine the way in which the filaments are synthesized (Asakura, 1970; Wakabayashi, 1974; Gerber, 1973). It is clear that bacterial filaments do not resemble the complex structure of protozoan and algal flagellae (Sleigh, 1974). It does, however, have an ultrastructure which has been detected by both negative and positive staining (Lowy and Hanson, 1965; Smith and Koffler, 1971; O'Brien, 1972). The filament of *Clostridium thermohydrosulphuricum* has been shown to have a hollow core (Sleytr and Glauert, 1973) but it is not known if this is true for all bacterial flagella. In some organisms, particularly *Vibrio* species, the filament may be encased in a sheath (Follett and Gordon, 1963; Glauert, 1963; Fuerst and Hayward, 1969). The function of this sheath which may make the overall diameter of the filament up to 35 nm has not been determined.

The bacterial flagellum is anchored to the basal granule by means of the proximal hook. It has been possible to isolate these hooks, (Abram, Mitchen, Koffler and Vatter, 1970; Dimmitt and Simon, 1971) and to show that they have specific antigenic differences from the flagella filaments, indicating they are composed of different proteins (Lawn, 1967; Dimmitt and Simon, 1970). More recently Kagawa, Asakura and Iino (1973) have shown that there are even antigenic differences in isolated hooks from different bacterial species.

The basal granule is considered to be the locus of flagella synthesis and to be associated with control and action of the flagellum (Figs 1 and 2). The ultrastructure has been most adequately depicted for both Gram-positive and Gram-negative bacteria by De Pamphilis and Adler (1970*a*, *b*).

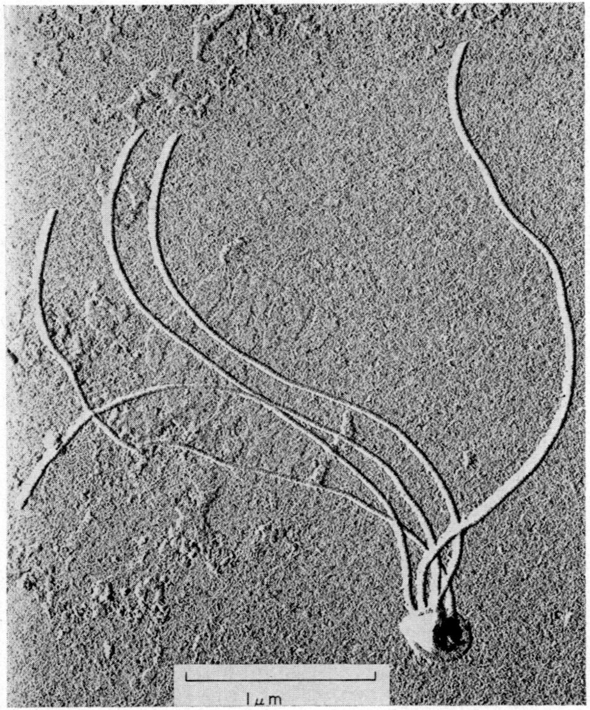

FIG. 1. Shadowed preparation showing detached basal granule and flagella of a *Pseudomonas* sp.

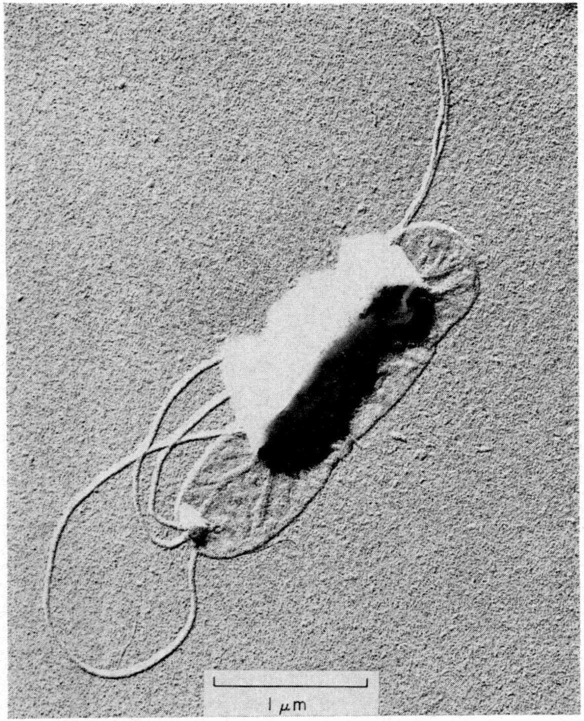

FIG. 2. Shadowed preparation of a *Pseudomonas* sp. showing basal granule and flagella.

Identification and taxonomy

Studies on flagellation in relation to identification and taxonomy were reviewed by Rhodes (1965), and Leifson (1966) published a critique of bacterial taxonomy which appealed for more careful studies on what he considers to be a neglected aspect of *taxonomy*. Both these authors quote lists of erroneous results which have appeared in the literature and Leifson deplores specific inaccuracies and inadequacies quoted in his paper. That flagellation *is* a neglected feature is not due solely to the difficulty of either flagella-staining techniques and the dubious results thereof (Hodgkiss 1960, 1964) or of obtaining access to an electron microscope. In general the bacteriologist can use a series of well-established tests which serve to identify isolates without recourse to examining this feature (Cowan, 1974). In Gram-positive organisms such as *Bacillus* spp. and *Clostridium* spp. and in the few motile strains which occur in the Lactobacillaceae and the coryneform group, the type of flagellation is not a diagnostic feature. Problems of identification arise in aquatic and food microbiology because the psychrophilic Gram-negative rods encountered therein have not been studied as intensively as the mesophilic enterobacteria. Few rapid diagnostic tests have been developed and the serology of these organisms is relatively unexplored.

The use of the electron microscope has not altered the original concepts relating to flagella and taxonomy. Because of the much greater resolution compared with the optical instrument it produces more accurate images and these images preserve some of the three-dimensional structure of the subject. However, more accurate images do not necessarily mean the solution of problems of identification. In the group of organisms recognized as *Vibrio parahaemolyticus* and *V. alginolyticus* (Fujino, *et al.*, 1974) which is currently of great interest to food hygiene microbiologists, some cultures consist of cells with polar flagella (Fig. 3), some consist of cells with peritrichous flagella and yet others show cells of both types in the same preparation (Fig. 4). The cells of the peritrichate strains also possess a polar, sheathed flagellum and closely resemble the morphology of some marine luminous bacteria assigned to the genus *Lucibacterium* by Hendrie, Hodgkiss and Shewan (1970). (In their paper Houwink and van Iterson (1950) describe this type of morphology in an organism isolated from a freshwater stream.)

The cells of the peritrichate *V. parahaemolyticus* strains and those of the luminous strains mentioned, when suspended in 2·5% (w/v) sodium chloride solution shed many flagella some of which coalesce to form large bundles (Fig. 5). Such bundles have been reported in a number of organisms and their formation in a strain of *V. alginolyticus* has been described

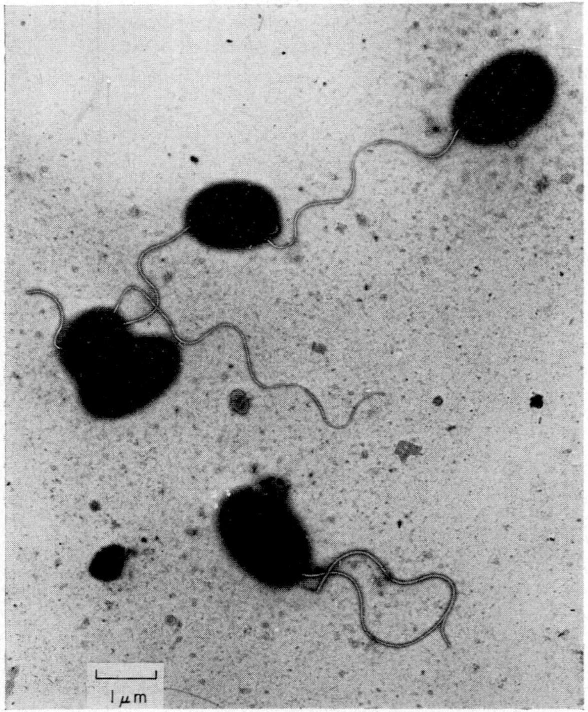

FIG. 3. Negatively-stained preparation of *V. parahaemolyticus* showing polar flagellation.

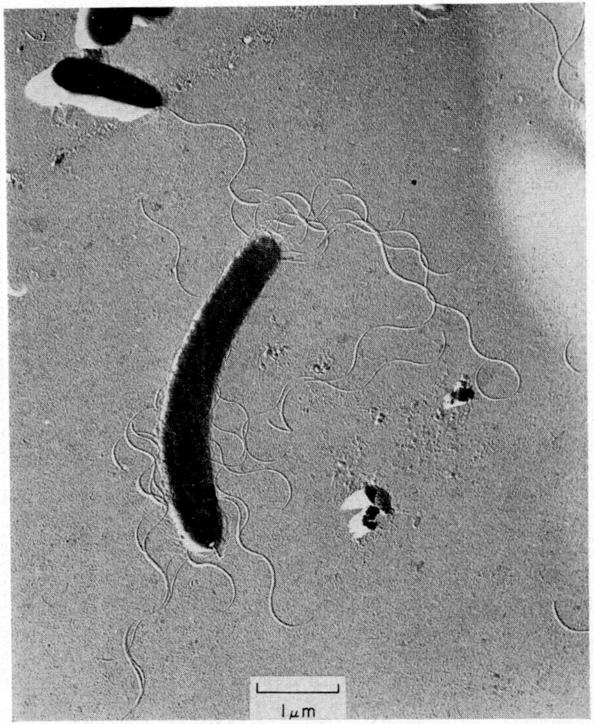

FIG. 4. Shadowed preparation of *V. parahaemolyticus* showing mixed flagellation; one cell polar, one cell peritrichous.

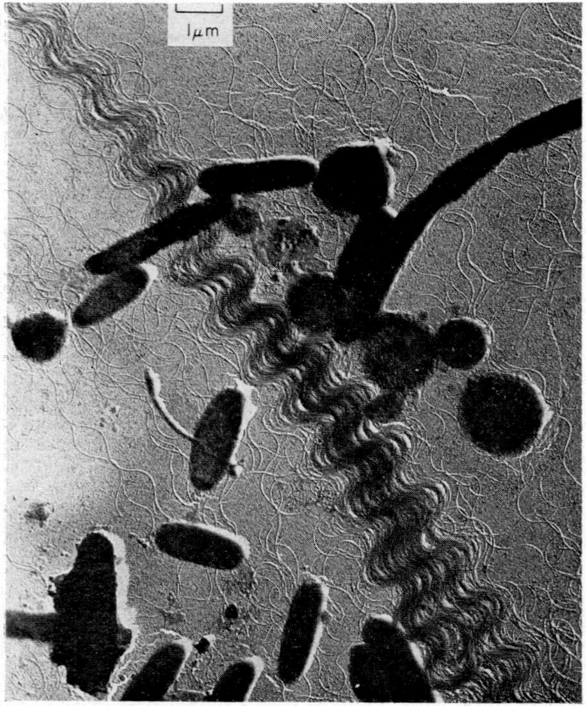

FIG. 5. Shadowed preparation of *V. parahaemolyticus* showing early stages of flagella-fasicle formation.

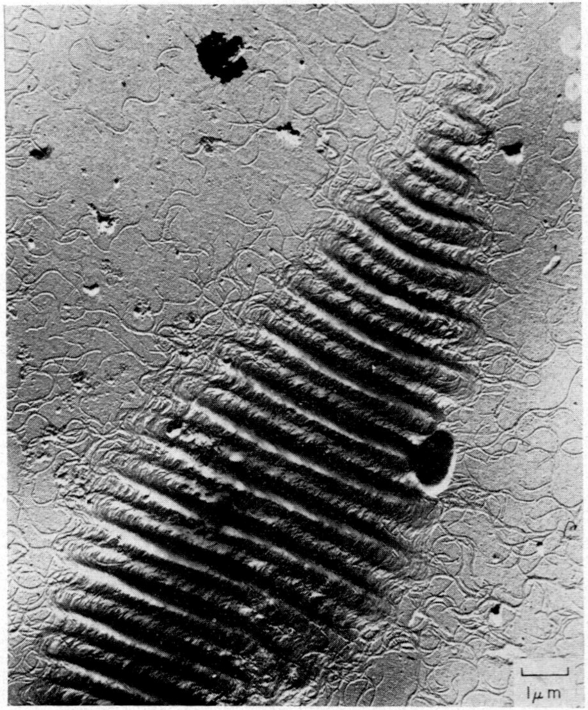

FIG. 6. Shadowed preparation of *V. alginolyticus* showing flagella fasicle.

by Ulitzur and Kessel (1973) (Fig. 6). This shedding of flagella is obviously important in the examination of cultures of these organisms and must be considered in some detail. However, we are sure that the single polarly flagellated cells described above have not arisen as a result of this process.

Mixed flagellation has been described in marine bacteria by Leifson and Hugh (1953) in *Chromobacterium* by Sneath (1956, 1960) and by Leifson (1963) who also claimed (Leifson, 1960) that it occurred in *Aeromonas* spp. although Hodgkiss and Shewan (1968) in electron microscope studies did not find this. A change from polar flagellation in liquid media to predominantly lateral or frankly peritrichous on agar is found in some strains of *P. putrefaciens* (Figs 7 and 8). The only distinction between *Alcaligenes* spp. and some of the "inactive" *Pseudomonas* spp. (Groups III and IV of Shewab, Hobbs and Hodgkiss, 1960) is in flagellation. There may well be a case for grouping these organisms together rather than dividing them upon a character which is rarely accurately investigated. To be deliberately controversial upon this point we would suggest that a long hard look be taken into the future application of this feature in taxonomy with the view that to the diagnostic bacteriologist it is not a practical proposition (*vide V. parahaemolyticus*). The day has passed when the optical microscope with a resolving power of 200 nm should be applied to the study of structures 20 nm or less in diameter and the results accepted as accurate.

At the SAB demonstration meeting in October 1974, as well as "normal" flagellation, some oddities were demonstrated. The illustrations, therefore, include micrographs of motile marine cocci (Fig. 9) of the genus *Planococcus* (Kocur, *et al.*, 1970), *Sarcina ureae* (Fig. 10)—an obvious choice as a model to refute Pijper's theory—and a *Selenomonas* spp. with its characteristic flagella (Fig. 11).

It is considered appropriate to point out that other forms of locomotion occur in bacteria and have not been dealt with here. These include the specialized locomotion of the spirochaetes (Fig. 12) (Jahn and Bovee, 1965), the gliding motility of the *Flexibacteriales* and the surface movement of some *Moraxella* and related species (Henrichsen, 1972).

Bacterial Fimbriae or Pili

Fimbriae or pili are filamentous surface appendages of bacteria morphologically distinct from flagella (Figs 13, 14 and 15). These appendages were first described but not named by Anderson (1949) and Houwink and van Iterson (1950). It was Duguid *et al.* (1955) who first used the term "fimbriae" to describe them, this being taken from the Latin meaning

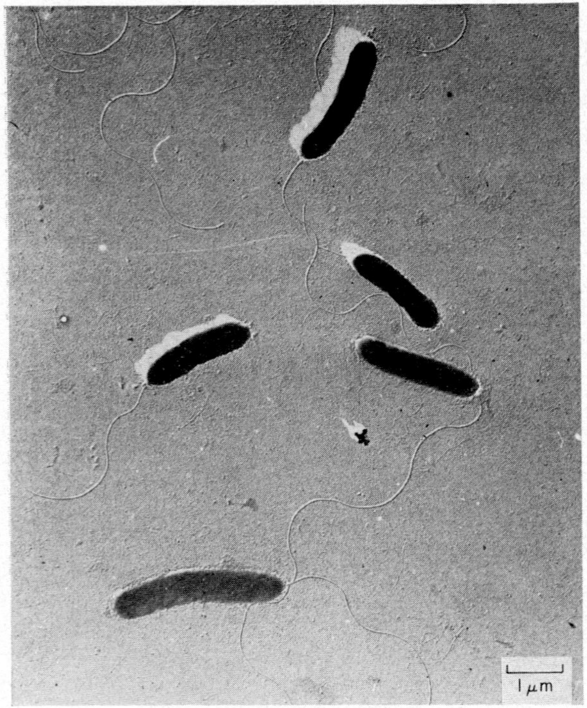

FIG. 7. Shadowed preparation of *Ps. putrefaciens* grown in tryptone water at 20° for 48 h.

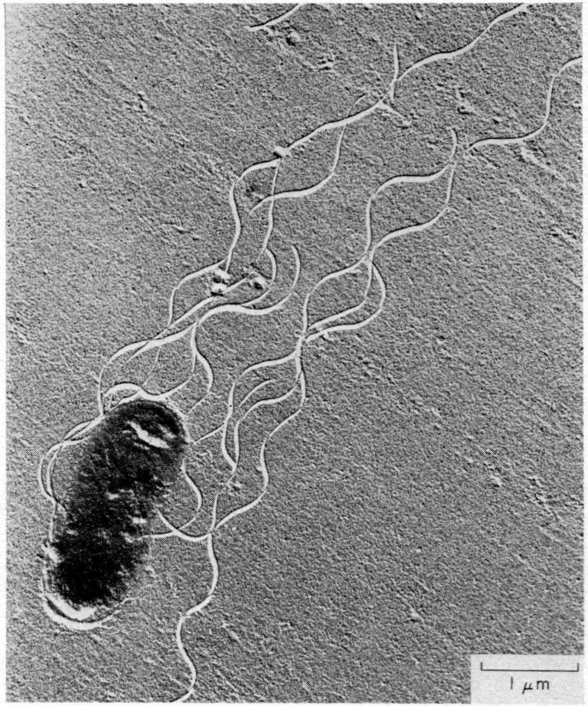

FIG. 8. Shadowed preparation of *Ps. putrefaciens* grown on nutrient agar at 20° for 24 h.

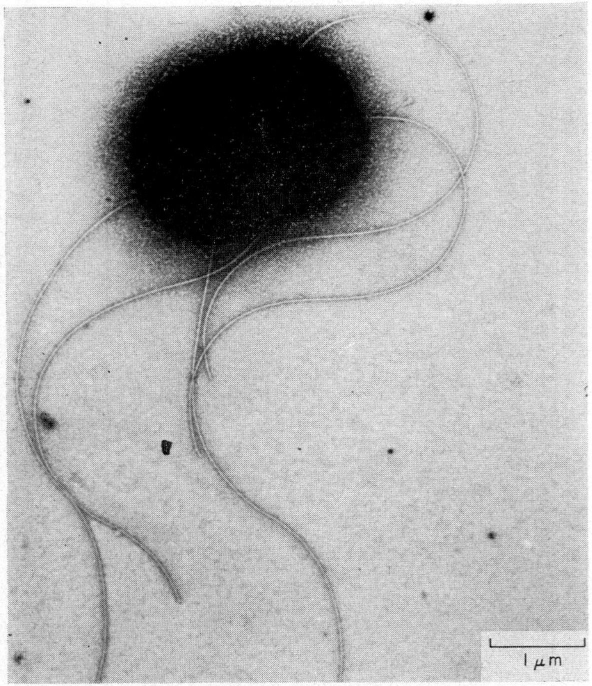

FIG. 9. Negatively-stained preparation of motile marine *Planococcus* species.

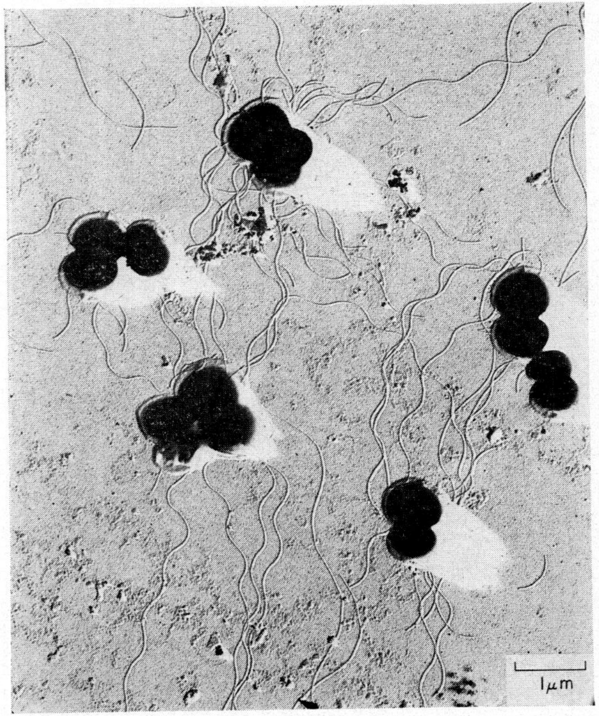

FIG. 10. Shadowed preparation of *Sarcina ureae*.

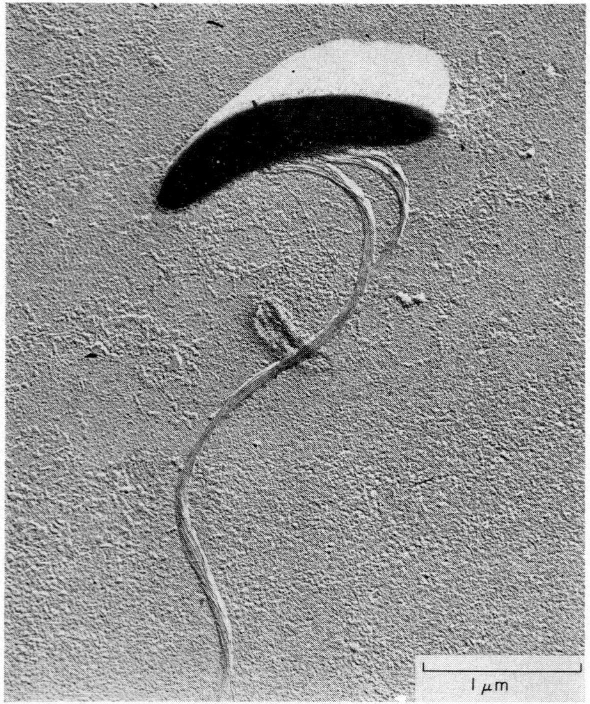

FIG. 11. Shadowed preparation of *Selenomonas* species from cow rumen.

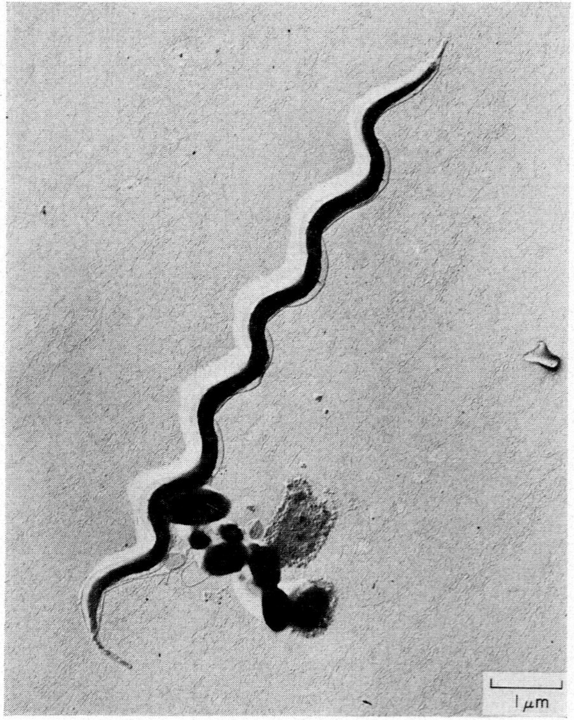

FIG. 12. Shadowed preparation of giant spirochaete from cow rumen.

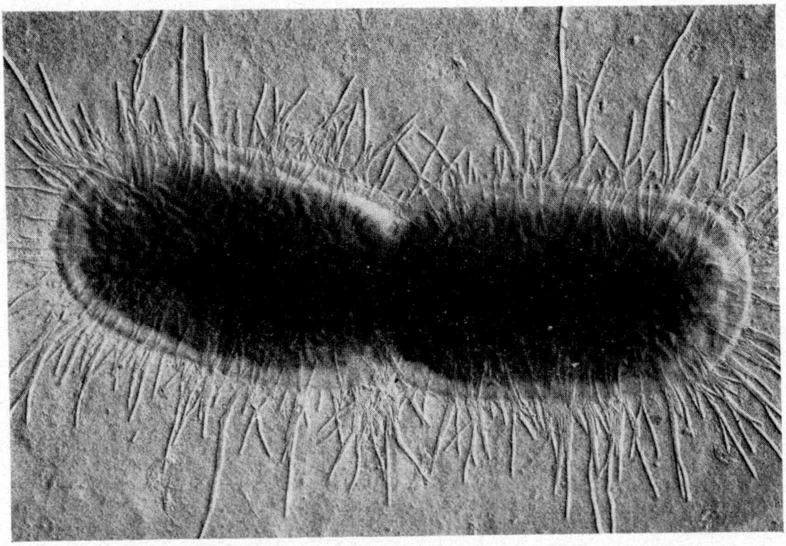

Fig. 13. Shadowed preparation of *Salmonella* pili (×35 200).

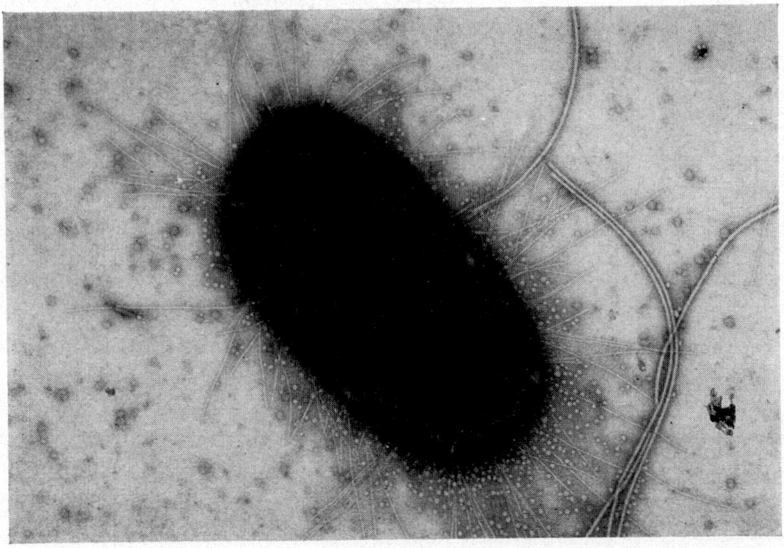

Fig. 14. Negatively-stained preparation of *Salmonella* pili (×27 600).

"threads, fibres or a fringe". Later Brinton (1959) in describing identical structures used the term "pili" taken from the Latin for "hair or fur". Despite discussion in the literature on the merits of the two terms the use of this dual terminology still exists in the literature today. Although in describing these appendages the term "fimbriae" obviously has a prior claim over that of "pili" the latter is probably in more general use. For the sake of conciseness and clarity it will, therefore, be used throughout this article.

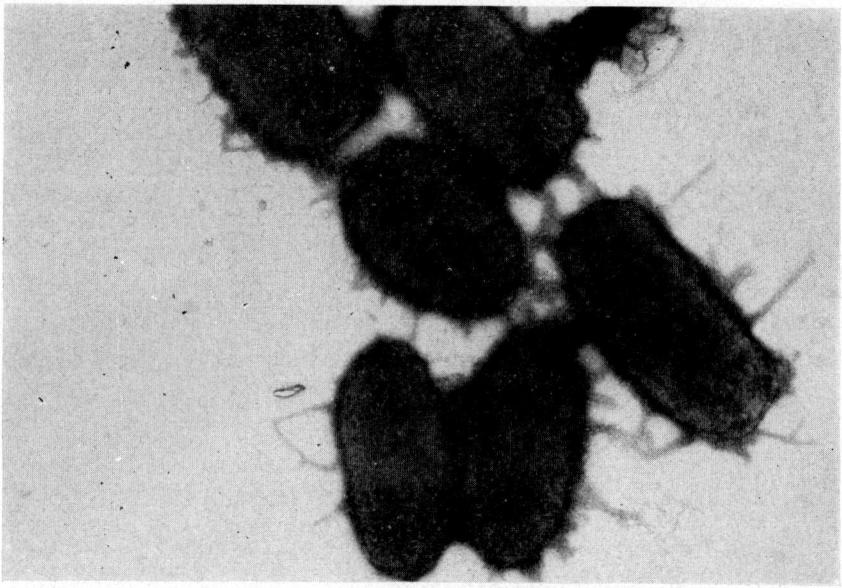

FIG. 15. Negatively-stained preparation of specifically agglutinated *Salmonella* pili (×22 440).

Bacterial pili due to their small dimensions can be visually detected only by electron microscopy. The first studies on the subject were carried out using metal-shadowing techniques (Duguid *et al.*, 1955) but with the introduction of negative staining, it was quickly realized that this technique could be used to advantage (Thornley and Horne, 1962). Since that time pili have been recognized in replicas (Swanson, 1973) and in thin-sectioning studies (Swanson, Kraus and Gotschlich, 1971; Novotny, Short and Walker, 1975).

Leaving aside sex pili, which are the subject of a separate article in this volume, Brinton (1965) listed five types of bacterial pili, these being found principally in the Enterobacteriaceae. Of these, he studied in detail those designated Type I (analagous to the Type I later described by Duguid, Anderson and Campbell, 1966). After establishing a technique for their purification he was able to demonstrate that they consisted of a protein (pilin) having a minimum molecular weight of 17 000. A study of their fine structure by electron microscopy, crystallography and X-ray diffraction demonstrated that they occurred in large numbers evenly distributed over the bacterial cell surface, were approximately 7·0 nm in diameter, 0·5–2 μm in length and had an empty core 2–2·5 nm. The pilin sub-units were arranged in a helical manner, the pitch of this helix being approximately 2·3 nm.

Duguid, Anderson and Campbell (1966) listed six types of bacterial pili which had previously been described by them and others, again principally in the Enterobacteriaceae. These types could be distinguished by their morphology, the presence on them of a mannose sensitive or resistant haemagglutinin or a combination of these factors. Duguid and his co-workers also raised the questions of the importance of pili with regard to taxonomy and had been instrumental in demonstrating that they conferred adhesive properties on some organisms. They also had shown that piliation could be varied by cultural conditions and discussed their finding in relation to infection and the life processes of the organisms. In an extension of their work Duguid and Campbell (1969) also investigated the antigenicity of the Type I pili of *Salmonella* concluding that crude or purified pilial antisera agglutinated the piliated-phase bacteria of a wide variety of heterologous serotypes.

The characterization of the physical, chemical and adhesive properties of the Type I pilus in the Enterobacteriacea is well documented. However, it is now recognized that pili or morphologically similar structures occur in a wide range of Gram-negative organisms especially in strains freshly isolated from natural sources—*Salmonella* spp. (Duguid *et al.* 1966), *Shigella* spp. (Duguid and Gillies, 1957), *Klebsiella aerogenes* (Duguid, 1959; Thornley and Horne, 1962), *Enterobacter* spp. (Constable, 1956; Duguid, 1959), *Serratia marcescens* (Duguid, 1959), *Proteus* spp. (Duguid and Gillies 1958; Coetzee, Pernet and Theron, 1962; Shedden, 1962; Hashimoto *et al.* 1963*a*, *b*, 1966; Hoeniger, 1965), *Ps. echinoides* (Heumann and Marx, 1964), *Pseudomonas* spp. (Fuerst and Hayward 1959), *Caulobacter* spp. (Schmidt 1966), *Agrobacterium* spp. (De Ley *et al.*, 1966), *Vibrio* spp. (Tweedy *et al.*, 1968), *Aeromonas liquefaciens, Photobacterium harveyi* (Hodgkiss and Shewan 1968).

In some strains of *Pseudomonas* the fimbriae are polar and Heumann and Marx (1964) showed that they are involved in the formation of typical cell aggregates—"Stars" (Fig. 16) and this was also found in an organism isolated from the Baltic Sea (Ahrens, Moll and Rheinheimer, 1968).

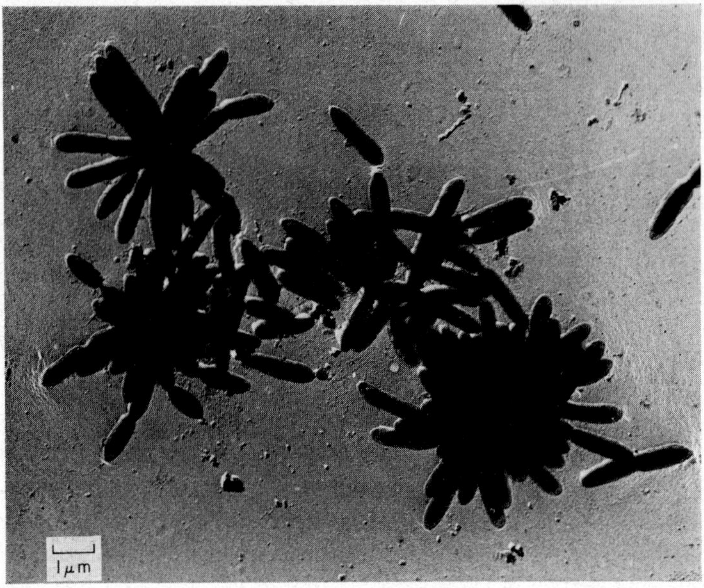

FIG. 16. Shadowed preparation showing star formation in *Pseudomonas* species.

A correlation between fimbriation and colony form in *Moraxella nonliquefaciens* was found by Bøvre *et al.* (1970)—a feature Brinton (1965) had noted in *Escherichia coli*. More recently two other examples of piliated organisms *Fusiformis nodosus* and *Neisseria gonorrhoeae* have been the subject of research at this and other laboratories. Whilst seeking to elucidate the physical, chemical and antigenic properties of their pili, all results obtained have been looked at in the light of the possible role pili may play in the pathogenicity of bacteria. (Analogies have been drawn between them and the pili types previously described, but it is difficult to say at this time if all such analogies are true.)

Fusiformis nodosus is an anaerobic organism which in conjunction with *F. necrophorus* causes the highly contagious disease known as ovine foot-rot. All primary isolates of this organism appear to be highly piliated (Short and Thorley, unpublished observations). When cultured further the cells remain in this highly piliated state during the logarithmic phase of growth. Negative staining shows the pili to arise as clusters from the poles of the cells, to be approximately 6 nm in diameter and up to 5 μm in length (Fig. 17). During the stationary and decline phases of growth the number of pili declines dramatically. Walker *et al.* (1973) showed that antibody in sera produced by caccination of animals with highly piliated cells was directed primarily against the pili (Fig. 18). This finding was important

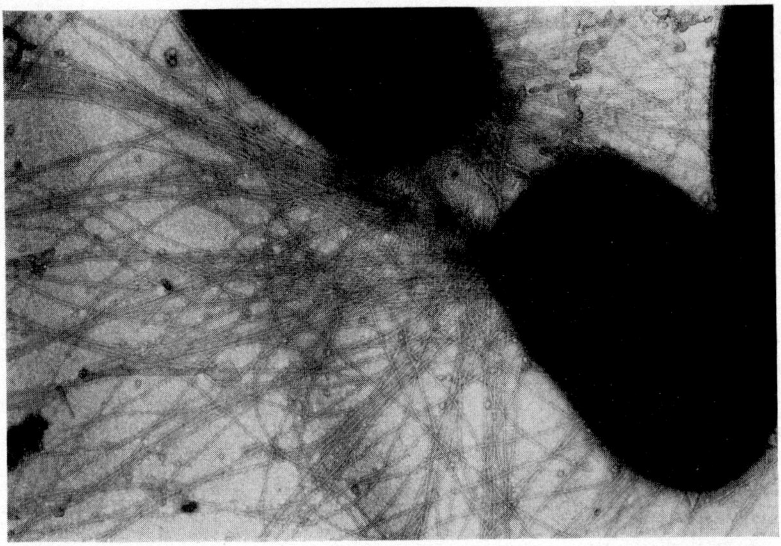

FIG. 17. Negatively-stained preparation of young cells of *F. nodosus* showing pili (×20 400).

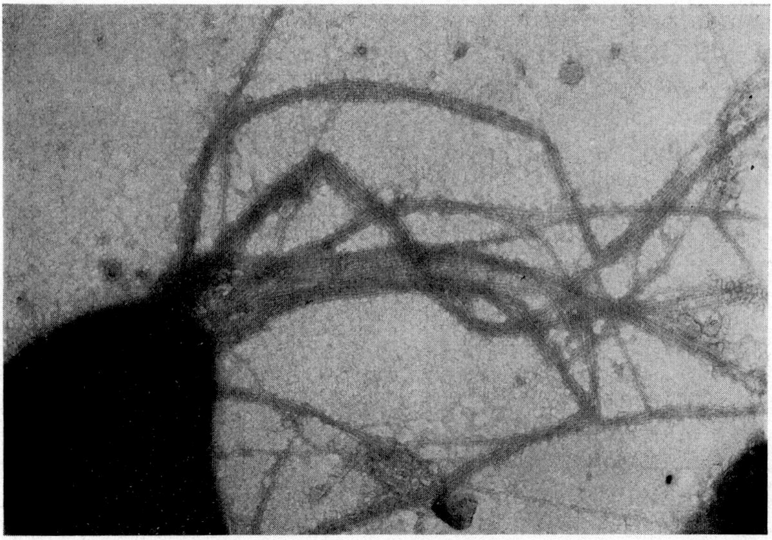

FIG. 18. Negatively-stained preparation of cells of *F. nodosus* agglutinated with antiserum from a vaccinated sheep. Note specific agglutination of the pili (×30 000).

because a definite correlation could then be established between agglutination titres found in vaccinated animals and protection against the disease. Research since that time has indicated, however, that perhaps other surface carried antigens may be important and that the pilial agglutinogen is type specific. The latter point is in direct contrast to that found in salmonellae by Duguid *et al.* (1969). Further work is, therefore, necessary before a full understanding of the role of pili in protection against ovine foot-rot is understood.

The discovery, by Jephcott *et al.* (1971) and independently by Swanson *et al.* (1971), that the virulent colony Types 1 and 2 of *N. gonorrhoeae* (Kellogg *et al.*, 1963) were piliated whereas the non-virulent Types 3 and 4 were not stimulated research into the possible role of pili in the virulence of this organism. Swanson *et al.* (1971) describe these pili as variable in number per cell, 0·5–2 μm in length and between 8·0–10·0 nm in diameter. It was noted that they have a marked tendency towards lateral aggregation at an acid pH and that sometimes pili aggregates show a faint periodicity amounting to 5·0 nm when examined by thin sectioning or negative staining. Jephcott *et al.* (1971) showed the pili to have a diameter of only 5–5·5 nm and in preliminary analysis of unpure aggregates obtained by lowering the pH concluded that lipid and protein were not major components.

Observations on gonococci have demonstrated three types of filamentous appendage (Novotny, Short and Walker, 1975). The first, designated pili *a*, are analagous to those described by Swanson *et al.* (1971) and Jephcott *et al.* (1971) and would be classified as Type 2 or because of their length Type 3 using Brinton's nomenclature (1965) (Fig. 19). The second, pili *b*, are fibres, usually bent and segmented, approximately only 70% the thickness of pili *a* but on some occasions they appear to be continuous with them (Fig. 20). It is not clear if they are structurally related. The third type, pili *c*, are morphologically similar to pili *a* with the exception that invariably they have a small knob at one end (Fig. 21). They never occur in aggregates, however, and can be antigenically distinguished when in mixtures with pili *a* (Novotny and Turner, 1975).

In order to explain the differences in the degree and type of piliation of different strains and even between colonies of the same strain, the following hypothesis was put forward. Pili *a* could be wrapped around the cells as proposed by Swanson (1973) and only seen on those occasions when the cell surface has been altered. This would occur during electron microscope preparative procedures. Alternatively, such piliation could be associated with organisms having disorganized cell surface layers. Pili *a* could, therefore, represent an integral cell wall moeity arising in an atypical form because incorporation into its normal functional position is lost.

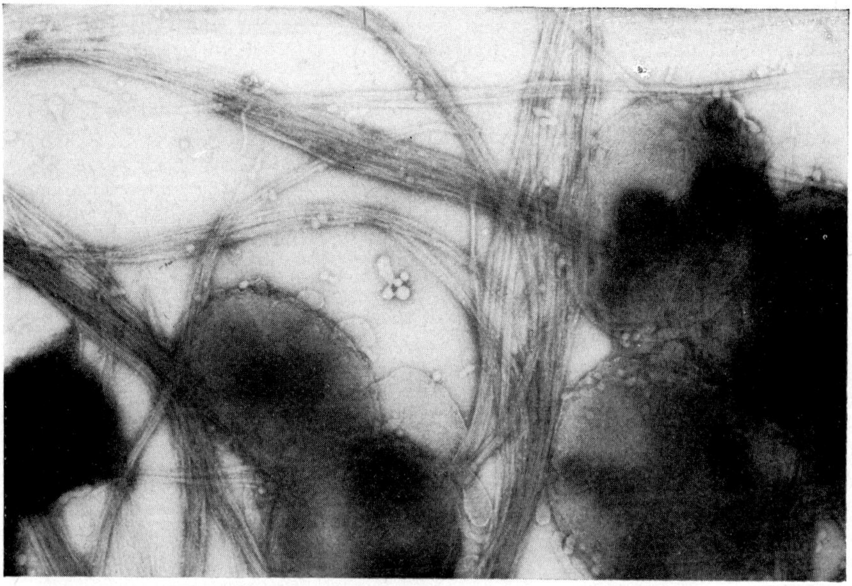

FIG. 19. Negatively-stained preparation of gonococcal pili type A (× 10 710).

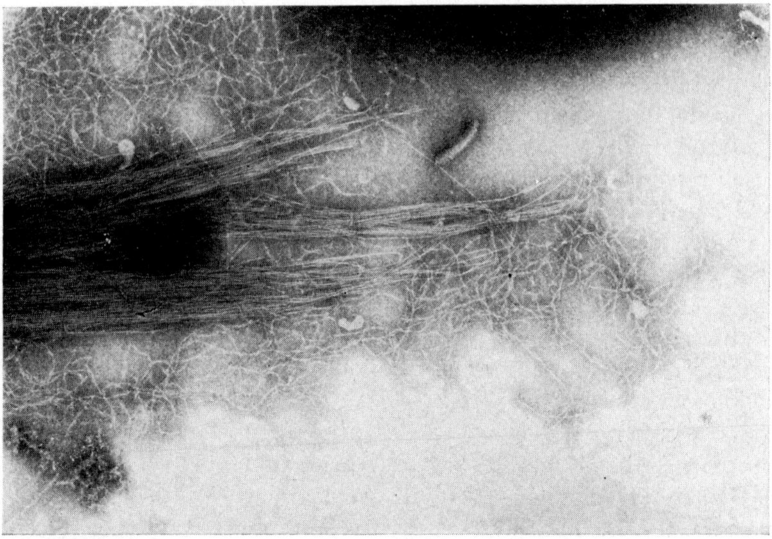

FIG. 20. Negatively-stained preparation of gonococcal pili type B (× 48 000).

This latter hypothesis is supported by evidence that Type 3 cells which are non-piliated give rise to antibodies which react with pili *a* of homologous Type 2 cells.

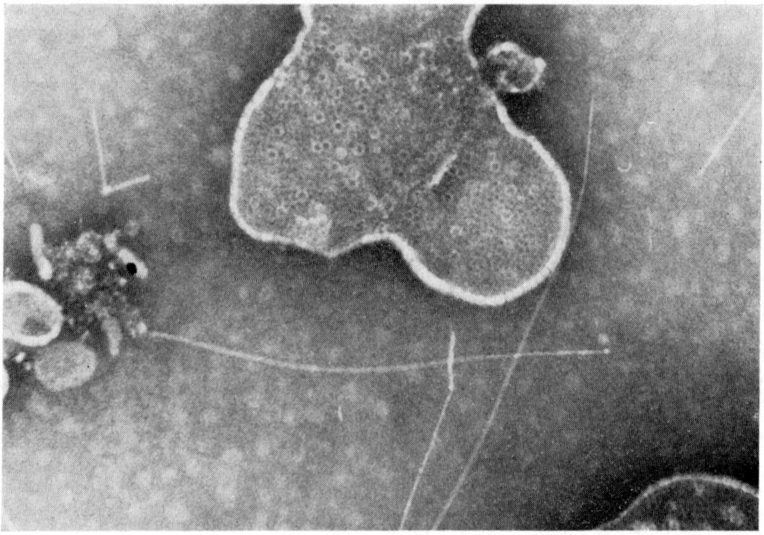

FIG. 21. Negatively-stained preparation of gonococcal pili type C (× 108 800).

Pili *c* morphologically resemble sex pili of *E. coli* (Lawn and Meynell, 1970). If they are sex pili then piliation could be determined by a regulatory mechanism. Pili *b* have not yet been well defined and it is difficult to attribute function or structure to them. Of interest is that similiar structures have been seen in the urine samples from gonorrhoeic males.

Swanson *et al.* (1971) postulated mechanisms whereby pili may alter host parasite relationships either by hindering phagocytosis, in which case they would be similar to other bacterial structures such as the M-protein of streptococci or by promoting adherence of the organism to epithelial cells. In this case they would be similar to the K88 antigen described in *E. coli* (Ørskov *et al.* 1961). Many reports have subsequently appeared in the literature aimed at proving one or both these points for *N. gonorrhoeae*. However, at the present time many of the properties, structure and function still await clarification.

More recently pili have been described on *Corynebacterium renale* a Gram-positive bacterium (Yanagawa, Otsuki and Tokui, 1968; Yanagawa and Otsuki, 1970; Kumazawa and Yanagawa, 1972, 1973; Hondo and Yanagawa, 1974). The diversity of the structures is further illustrated by the appearance of morphologically similar threads on some strains of staphylococci assocated with bovine mastitis (Fig. 22). These threads can be demonstrated during part of the logarithmic growth phase of staphylococci in fermenter cultures and later disappear. This adds further

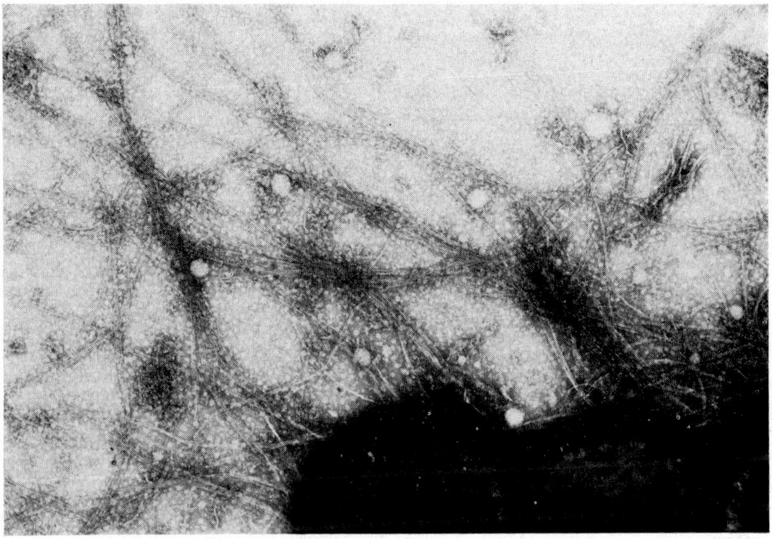

FIG. 22. Negatively-stained preparation of staphylococcus showing pili type structures (× 40 000).

support to the view that such structures could represent an integral cell wall moeity arising on atypical form because incorporation into its normal functional position is lost (Short, Thorley and Walker, unpublished).

References

ABRAM, D., MITCHEN, J., KOFFLER, H. & VATTER, A. (1970). Differentiation within the bacterial flagellum and isolation of the proximal hook. *J. Bact.*, **101**, 250.

ANDERSON, T. F. (1949). On the mechanism of adsorption of bacteriophages on host cells. In *The nature of the bacterial surface*. (A. A. Miles and N. W. Pirie, eds). Blackwell Scientific Publications, Oxford.

AHRENS, R., MOLL, G. & RHEINHEIMER, G. (1968). Die Rolle der Fimbrien bei der eigenartigen Sternbildung von *Agrobacterium luteim*. *Arch. Mikrobiol.*, **63**, 321.

ASAKURA, S. (1970). Polymerisation of flagellin and polymorphism of flagella. *Adv. Biophys.*, **1**, 99.

BØVRE, K., BERGAN, T. & FROHOLM, L. O. (1970). Electron microscopical and serological characteristics associated with colony type in *Moraxella non-liquefaciens*. *Acta path. microbiol. scand.*, **78**, 765.

BRINTON, C. C. (1959). Non-flagellar appendages of bacteria. *Nature, Lond.*, **183**, 782.

BRINTON, C. C. (1965). The structure, function, synthesis and genetic control of bacterial pili and a model for DNA and RNA transport in Gram negative bacteria. *Trans. N.Y. Acad. Sci.*, **27,** 1003.

COETZEE, J. N., PERNET, G. & THERON, J. J. (1962). Fimbriae and haemagglutinating properties in strains of *Proteus. Nature, Lond.*, **196**, 497.

CONSTABLE, D. (1956). Fimbriae and haemagglutinating activity in strains of *Bacterium cloacae. J. Path. Bact.*, **72**, 133.

COWAN, S. T. (1974). *Cowan and Steel's Manual for the Identification of Medical Bacteria.* 2nd ed. Cambridge University Press, Cambridge.

DE PAMPHILIS, M. & ADLER, J. (1970*a*). Fine structure and isolation of the hook-basal body complex of flagella from *Escherichia coli* and *Bacillus subtilis. J. Bact.*, **105,** 384.

DE PAMPHILIS, M. & ADLER, J. (1970*b*). Attachment of flagellar basal bodies to the cell envelope: specific attachment to the outer lipopolysaccharide membrane and the cytoplasmic membrane. *J. Bact.*, **105,** 396.

DIMMITT, K. & SIMON, M. (1971). Purification and partial characterisation of *Bacillus subtilis* flagellar hooks. *J. Bact.*, **108,** 282.

DE LEY, J., BERNAERTS, M., RASSEL, A. & GUILMOT, J. (1966). Approach to an improved taxomony of the genus *Agrobacterium. J. gen. Microbiol.*, **43,** 7.

DUGUID, J. P. (1955). Non-flagellar filamentous appendages (fimbriae) in *Bacterium coli. J. Path. Bact.*, **70,** 335.

DUGUID, J. P. (1959). Fimbriae and adhesive properties in *Klebsiella* strains. *J. gen. Microbiol.*, **21,** 271.

DUGUID, J. P., ANDERSON, E. S. & CAMPBELL, I. (1966). Fimbriae and adhe sive properties in salmonellae. *J. Path. Bact.*, **92,** 107.

DUGUID, J. P. & CAMPBELL, I. (1969). Antigens of the type 1 fimbriae of salmonellae and other enterobacteria. *J. med. Microbiol.*, **2,** 535.

DUGUID, J. P. & GILLIES, R. R. (1957). Fimbriae and adhesive properties in dysentery bacilli. *J. Path. Bact.*, **74,** 397.

DUGUID, J. P. & GILLIES, R. R. (1958). Fimbriae and haemagglutinating activity in *Salmonella, Klebsiella, Proteus* and *Chromobacterium. J. Path. Bact.*, **75,** 519.

DUGUID, J. P., SMITH, I. W., DEMPSTER, G. & EDMUNDS, P. N. (1955). Non-flagellar filamentous appendages (fimbriae) and haemagglutinating activity of *Bacterium coli. J. Path. Bact.*, **70,** 335.

FOLLETT, E. A. C. & GORDON, J. (1963). An electron microscope study of Vibrio flagella. *J. gen. Microbiol.*, **32**, 235.

FUERST, J. A. & HAYWARD, A. C. (1969). Surface appendages similar to fimbriae (pili) on *Pseudomonas* species. *J. gen. Microbiol.*, **58,** 227.

FUJINO, T., SAKAGUCHI, G., SAKAZAKI, R. & TAKEDA, Y. (1974). *International symposium on* Vibrio parahaemolyticus. Tokyo, Saikon Publishing Co.

GERBER, B. (1973). Effect of temperature on the *in vitro* assembly of bacterial flagella. *J. mol. Biol.*, **74,** 467.

GLAUERT, A. M. (1963). Fine structure and mode of attachment of the sheathed flagellum of *Vibrio metschnikovii. J. Cell Biol.*, **18,** 237.

HASHIMOTO, M., TSCUCHMOTO, S., KATO, A., TAKEMOTO, T., YOSHINO, T., IKUTA, M. & NAKAJIMA, M. (1963). Studies on structure and function of bacterial fimbriae. I. Demonstration of fimbriae in Gram negative enteric bacteria. *Bull. Tokyo Med. Dental Univ.*, **10,** 181.

HASHIMOTO, M., TANIGUCHI, H., TAKEMOTO, T., YOSHINO, T., NAKAJIMA, M., HAGIWARA, I., TUBA, M. & NISHIDO, T. (1963). Studies on structure and function of bacterial fimbriae. II. Adhesive properties of the fimbriae. *Bull. Tokyo Med. Dental Univ.*, **10,** 493.

HASHIMOTO, M., NAKAJIMA, M., YOSHINO, T., YAOI, S. & TAKADA, R. (1966). Studies on structure and function of bacterial fimbriae. III. Fimbriation and acid agglutination. *Tokyo Med. Dent. Univ. Bull.*, **12,** 294.

HENDRIE, M. S., HODGKISS, W. & SHEWAN, J. M. (1970). The identification, taxonomy and classification of luminous bacteria. *J. gen. Microbiol.*, **64,** 151.

HENRICHSEN, J. (1972). Bacterial surface translocation: a survey and a classification. *Bact. Rev.*, **36,** 478.

HEUMANN, W. & MARX, R. (1964). Feinstruktur und Funktion der Fimbrien bei dem sternbildenden Bakterium, *Pseudomonas echinoides. Arch. Mikrobiol.*, **47,** 325.

HODGKISS, W. (1960). The interpretation of flagella stains. *J. appl. Bact.*, **23,** 398.

HODGKISS, W. (1964). The flagella of *Pseudomonas solanacearum. J. appl. Bact.*, **27,** 278.

HODGKISS, W. & SHEWAN, J. M. (1968). Problems and modern principles in the taxonomy of marine bacteria. In *Advances in Microbiology of the sea.* (M. R. Droop and E. J. Ferguson Wood, eds). Academic Press, London and New York.

HOENIGER, J. F. M. (1965). Development of flagella by *Proteus mirabilis. J. gen. Microbiol.*, **40,** 29.

HONDO, E. & YANAGAWA, R. (1974). Agglutination of trypsinised sheep erythrocytes by the pili of *Corynebacterium renale. Infect. Immun.*, **10,** 1426.

HOUWINK, A. C. & VAN ITERSON, W. (1950). Electron microscopical observations on bacterial cytology. II. A study of flagellation. *Biochim. biophys. Acta.*, **5,** 10.

JAHN, T. L. & BOVEE, E. C. (1965). Movement and locomotion of microorganisms. *A. Rev. Microbiol.*, **19,** 21.

JEPHCOTT, A. E., REYN, A. & BIRCH-ANDERSON, A. (1971). *Neisseria gonorrhoeae.* III. Demonstration of presumed appendages to cells from different colony types. *Acta path. microbiol. scand.* (B), **79,** 437.

KAGAWA, H., ASAKURA, S. & IINO, T. (1973). Serological study of bacterial flagella hooks. *J. Bact.*, **113,** 1474.

KELLOGG, D. S., PEACOCK, W. L., DEACON, W. E., BROWN, L. & PIRKLE, C. I. (1963). *Neisseria gonorrhoeae.* I. Virulence genetically linked to clonal variation. *J. Bact.*, **85,** 1274.

KINGMA BOLTJES, T. Y. (1948). Function and arrangement of bacterial flagella. *J. Path. Bact.*, **60,** 275.

KOCUR, M., PÁČOVÁ, Z., HODGKISS, W. & MARTINEC, T. (1970). The taxonomic status of the genus *Planococcus* Migula 1894. *Int. J. Syst. Bact.*, **20,** 241.

KUMAZAWA, N. & YANAGAWA, R. (1972). Chemical properties of the pili of *Corynebacterium renale. Infec. Immunity*, **5,** 27.

KUMAZAWA, N. & YANAGAWA, R. (1973). Comparison of the chemical and immunological properties of the pili of three types of *Corynebacterium renale. Jap. J. Microbiol.*, **17,** 13.

LAUTROP, H. & JESSEN, O. (1964). On the distinction between polar monotrichous and lophotrichous flagellation in green fluorescent pseudomonads. *Acta path. microbiol. scand.*, **60,** 580.

LAWN, A. M. & MEYNELL, E. (1970). Serotypes of sex pili. *J. Hygiene*, **68,** 683.

LEIFSON, E. (1956). Morphological and physiological characteristics of the genus *Chromobacterium*. *J. Bact.*, **71**, 393.

LEIFSON, E. (1960). *Atlas of Bacterial Flagellation*. Academic Press, New York and London.

LEIFSON, E. (1963). Mixed polar and peritrichous flagellation of marine bacteria. *J. Bact.*, **86,** 73.

LEIFSON, E. (1966). Bacterial taxonomy: a critique. *Bact. Rev.*, **30,** 257.

LEIFSON, E. & HUGH, R. (1953). Variation in shape and arrangement of bacterial flagella. *J. Bact.*, **65,** 263.

LOWY, J. & HANSON, J. (1965). Electron microscope studies of bacterial flagella. *J. molec. Biol.*, **11**, 293.

NOVOTNY, P., SHORT, J. A. & WALKER, P. D. (1975). An electron microscope study of naturally occurring and cultured cells of *Neisseria gonorrhoeae*, *J. med. Microbiol.*, **89**, 87, 413.

NOVOTNY, P. & TURNER, W. H. (1975). Immunological heterogeneity of pili of *Neisseria gonorrhoeae*. *J. gen. Microbiol*, **89,** 87.

O'BRIEN, E. J. (1972). Structure of straight flagella from a mutant *Salmonella*. *J. mol. Biol.*, **70,** 133.

PIJPER, A. (1946). Shape and motility of bacteria. *J. Path. Bact.*, **58,** 325.

RHODES, M. E. (1965). Flagellation as a criterion for the classification of bacteria. *Bact. Rev.*, **29,** 442.

RINKER, J. N. & KOFFLER, M. (1951). Preliminary evidence that bacterial flagella are not 'polysaccharide twirls'. *J. Bact.*, **61,** 421.

SCHMIDT, J. M. (1966). Observations on the adsorption of caulobacter bacteriophages containing ribonucleic acid. *J. gen. Microbiol.*, **45,** 347.

SHEDDEN, W. I. M. (1962). Fimbriae and haemagglutinating activity in strains of *Proteus hauseri*. *J. gen. Microbiol.*, **28,** 1.

SHEWAN, J. M., HOBBS, G. & HODGKISS, W. (1960). A determinative scheme for the identification of certain genera of Gram negative bacteria, with special reference to the *Pseudomonadaceae*. *J. appl. Bact.*, **23,** 379.

SLEIGH, M. A. (ed.), (1974). *Cilia and flagella*. Academic Press, London and New York.

SLEYTR, U. B. & GLAUERT, A. M. (1973). Evidence for an empty core in a bacterial flagellum. *Nature, Lond.*, **241**, 542.

SMITH, R. W. & KOFFLER, M. (1971). Bacterial flagella. In *Adv. Microb. Physiol.* Volume 6. (A. H. Rose and J. F. Wilkinson, eds). Academic Press, London and New York, p. 219.

SNEATH, P. H. A. (1956). The change from polar to peritrichous flagellation in *Chromobacterium* spp. *J. gen. Microbiol.*, **15,** 99.

SNEATH, P. H. A. (1960). A study of the bacterial genus *Chromobacterium*. *Iowa State J. Sci.*, **34,** 243.

SWANSON, J., KRAUS, S. J. & GOTSCHLICH, E. C. (1971). Studies on gonococcus infection. I. Pili and zones of adhesion: their relation to gonococcal growth patterns. *J. exp. Med.*, **134,** 886.

THORNLEY, M. J. & HORNE, R. W. (1962). Electron microscope observations on the structure of fimbriae, with particular reference to *Klebsiella* strains, by the use of the negative staining technique. *J. gen. Microbiol.*, **28,** 51.

TWEEDY, J. M., PARK, R. W. A. & HODGKISS, W. (1968). Evidence for the presence of fimbriae (pili) on *Vibrio* species. *J. gen. Microbiol.*, **51,** 235.

ULITZUR, S. & KESSEL, M. (1973). Giant flagellar bundles of *Vibrio alginolyticus* (NCMB 1803). *Arch. Mikrobiol.*, **94,** 331.

WAKABAYASHI, K. (1974). Structural studies of P. filaments produced from *Salmonella* species. *J. molec. Biol.*, **83,** 545.

WALKER, P. D., SHORT, J., THOMSON, R. O. & ROBERTS, D. S. (1973). The fine structure of *Fusiformis nodosus* with special reference to the location of antigens associated with immunogenicity. *J. gen. Microbiol.*, **77,** 351.

YANAGAWA, R. & OTSUKI, K. (1970). Some properties of the pili of *Corynebacterium renale*. *J. Bact.*, **101,** 1063.

YANAGAWA, R., OTSUKI, K. & TOKUI, T. (1968). Electron microscopy of fine structure of *Corynebacterium renale* with special reference to pili. *Jap. J. Vet. Res.*, **16,** 31.

Sex Pili of Enterobacteria

A. M. Lawn

Lister Institute of Preventive Medicine, Chelsea Bridge Road, London, England

In 1964 Crawford and Gesteland demonstrated that a special hair-like surface structure was the receptor for the phage R17 which infects only donor (F) bacteria of the *Escherichia coli* conjugation system. An extensive series of investigations has since shown that these protein appendages, now called sex pili or sex fimbriae, are essential for F-mediated conjugation and that their synthesis is directed by a DNA segment called an infectious plasmid or sex factor. This is capable of autonomous replication and confers donor properties on its host. Similar sex factors are responsible for transmissible drug resistance (R factors) and for other transmissible characters (e.g. colicine factors), and in many instances sex pili are an essential part of the transfer mechanisms of these sex factors. (For recent reviews see Brinton, 1971; Meynell 1972).

Sex pili are usually examined in negative-contrast preparations of whole organisms. Uranyl acetate gives good contrast and the known sex pili, unlike flagella, are not dissociated by its acidity. As this contrast agent is intolerant of small molecular weight contamination, the bacteria must be well washed with water before its application. A method using small discs of membrane filter has been described (Lawn and Meynell, 1970) which is useful as a means of concentrating and washing bacteria and for transferring them to specimen support grids for examination in the electron microscope (Fig. 1). Ionic bombardment of specimen grids aids uniform spreading of the contrast agent.

In negative-contrast preparations the sex pili of the F sex factor appear as filaments of uniform diameter (about 9 nm) varying in length from less than 1 to more than 20 μm. They are slightly wider and more flexible than the common pili or fimbriae present on many enterobacteria (Brinton, Gemski and Carnahan, 1964; Brinton, 1971; Lawn 1966). Labelling them with an isometric, donor-specific phage makes them more conspicuous. A number of sex factors determine sex pili which are morphologically indistinguishable from F pili in conventional preparations (Figs 2 and 3);

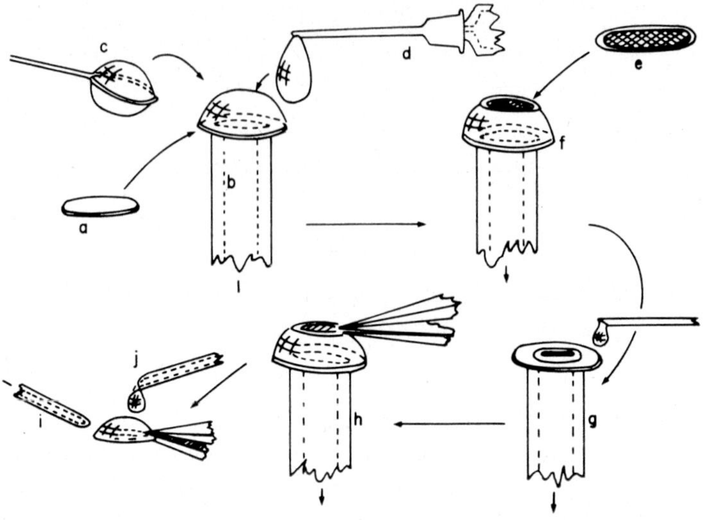

FIG. 1. General method of preparation of bacteria for examination of their appendages in the electron microscope. A circle cut from membrane filter (a) is positioned over the end of a glass tube (b) to which constant gentle suction is applied. A drop of the bacterial culture (c), or a preincubated mixture of culture and specific antiserum, is applied to the membrane. When the fluid has passed through the filter the bacteria are washed with several drops of distilled water from a disposable syringe and needle (d). The bacteria are transferred to a specimen support grid (e, formvar and carbon coated and subjected to ionic bombardment) by sucking the grid onto the filter surface and refloating it (f, g, h). The grid is then picked up, surplus water removed by suction (i), a drop of uranyl acetate solution is added (j) and the surplus stain removed as before.

these form the F-like class of sex pili (Lawn *et al.*, 1967). Another group of sex factors determine sex pili which are narrower, shorter and more fragile than F-like pili (Fig. 8, *a* and *b*) and which are the receptor for the filamentous phages If1 and If2 which do not infect bacteria carrying only F-like sex factors (G. G. Meynell and Lawn, 1968). This I-like group of sex pili (Lawn *et al.*, 1967) are almost indistinguishable from common pili without specific labelling. As an isometric phage for I-like pili has not yet been isolated, antibody must be used to confirm their identity in the presence of common pili. The size relationships between F-like, I-like and common pili are demonstrated in Fig. 4.

Recently an isometric phage specific for donor bacteria carrying the R factor R 1822 has been shown to adsorb to a short plasmid-specific pilus of 9 nm width (Bradley, 1974; Olsen and Thomas, 1973) which is presumably a sex pilus. R 1822 belongs to the P incompatability group and is unrelated

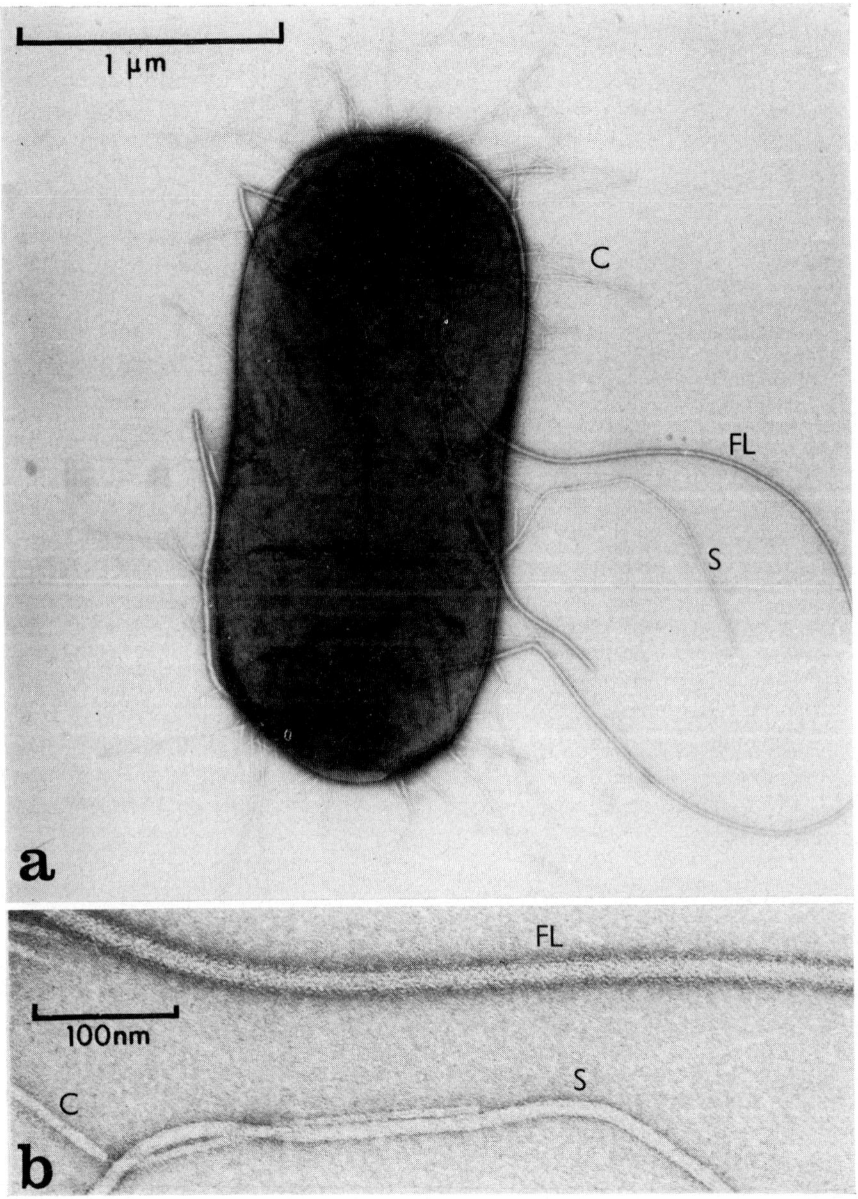

FIG. 2. *E. coli* infected with the sex factor R538Fdrd1 (a) showing flagella (FL), common pili (C) and F-like sex pili (S). A portion of this micrograph at higher magnification (b) demonstrates that these three types of appendage are readily distinguishable from each other.

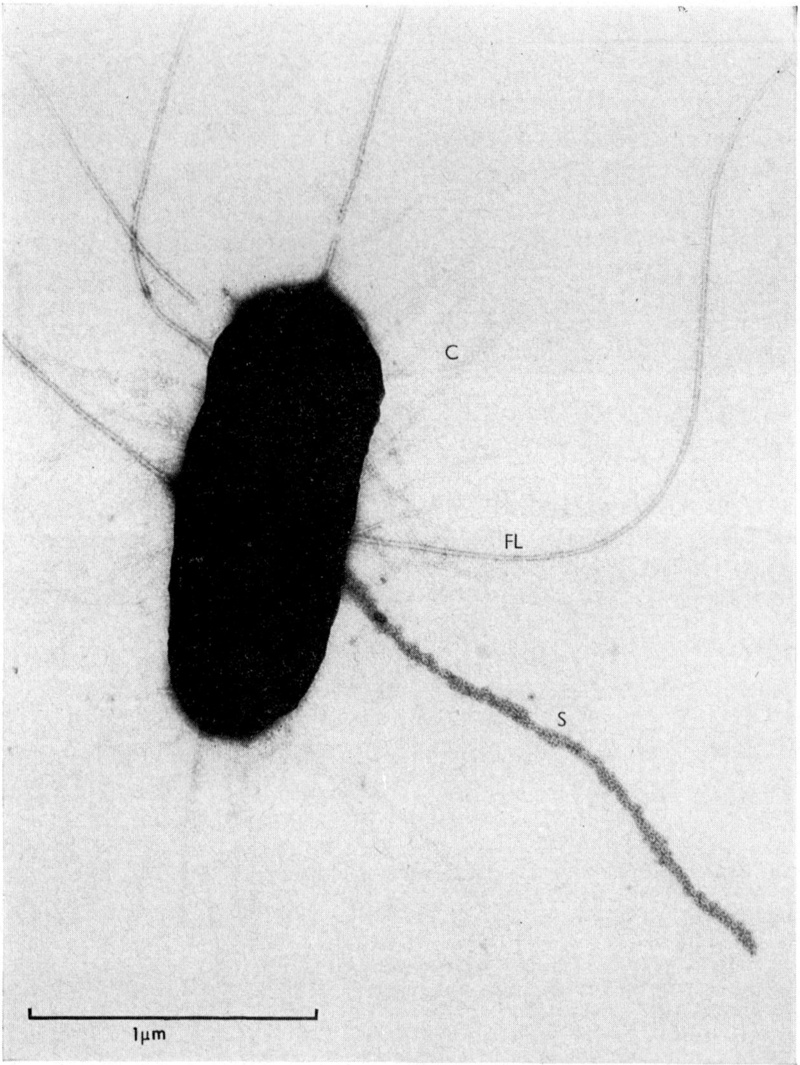

FIG. 3. *E. coli* infected with the sex factor R237 showing flagella (FL), common pili (C) and an F-like sex pilus (S). The sex pilus is conspicuous because it is labelled by adsorption of the bacteriophage MS2.

to F-like and I-like sex factors (Datta, *et al.*, 1971). Other phages specific for bacteria carrying R 1822 adsorb randomly to the cell wall as does the filamentous phage IKe (Brodt, Leggett and Iyer, 1974) specific for bacteria carrying sex factors of the N compatibility group (Datta and Hedges, 1971), for which no sex pilus has yet been found.

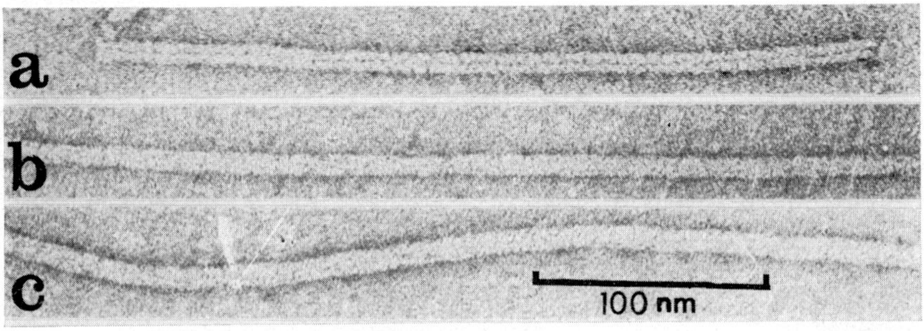

FIG. 4. This figure demonstrates the size differences between a common pilus of *E. coli* (a), an I-like sex pilus (b) of the sex factor R538Idrd2 and an F sex pilus (c).

In many instances sex factors are naturally repressed and only 0·1% or less of the bacteria in a culture carry a sex pilus. In this case direct search for sex pili may be impracticable but they can easily be discovered by using an enrichment technique (author's unpublished experiments). Confluent growth on a nutrient agar plate is suspended in a few drops of 0·9% sodium chloride (less than 0·5 ml) and centrifuged in a bench centrifuge to sediment the bacteria. A drop of the supernatant is washed on a membrane filter of 0·05 μm pore diameter, transferred to a support grid as for bacteria and stained with uranyl acetate. If flagella are numerous they may be removed by acidifying the drop of supernatant before filtration and if a suitable phage or antiserum is available the pili may be specifically labelled (Fig. 5). This technique has confirmed the serological identity between the sex pili of wild type repressed R factors and those of their derepressed mutants, and has also confirmed the serological specificity of the pili of some repressed sex factors for which no such mutants are available. The specificity of these sex pili could previously be deduced only indirectly by the inhibition of transmission of the sex factor in the presence of specific sera (Harden and Meynell, 1972). When the above enrichment technique was applied to strains carrying R27 and R28, members of the N compatibility group, (Datta and Hedges, 1971) no sex pili were discovered (unpublished observations).

Ørskov and Ørskov (1960) first demonstrated that donor (F) *E. coli*

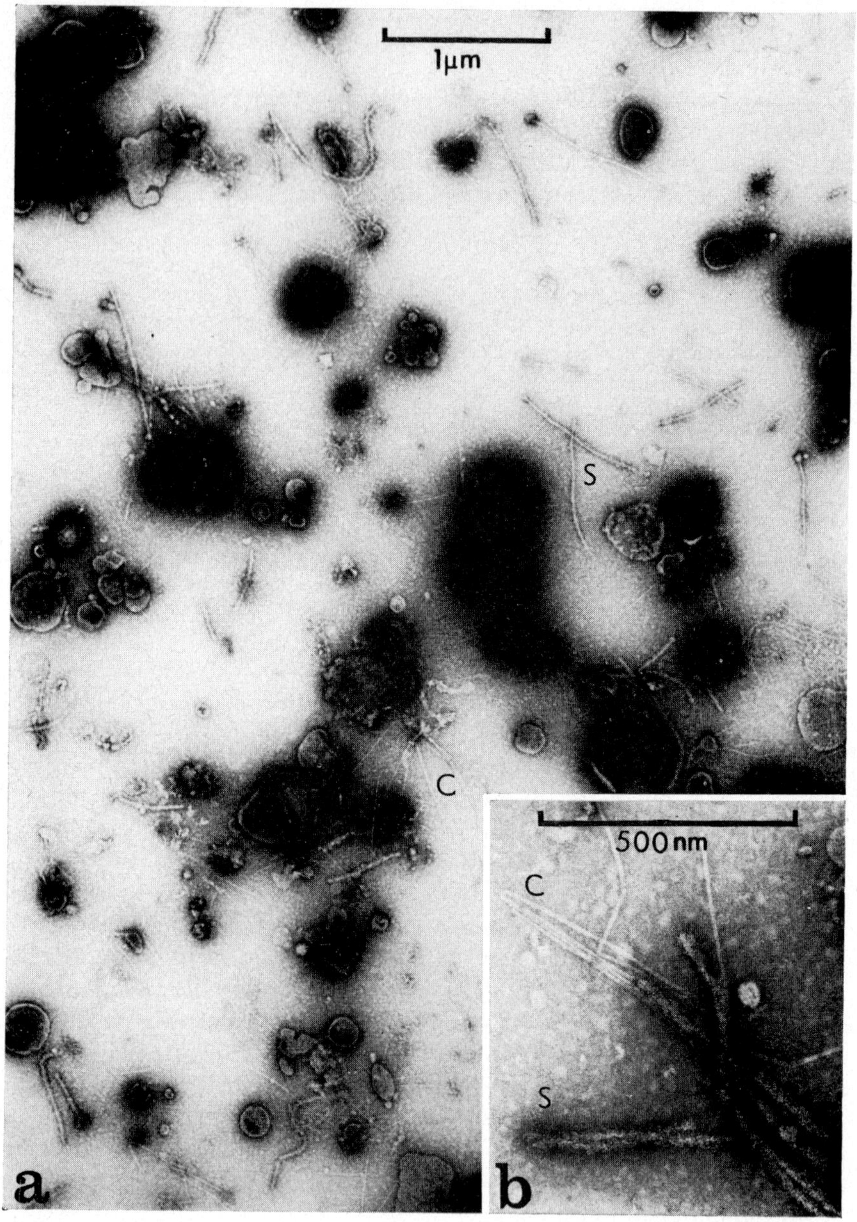

FIG. 5. A sample prepared from *E. coli* carrying the repressed sex factor R502, using a special technique described in the text. The micrograph (a) shows abundant sex pili (S), which have been labelled with specific antibody. Common pili are unlabelled. In the micrograph at higher magnification (b), from another part of the same sample, the two types of pili can be clearly distinguished.

strains possessed a specific surface antigen which was later shown to be the F pilus (Ishibashi, 1967). Specific antibody is a valuable tool for the investigation of sex pili. It can be used for labelling sex pili to aid in their recognition, particularly I-like pili for which there is no suitable isometric phage (Fig. 6) and also for distinguishing between pili of different sex factors. Conventional serological typing of sex pili by bacterial agglutination is possible (Kétyi and Ørskov, 1969) but its validity may be doubtful because of the multiplicity of surface antigens of enterobacteria. With the electron microscope, specific binding of antibody molecules to morphologically identifiable sex pili can be directly observed and quantified, so avoiding confusion with other antigenic specificities. If minimal binding is to be identified unambiguously, efficient removal of unbound antibody is essential; the filter disc method (Fig. 1) gives good results. By this method a number of serotypes can be distinguished within both the F- and I-like classes of sex pili (Lawn and Meynell, 1970). Absorbed sera may be necessary as cross reactions are common between members of the same class although I-like and F-like pili do not cross-react with each other. With specific sera prepared by suitable absorption it has been demonstrated that when two F-like plasmids, neither of which is repressed, are present in one bacterium all the sex pili possess both types of antigenic determinant; they appear to be co-polymers of the two pilus proteins (Fig. 6, Lawn, Meynell and Cooke, 1971). If one of a pair is an F-like factor and the other I-like, then the two types of pili are produced independently (Fig. 7).

Although close proximity between donor and recipient is essential for genetic transfer by bacterial conjugation, actual surface contact may not be necessary (Ou and Anderson, 1970). The exact function of pili in conjugation and the conditions which influence their synthesis and assembly are as yet uncertain. Either or both extrusion and retraction of sex pili in response to some external stimulus may play a part; it has been reported that chemical stimuli (O'Callahan, *et al.*, 1973; Novotny and Fives-Taylor, 1974) or the attachment of filamentous phage (Jacobsen, 1972) cause retraction of sex pili. Abrupt extrusion of I-like pili may occur in the presence of specific antibody (Fig. 8, Lawn and E. Meynell, 1972) or during rapid washing on a membrane filter (Fig. 9, Lawn and Meynell, 1975). The labile nature of sex pili must therefore be considered when interpreting quantitative studies of them.

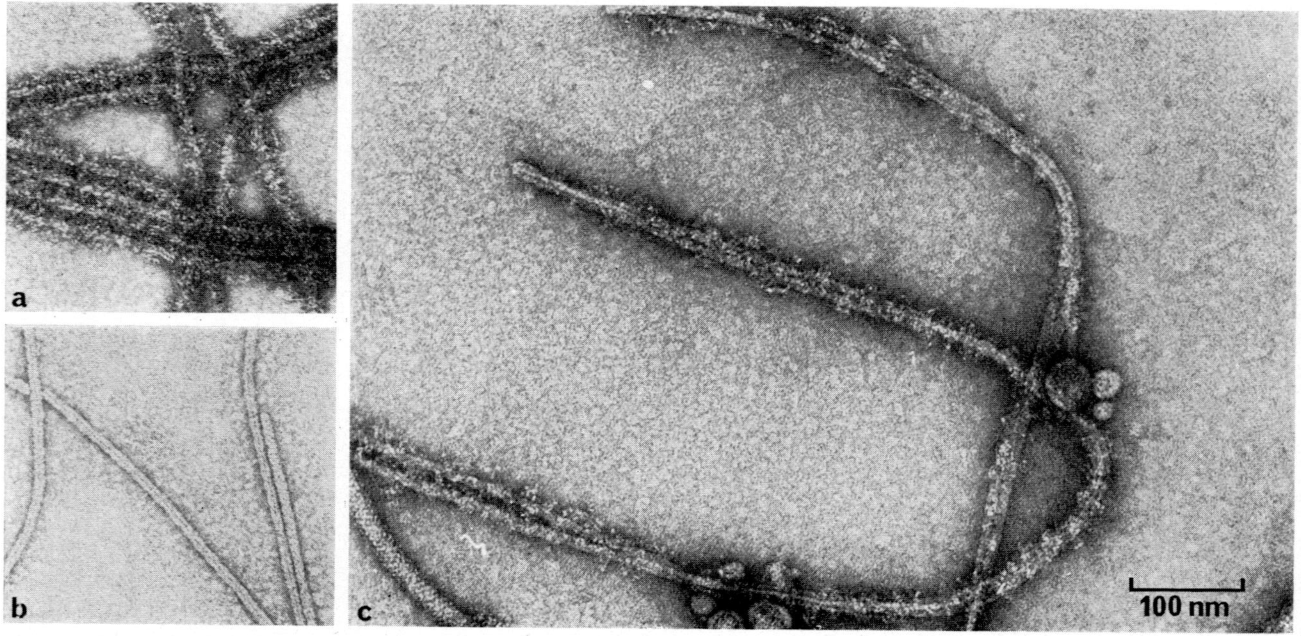
a
b
c
100 nm

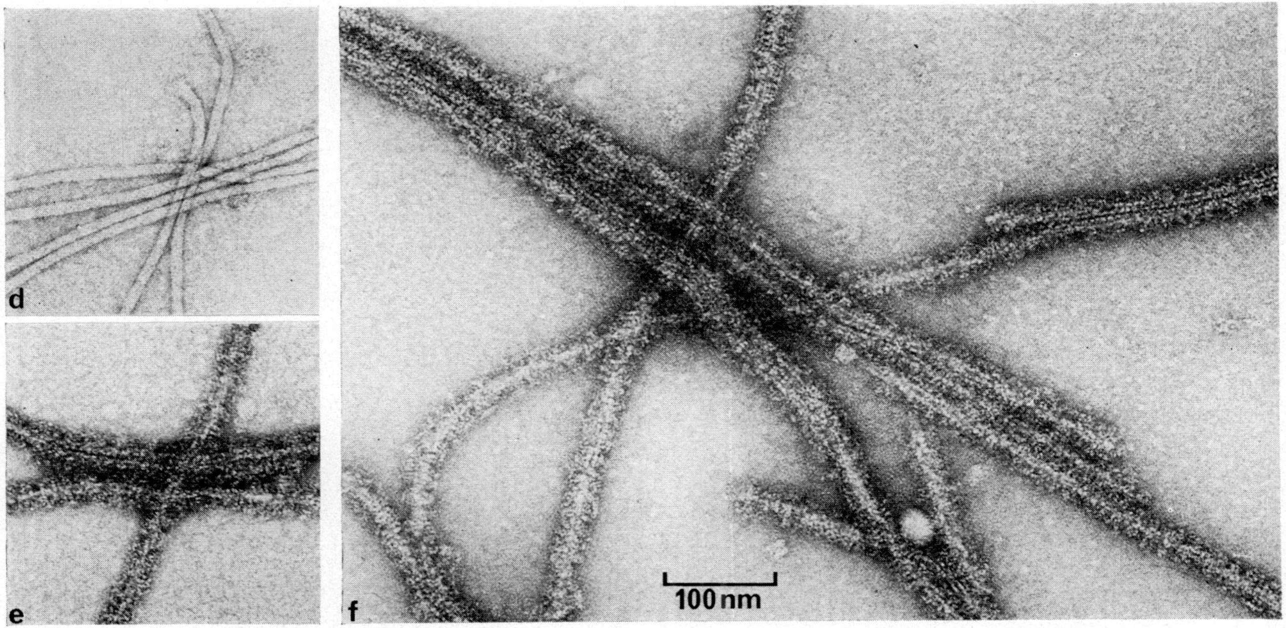

FIG. 6. Each sex pilus of *E. coli* which is infected with both the F sex factor and another F-like sex factor (R1drd19) appears to be a mixture of two types of pilus protein. Specific anti-F serum, prepared by absorption, reacts with F pili (a) and the mixed pili (c) but not with R1 pili (b). Specific anti-R1 antiserum reacts with R1 pili (e) and the mixed pili (f) but not with F pili (d).

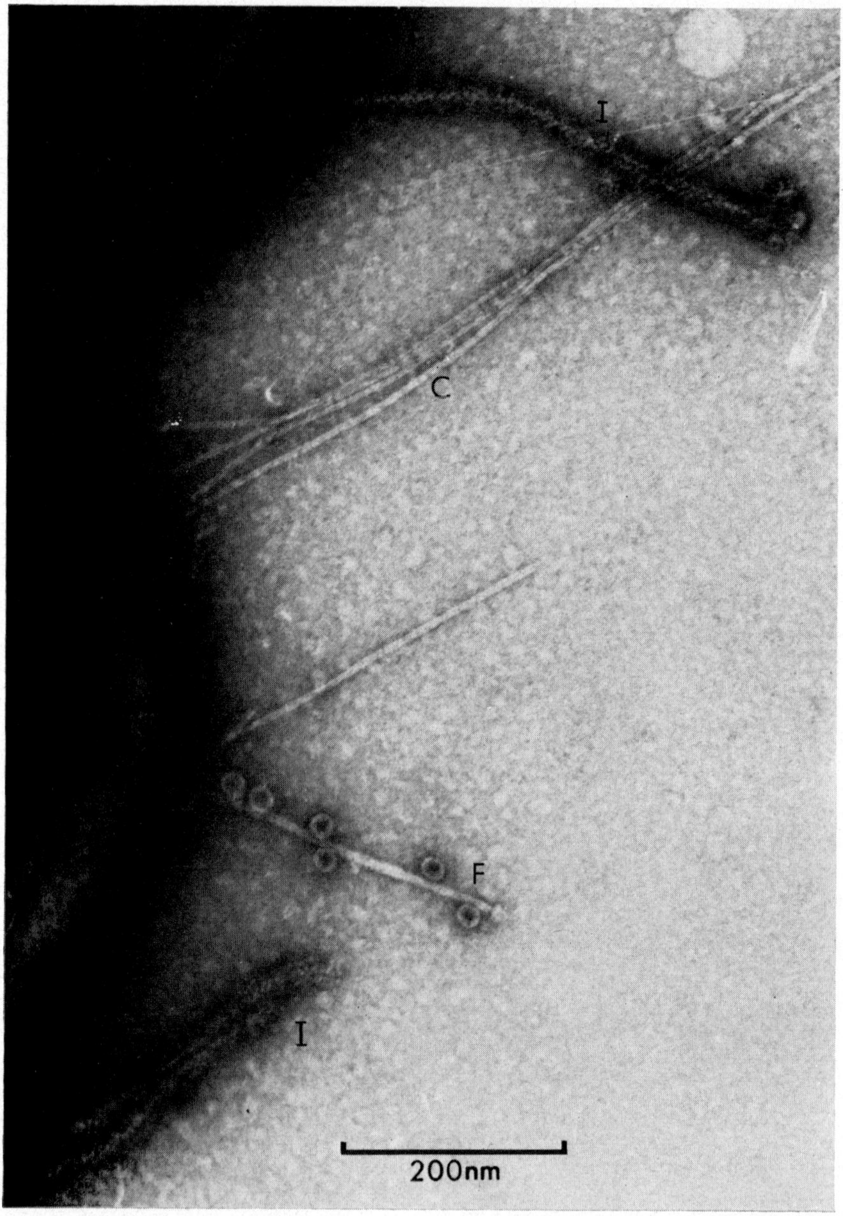

FIG. 7. This Hfr bacterium carries the I-like sex factor R538Idrd2 as well as F. It has two distinct types of sex pili, I-like pili (I), which have been labelled with specific antiserum, and F pili (F) which have been labelled with the bacteriophage MS2. Common pili (c) are unlabelled.

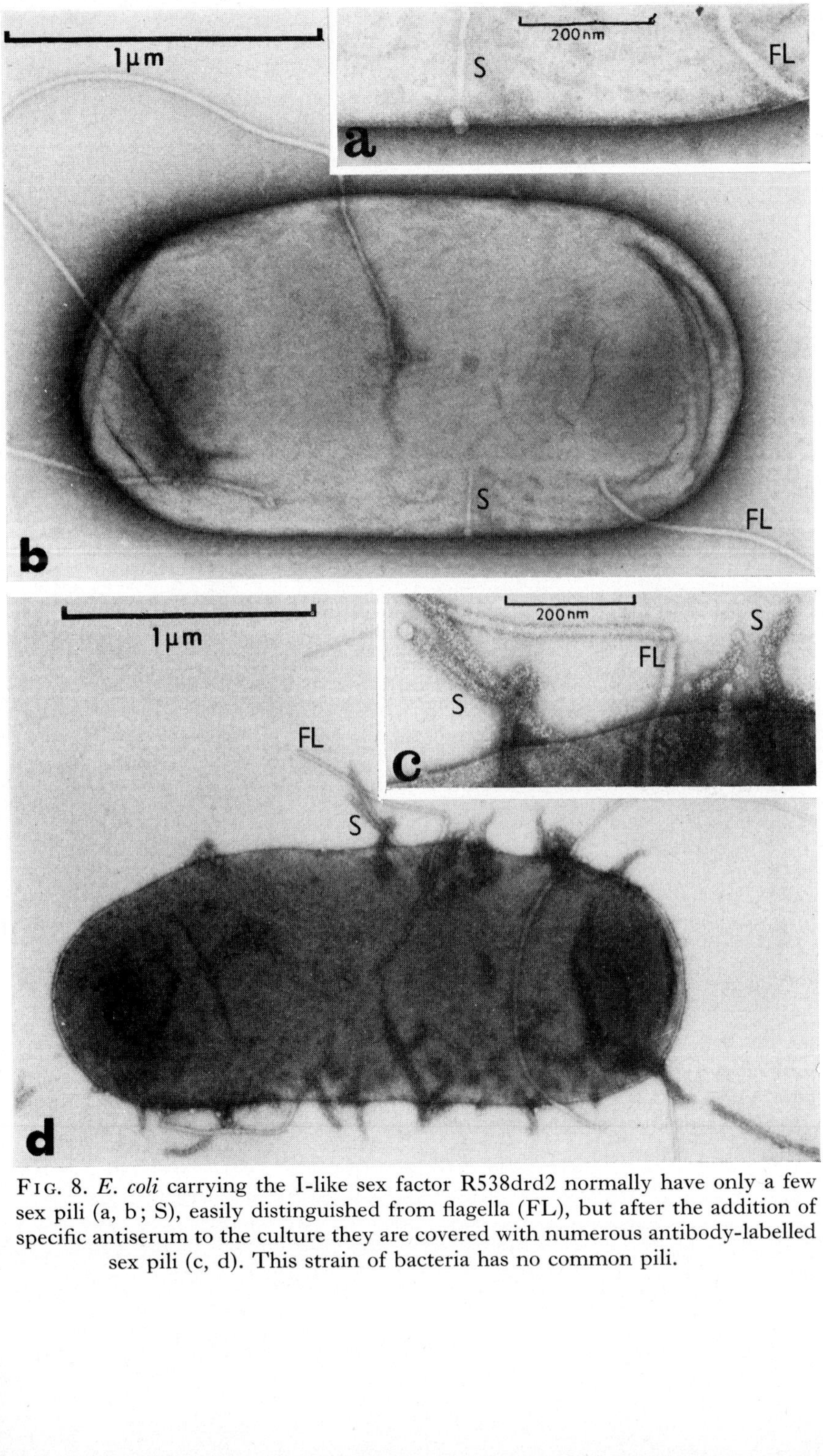

FIG. 8. *E. coli* carrying the I-like sex factor R538drd2 normally have only a few sex pili (a, b; S), easily distinguished from flagella (FL), but after the addition of specific antiserum to the culture they are covered with numerous antibody-labelled sex pili (c, d). This strain of bacteria has no common pili.

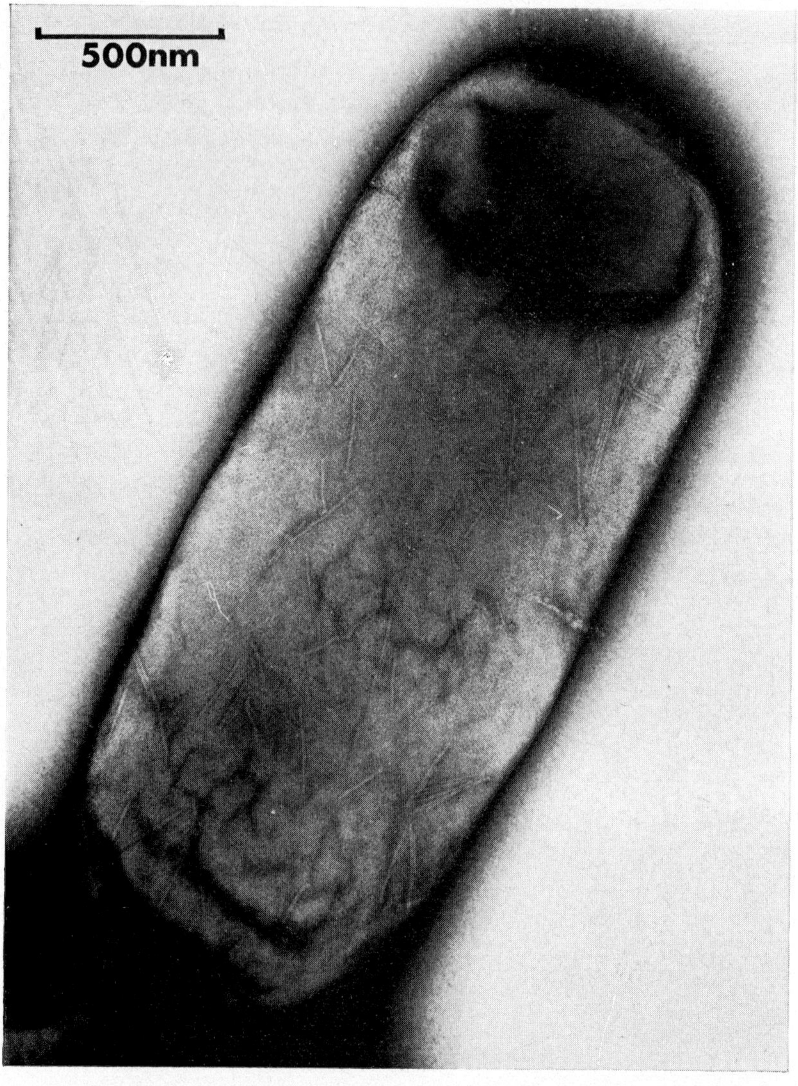

FIG. 9. This bacterium infected with the sex factor R538Idrd2 is from a sample washed on a membrane filter by rapid flow of warm broth before it was prepared for electron microscopy. It is covered with over 50 short sex pili in contrast to the 1 or 2 normally found on bacteria of this strain (see Fig. 8a).

References

BRADLEY, D. E. (1974). Adsorption of bacteriophages specific for *Pseudomonas aerogenosa* R factors RP1 and R1822. *Biochem. Biophys. Res. Commun.*, **57,** 893.

BRODT, P., LEGGETT, F. & IYER, R. (1974). Absence of a pilus receptor for filamentous phage IKe. *Nature, Lond.*, **249,** 856.

BRINTON, C. C. (1971). The properties of sex pili, the viral nature of "conjugal" genetic transfer systems, and some possible approaches to the control of bacterial drug resistance. *CRC crit. Rev. Microbiol.*, **1,** 105.

BRINTON, C. C., GEMSKI, P. & CARNAHAN, J. (1964). A new type of bacterial pilus genetically controlled by the fertility factor of E. coli K12 and its role in chromosome transfer. *Proc. natn. Acad. Sci. USA.*, **52,** 776.

CRAWFORD, E. M. & GESTELAND, R. F. (1964). The adsorption of bacteriophage R-17. *Virology*, **22,** 165.

DATTA, N. & HEDGES, R. W. (1971). Compatibility groups among fi^- R factors. *Nature, Lond.*, **234,** 222.

DATTA, N., HEDGES, R. W., SHAW, E. J., SYKES, R. B. & RICHMOND, M. H. (1971). Properties of an R factor from *Pseudomonas aerogenosa*. *J. Bact.*, **108,** 1244.

HARDEN, V. & MEYNELL, E. (1972). Inhibition of gene transfer by antiserum and identification of serotypes of sex pili. *J. Bact.*, **109,** 1067.

ISHIBASHI, M. (1967). F plius as f^+ antigen. *J. Bact.*, **93,** 379.

JACOBSEN, A. (1972). Role of F pili in the penetration of bacteriophage f1. *J. Virol.*, **10,** 835.

KÉTYI, I. & ØRSKOV, I. (1969). Studies on the antigenic structure of sex fimbriae carried by a strain of *Shigella flexneri* 4b. *Acta. path. microbiol. scand.*, **77,** 229.

LAWN, A. M. (1966). Morphological features of the pili associated with $R^+ F^-$ and $R^- F^+$ bacteria. *J. gen. Microbiol.*, **45,** 377.

LAWN, A. M. & MEYNELL, E. (1970). Serotypes of sex pili. *J. Hyg., Camb.*, **68,** 683.

LAWN, A. M. & MEYNELL, E. (1972). Antibody-stimulated increase in sex pili in R^+ enterobacteria. *Nature, Lond.*, **255,** 441.

LAWN, A. M. & MEYNELL, E. (1975). Extrusion of sex pili by rapidly washed R^+ *Escherichia coli*. *J. gen. Microbiol.*, **86,** 188.

LAWN, A. M., MEYNELL, E. & COOKE, M. (1971). Mixed infections with bacterial sex factors: sex pili of pure and mixed phenotype. *Annls Inst. Pasteur, Paris,* **120,** 3.

LAWN, A. M., MEYNELL, E., MEYNELL, G. G. & DATTA, N. (1967). Sex pili and the classification of sex factors in the Enterobacteriaceae. *Nature, Lond.*, **216,** 343.

MEYNELL, G. G. (1972). *Bacterial plasmids*. London: Macmillan.

MEYNELL, G. G. & LAWN, A. M. (1968). Filamentous phages specific for the I sex factor. *Nature, Lond.*, **217,** 1184.

NOVOTNY, C. P. & FIVES-TAYLOR, P. (1974). Retraction of F pili. *J. Bact.*, **117,** 1306.

O'CALLAGHAN, R. J., BUNDY, L., BRADLEY, R. & PARANCHYCH, W. (1973). Unusual arsenate poisoning of the F pili of *Escherichia coli*. *J. Bact.*, **115,** 76.

OLSEN, R. H. & THOMAS, D. D. (1973). Characteristics and purification of PRR1, an RNA phage specific for the broad host range *Pseudomonas* R 1822 drug resistant plasmid. *J. Virol.*, **12,** 1560.

ØRSKOV, I. & ØRSKOV, F. (1960). An antigen termed f^+ occurring in F^+ *E. coli*. *Acta path. microbiol. scand.*, **48,** 37.

OU, J. T. & ANDERSON, T. F. (1970). Role of pili in bacterial conjugation. *J. Bact.*, **102,** 648.

Demonstration of a Carbohydrate Layer Involved in the Attachment of Lactobacilli to the Chicken Crop Epithelium

B. E. Brooker and R. Fuller

National Institute for Research in Dairying, Shinfield, Reading, England

The alimentary canal of vertebrates provides a number of distinct habitats for microbial colonization. Each of these habitats is normally populated by a characteristic bacterial flora of which some species are constantly associated with the lining epithelium. The resulting association of different species of bacteria with the epithelia of different regions of the gut is probably a reflection of the specific nature of bacterial adhesion (Dubos, *et al*, 1965; Fuller and Turvey, 1971; Gibbons and van Houte, 1971; Savage, 1972). Since such selective adhesion appears to be a property not only of those bacteria which constitute the normal flora but also of some pathogens, whose ability to adhere to the gut epithelium is believed to contribute to their virulence (Freter, 1969; Jones and Rutter, 1972) the mechanism of adhesion and of its attendant specificity is of considerable interest. In some species, adhesion has been attributed to a surface coat of protein (Gibbons, van Houte and Liljemark, 1972; Jones and Rutter, 1972) but the postulated presence of epithelial receptors as co-determinants of specificity (Jones and Rutter, 1972) has not been demonstrated.

The surface of the chicken crop is partially covered by a layer of lactobacilli attached to the superficial cells of the stratified, squamous epithelium (Fuller and Turvey, 1971). Previous work suggests that this population of bacteria is of importance in regulating the composition of the intestinal flora (Fuller, 1973). Since the ability to attach to crop cells is restricted to certain biotypes of lactobacilli found only in the gut of birds, their adhesion is considered to be specific (Fuller, 1973). Experimental evidence for the involvement of carbohydrate in bacterial adhesion has been obtained by Fuller (1975) who, using an *in vitro* test, found that pretreatment of lactobacilli with sodium periodate or monovalent concanavalin A significantly reduced their ability to adhere to crop cells *in vitro*. The results of the present study, in which attempts were made to visualize this carbohydrate, suggest that the attachment of lactobacilli to the crop epithelium is effected by the interaction of carbohydrate-rich coats located at the surface of bacterial and host cells and that the observed specificity of adhesion is a function of this interaction. Some aspects of this work have been discussed at greater length elsewhere (Brooker and Fuller, 1975).

Materials and Methods

Bacteria

Two strains of lactobacilli were used. One of these (strain 59 belonging to biotype B, see Fuller, 1973) was known to adhere to chicken crop cells and was isolated from the washed crop wall macerates of Light Sussex hens from the Institute breeding flock (Fuller, 1973). The other organism, *Lactobacillus acidophilus* NCTC 1723 isolated from rat faeces, was known from a previous study (Fuller, 1973) not to adhere to crop cells. Both were maintained in MRS broth (de Man, Rogosa and Sharpe, 1960) at 37°.

Chickens

All birds used in this study were the progeny of Light Sussex hens crossed with Rhode Island Red cocks. Germ-free chicks (3–4 weeks old) which had been hatched and reared in stainless steel isolators by the method of Coates *et al.* (1963) were monoassociated with a single strain of lactobacillus (strain 59) by the introduction of the bacteria into their drinking water. Chicks were killed 3 days later by dislocation of the neck.

Electron microscopy

Transmission electron microscopy—The general procedure for the preparation of chicken crop for electron microscopy was as follows. After removal from the chick, the crop was cut open and washed free of its contents with saline. A small piece of crop was then transferred to a Petri dish containing the primary fixative and chopped into 1 mm^3 pieces using a razor blade. Fixation was continued for 90 min at room temperature. The samples were then washed for several hours in the same buffer as that used in primary fixation and then postfixed for 1–2 h in an osmium tetroxide-based fixative at room temperature. After dehydration in a series of ethanol-water mixtures and absolute ethanol, specimens were transferred to propylene oxide and a number of propylene oxide-Araldite mixtures before finally embedding in Araldite. To examine lactobacilli which had been grown *in vitro*, a log. phase culture was centrifuged at

1000 *g* for 15–20 min to sediment the bacteria and the resulting pellet fixed, washed and postfixed in the appropriate reagents (see below). The bacteria were then embedded in agar, the latter chopped into small pieces and processed as above.

For the conventional morphological study of crop tissue and cultured lactobacilli, the primary fixative was 0·1M cacodylate-buffered 2·5% (v/v) glutaraldehyde (pH 7·2). After a thorough washing in 0·1M cacodylate buffer, specimens were postfixed in 0·1M cacodylate-buffered 1% osmium tetroxide. They were then treated with 1% (w/v) uranyl acetate in 25% (v/v) ethanol for 30 min, dehydrated and embedded.

Each of the following cytochemical procedures were carried out using (i) crop tissue from a germ-free chick, (ii) crop tissue from a gnotobiotic chick which had been monoassociated with lactobacillus strain 59, (iii) *in vitro* grown lactobacillus strain 59 and (iv) *in vitro* grown *L. acidophilus*.

(a) *Colloidal iron*. After primary fixation for 90 min at room temperature in buffered glutaraldehyde, specimens were treated for 1 h at pH 1·8 with a mixture of distilled water (18 volumes), glacial acetic acid (12 volumes) and stock colloidal iron (10 volumes). The latter was prepared by the method given by Mowry (1963). Specimens were then rinsed twice in 30% (v/v) acetic acid and postfixed for 2 h in buffered 1% (w/v) osmium tetroxide.

(b) *Ruthenium red*. The method used was essentially that described by Luft (1971). Specimens were fixed for 1 h at room temperature in a solution prepared by the addition of equal volumes of 4% (v/v) aqueous glutaraldehyde, 0·2M cacodylate buffer (pH 7·2) and 0·15% (w/v) aqueous ruthenium red (Taab Laboratories). After several brief washes in buffer, specimens were postfixed at room temperature for 2 h in a solution containing equal volumes of 2% osmium tetroxide, 0·2M cacodylate buffer (pH 7·2) and 0·15% aqueous ruthenium red.

(c) *Alcian blue—lanthanum nitrate*. The technique used was that described by Shea (1971). Primary fixation in 0·1M cacodylate-buffered 2·5% (v/v) glutaraldehyde (pH 7·2) containing 0·5% Alcian blue 8GX (ICI Ltd) was followed by several rinses in buffer. Postfixation was in 1% (w/v) osmium tetroxide and 1% (w/v) lanthanum nitrate in 0·1M s-collidine buffer (pH 8·0) for 2 h.

All sections were cut using a Reichert OmU2 ultramicrotome. Sections of material which had been treated by the cytochemical procedures described above were examined in an Hitachi HU 11E electron microscope without prior staining in lead citrate or uranyl acetate.

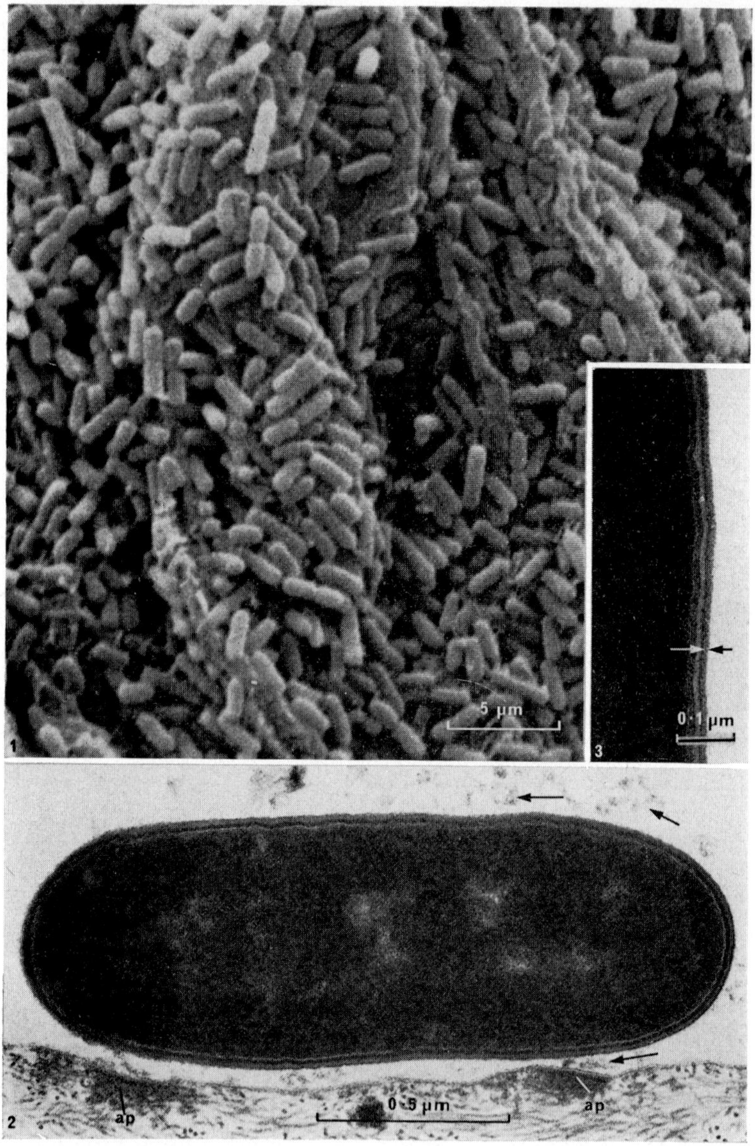
1
5 µm
3
0·1 µm
2
ap
0·5 µm
ap

Scanning electron microscopy—Pieces of fresh crop tissue from conventional and monoassociated birds were pinned to a sheet of dental wax and fixed by immersion in buffered 2·5% (v/v) glutaraldehyde containing 0·5% (w/v) Alcian blue 8GX for 2 h. After dehydration in a series of ethanol-water mixtures and absolute ethanol, the tissue was soaked for 1 h in amyl acetate and dried by the critical point method of Anderson (1951). The dried tissue was placed on a stub, coated evenly with 10–15 nm of gold in an Edwards 12E6 coating unit and examined in a Cambridge "Stereoscan" Mk. II.

Results

The entire surface of the stratified squamous epithelium lining the chicken crop was covered by cells in various stages of desquamation. In birds with a normal gut flora and in gnotobiotic chicks which had been monoassociated with strain 59, bacilli were found attached to the crop cells in large numbers and many showed central constrictions (Fig. 1). Scanning electron micrographs of monoassociated crop epithelium which had been fixed in the presence of Alcian blue showed sporadic filamentous connections between bacteria and between bacteria and the crop wall. The cell wall of strain 59 lactobacilli consisted of an inner and outer electron dense layer enclosing an intermediate zone of lower electron density (Fig. 2). Serial sections showed that adhesion to the crop wall did not involve direct contact between the exterior layer of the bacterial cell wall and the outer leaflet of the crop cell membrane but that a gap at least 7 nm wide usually separated the two at the point of their closest apposition (Fig. 2). The possibility that this gap was at least partially filled with material of bacterial origin was suggested by electron micrographs of conventionally fixed specimens in which a poorly defined filamentous or floccular layer was observed covering part of the cell surface (Fig. 2).

FIG. 1. Scanning electron micrograph of lactobacilli attached to the crop epithelium of a conventional bird. (× 3700)

FIG. 2. Longitudinal section of a lactobacillus (strain 59) lying adjacent to a crop epithelial cell showing the layered structure of the bacterial cell wall. There is no apparent direct contact between bacterium and epithelial cell. Loose floccular material associated with the bacterial cell wall is indicated by arrows. The two electron dense areas associated with the crop cell membrane (*ap*) are the remains of adhesion plaques (maculae adherentes) formed by this and an overlying cell which has been sloughed off. They are not induced by bacterial adhesion (×64 000).

FIG. 3. Longitudinal section of the cell wall of *L. acidophilus* showing its layered structure. In addition to the layers present in lactobacillus strain 59, this species possesses an outer component (between arrows). Lead citrate and uranyl acetate (×96 500).

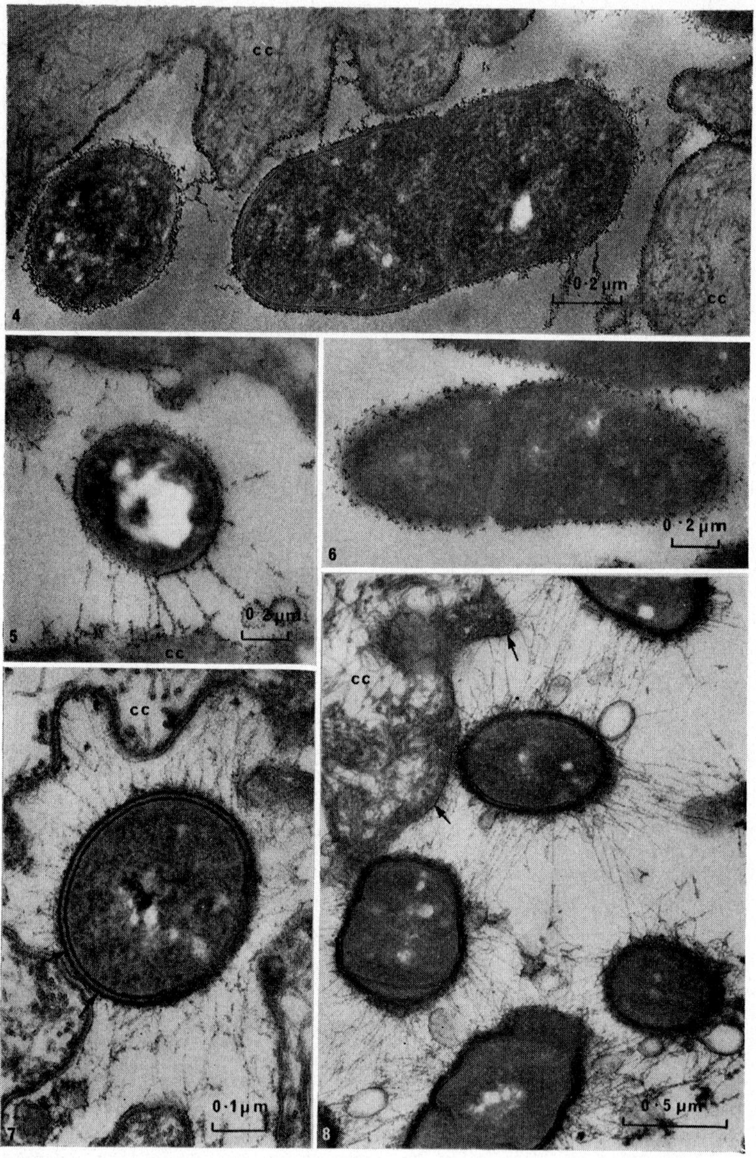

cc
0·2 µm
cc
4
0 ·2 µm
6
0·2 µm
5
cc
cc
7
0·1 µm
cc
8
0·5 µm

Indications of the same layer were also present in lactobacilli grown in culture. Presumed sites of adhesion were not marked by any change in the appearance of host or bacterial cells.

The cell wall of *in vitro* grown *L. acidophilus* closely resembled that of strain 59 but possessed in addition a uniform outer layer of low electron density (Fig. 3).

Colloidal iron

After treatment with colloidal iron at pH 1·8, a thin layer of colloidal particles was bound to the external surface of the cell wall of lactobacilli attached to the crop epithelium (Fig. 4). Filaments extending from this layer both to other bacteria and to the crop cells were similarly reactive (Fig. 5). A thin but often discontinuous layer of colloidal particles was bound to the cell membrane of crop cells from germ-free chicks and those which had been monoassociated with lactobacilli (Figs 4 and 5).

In strain 59 grown *in vitro*, a pronounced layer of colloidal iron was bound to the outside of the cell wall (Fig. 6). Reactive filaments arising from this layer were never seen. In the case of *L. acidophilus*, colloidal iron was never found bound to the cell wall.

Ruthenium red

In lactobacilli attached to the crop epithelium, three distinct layers of the cell wall were stained after fixation in the presence of ruthenium red (Fig. 7). Two of these corresponded approximately to the position of the inner and outer, electron-dense layers of the cell wall seen after lead and

---

FIG. 4. Lactobacilli attached to the crop wall treated with colloidal iron at pH 1·8 but otherwise unstained. Iron is bound to the surface of crop cells (*cc*) and to the bacterial cell wall (× 58 000).

FIG. 5. As for Fig. 4 but showing iron bound to filaments which arise from the cell wall and extend to a neighbouring crop cell (× 37 800).

FIG. 6. A lactobacillus (strain 59) from culture treated with colloidal iron at pH 1·8 but otherwise unstained (× 37 800).

FIG. 7. A lactobacillus attached to the crop epithelium treated with ruthenium red but otherwise unstained. In addition to two intensely staining layers of the cell wall, a superficial layer of material is visible from which thin filaments radiate and extend to nearby areas of crop cell membrane. A probable site of primary bacterial adhesion extends between the arrows (× 86 000).

FIG. 8. As for Fig. 7 but showing bacterial filaments extending to other organisms as well as to the crop cell membrane. Arrows indicate areas of the crop cell membrane where surface staining material can be seen clearly (× 36 500).

uranium staining as shown in Fig. 2. The third layer appeared as diffuse material on the outside of the cell wall from which numerous branching filaments radiated in all directions (Fig. 7). In germ-free chicks and those which had been monoassociated with lactobacilli, the plasmalemma of the superficial crop cells was covered with a thin layer of ruthenium red-positive material (Fig. 8). In those profiles where lactobacilli very closely approached the crop cell membrane, there was direct contact of the stained layers at the surface of bacteria and crop cells (Fig. 7). Thus, the almost featureless, narrow gap seen between bacteria and the crop cell membrane in conventionally treated specimens was either filled with diffuse material or bridged by short filaments after exposure to ruthenium red. The filamentous extensions of the bacterial cell wall made contact with one or more neighbouring organisms and with adjacent areas of the crop cell membrane (Fig. 8).

The staining reaction of strain 59 lactobacilli grown *in vitro* was identical to that of bacilli attached to the crop (Fig. 9). However, the filamentous extensions of the outer diffuse layer of the cell wall were generally fewer in number and much shorter than those observed *in vivo*. In the case of *L. acidophilus*, three layers of the cell wall gave a positive staining reaction after exposure to ruthenium red (Fig. 10). As with strain 59, two of these corresponded approximately to the inner and outer electron-dense layers of the cell wall seen after lead and uranium straining (see Fig. 3). The third component of the cell wall to stain was a very thin (4–5 nm thick) and clearly defined surface layer from which filamentous extensions were never seen to arise (Fig. 10).

Alcian blue—lanthanum nitrate

In lactobacilli attached to the crop epithelium which had been exposed to Alcian blue and lanthanum nitrate during fixation, lanthanum was bound to two components of the cell wall. One of these corresponded in position to the outer, electron-dense layer seen in conventionally fixed and stained material (arrows in Fig. 11); the other was a much thicker and less clearly defined layer on the outside of the cell wall (Figs 11 and 12). Fine filaments radiating from this outer layer to other bacteria and to adjacent areas of crop cell membrane were recorded but they were very rare compared with their incidence in material treated with ruthenium red. The cell membrane of superficial crop cells from germ-free and monoassociated chicks was covered with a layer of stained material (Fig. 12). However, it was frequently observed that where a bacillus lay very close to a crop cell there was no lanthanum staining of either of the apposed surfaces. This is seen clearly in Fig. 11.

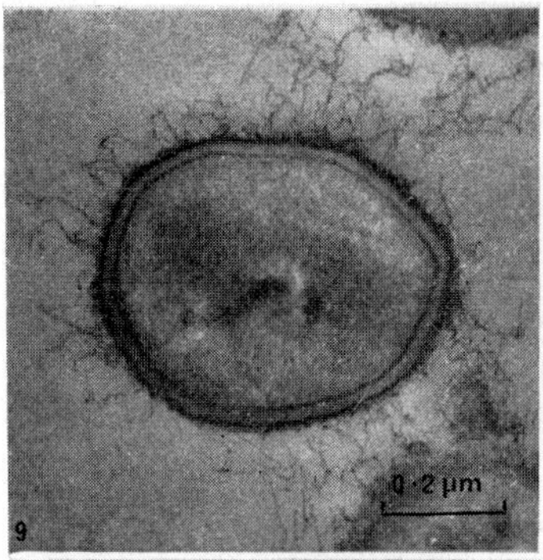

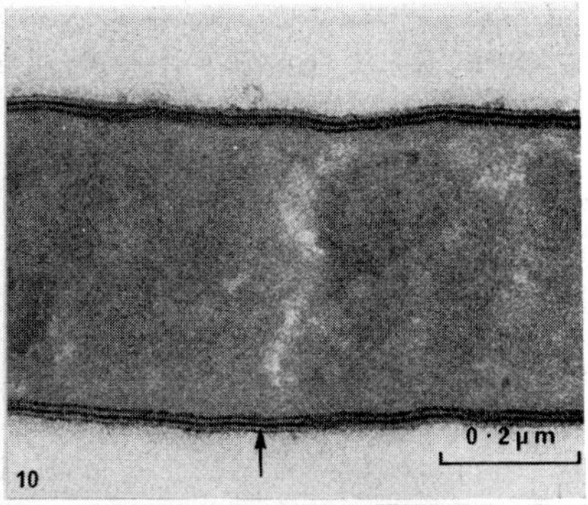

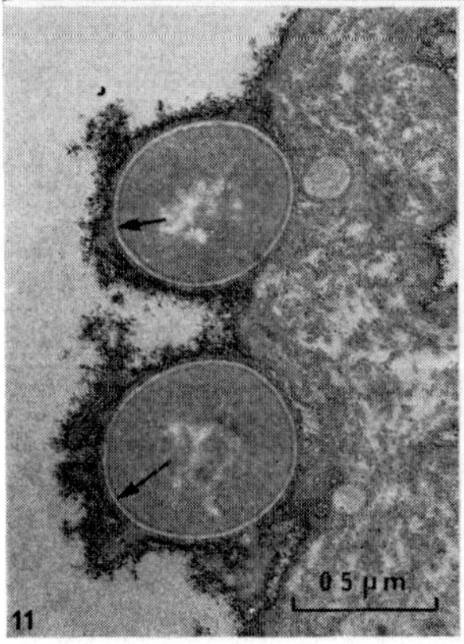

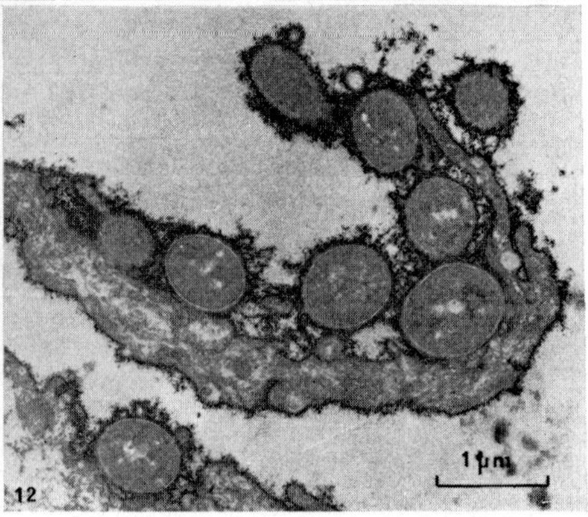

FIG. 9. Cultured lactobacillus (strain 59) treated with ruthenium red. The stained layers of the cell wall are identical to those in Fig. 7. (× 57 500).

FIG. 10. Longitudinal section of *L. acidophilus* treated with ruthenium red only. Note the very thin layer of strained material at the surface of the cell wall (arrowed) (×77 000).

FIG. 11. Two lactobacilli (strain 59) attached to a crop epithelial cell treated with Alcian blue-lanthanum nitrate only. In addition to the diffuse mass of surface material, a more clearly defined layer of the cell wall has stained (at arrows). Note the exclusion of lanthanum from two sites at which the bacteria are closely apposed to the epithelial cell. (× 28 500).

FIG. 12. Lactobacilli attached to a desquamating epithelial cell treated as in Fig. 11. The surfaces of the bacterial and crop cells are heavily stained. (×14 500).

The staining reaction of strain 59 from culture was similar to that seen *in vivo*, but filamentous projections from the cell wall were much better developed. The cell wall of *L. acidophilus* was never stained.

Discussion

The present study shows that the cell wall of lactobacillus strain 59 has a surface layer of material which is not rendered visible by the conventional methods of fixation and heavy-metal staining used in electron microscopy. The staining of this layer with cationic dyes indicates that it is composed of polyanionic material. More specifically, its affinity for colloidal iron at low pH suggests that it contains acidic carbohydrate because previous studies have shown that the positively charged colloidal iron hydroxide binds selectively to the negatively charged carboxyl and sulphate groups of these compounds (Benedetti and Emmelot, 1967; Wetzel, Wetzel and Spicer, 1966).

The ability of ruthenium red, as a cationic dye, to bind polyanions has been used by many previous authors to demonstrate what has usually been referred to as acid polysaccharide or mucopolysaccharide at the surface of a variety of prokaryotic cells including *Chondrococcus columnaris* (Pate and Ordal, 1967), *Anabaena* sp. (Leak, 1968), bacteria of the human oral cavity (Luft, 1971), Gram-negative marine bacteria (Fletcher and Floodgate, 1973), *Diplococcus pneumoniae* and *Klebsiella pneumoniae* (Springer and Roth, 1973) and *Mycoplasma dispar* (Howard and Gourlay, 1974). However, from the work of Luft (1971) it appears that ruthenium red will react strongly not only with acid polysaccharides but also with other polyanions of high charge density. Strong affinity of a material for ruthenium red *per se* cannot therefore be considered as evidence for its positive identification as an acid polysaccharide since other polyanions of a quite different nature might well be involved. In the present study, however, the ruthenium red-staining material of lactobacillus strain 59 corresponded in position to that demonstrated by the colloidal iron method and is considered to reinforce its identification as acidic polysaccharide.

The inclusion of Alcian blue in primary aldehyde fixatives has been shown by Behnke and Zelander (1970) to improve the fixation of the "cell coat" of a number of mammalian cell types and has been used by Fletcher and Floodgate (1973) to demonstrate material coating the cell wall of a marine bacterium. To improve the electron density of negatively charged mucosubstances precipitated by Alcian blue treatment, Shea (1971) has suggested the inclusion of lanthanum in postfixatives and has argued that although the precise chemical basis of lanthanum staining remains to be established, the resulting reaction product marks the site of mucopolysaccharides or mucopolysaccharide-protein complexes. The results obtained with strain 59 after treatment with Alcian blue-lanthanum are therefore consistent with the presence of an acidic carbohydrate-rich layer on the surface of the cell wall as suggested by the colloidal iron staining. Some aspects of the morphology of this superficial layer varied with the cationic dye used. Thus, the filaments demonstrated with ruthenium red which radiated from the cell wall were not nearly as conspicuous following colloidal iron and Alcian blue-lanthanum treatment. Assuming that this is not due to the demonstration of different components of the same surface layer and that with all three dyes the electron density of this layer is generated by their interaction with the same molecular species, the paucity of filaments is probably due to their loss caused either by the conditions of low pH in the case of colloidal iron or by their clumping and subsequent collapse in the case of Alcian blue-lanthanum.

In common with many other vertebrate cells (Rambourg, 1971; Cook and Stoddart, 1973), those of the chicken crop epithelium appear, from their affinity for cationic dyes, to possess a surface coat of acidic carbohydrate. The presence of this material on the crop cells from germ-free chicks as well as those monoassociated with strain 59 lactobacilli demonstrates that it is truly of host origin and does not represent carbohydrate from the bacterial cell wall which has been adsorbed by epithelial cells. Therefore, any direct contact of lactobacilli with crop cells occurs between their respective surface carbohydrate layers.

The adhesion of bacteria to gut epithelia usually results in two possible, morphologically distinct, changes in the appearance of the host cells in the vicinity of the attachment site (Hampton and Rosario, 1965; Papadimitriou, 1966; Staley, Jones and Corley, 1969; Erlandsen and Chase, 1974; Wagner and Barrnett, 1974). The plasma membrane of the host cell becomes invaginated to accommodate that part of the bacterium involved in the adhesion and a dense zone of filaments appears in the cytoplasm of the epithelial cell adjacent to the attachment site. In some cases, the bacterium may undergo changes both in shape and in the appearance of its cell wall in the region of attachment. In the present study it appears that neither host nor bacterial cells undergo morphological changes which enable the precise site of attachment to be determined. However, it seems likely that the primary site of bacterial adhesion to the crop epithelium is at that point where separation of the bacterial cell wall and crop cell membrane is minimal. The narrow gap which appears to exist at this point in conventionally stained preparations (Fig. 2) is, in reality, largely occupied by the ruthenium red-positive material on the outside of the crop cell membrane and bacterial cell wall. It is suggested therefore that bacterial adhesion to the crop wall depends on the mutual interaction of acidic carbohydrates of bacterial and host origin and that the specificity of adhesion reported by Fuller (1973) is a function of this interaction. The close association of bacterial and crop cell surfaces at the sites of primary adhesion is demonstrated by the apparent exclusion of lanthanum from these areas as shown in Fig. 11. At the sites of adhesion of other bacteria to gut epithelia, the existence of a narrow gap between bacterial and host cells has been noted by previous authors (Hampton and Rosario, 1965; Staley, Jones and Corley, 1969; Wagner and Barrnett, 1974). In some cases, this gap is traversed by filamentous material of bacterial and/or host cell origin which is readily visible without the use of cationic dyes and is probably of considerable importance in maintaining adhesion since it represents the only visible means of physical contact between the two cells.

The filamentous extensions of the surface carbohydrate layer are very similar to those demonstrated by Luft (1971) on bacteria associated with the squamous epithelium of the human oral cavity. In the case of lactobacilli, they probably provide an additional means of adhesion to the crop epithelium which reinforces the effect of the primary sites. It does not seem likely that the filaments are of major importance to the adhesion of individual lactobacilli since, although often numerous in ruthenium red preparations, their total area of contact with the crop appears to be smaller and the distance over which they would be expected to exert their effect much greater than the carbohydrate involved at primary sites of adhesion. However, filamentous connections between bacteria may afford a degree of stability to the population as a whole.

The production of an external layer of carbohydrate by lactobacilli grown *in vitro* indicates that the synthesis of this material is not dependent for its initiation on the conditions encountered in the crop or on contact with the crop cells. Its importance for the adhesion of lactobacilli is indicated by the results obtained using *L. acidophilus* grown *in vitro*. Although a very thin layer of material was detected on the surface of the cell wall with ruthenium red, it was far less obvious than that seen in strain 59. The inability of the cell wall of *L. acidophilus* to bind colloidal iron at pH 1·8 indicates either that the surface layer demonstrated with ruthenium red is not acid polysaccharide (but nevertheless polyanionic) or that the method is too insensitive to visualize any small amounts of acid polysaccharide present. Since *L. acidophilus* does not adhere to crop cells, it appears from these results that there may be quantitative and qualitative differences in the surface layer between "stickers" and "non-stickers" which are reflected by differences in their adhesive behaviour *in vivo*.

References

ANDERSON, T. F. (1951). Techniques for the preservation of three-dimensional structure in preparing specimens for the electron microscope. *Trans. N. Y. Acad. Sci.*, **13**, 130.

BEHNKE, O. & ZELANDER, T. (1970). Preservation of intercellular substances by the cationic dye Alcian blue in preparative procedures for electron microscopy. *J. Ultrastruct. Res.*, **31**, 424.

BENEDETTI, E. L. & EMMELOT, P. (1967). Studies on plasma membrane IV. The ultrastructural localization and content of sialic acid in plasma membranes isolated from rat liver and hepatoma. *J. Cell Sci.*, **2**, 499.

BROOKER, B. E. & FULLER, R. (1975). Adhesion of lactobacilli to the chicken crop epithelium. *J. Ultrastruct. Res.*, **52**, 21.

COATES, M. E., FULLER, R., HARRISON, G. F., LEV, M. & SUFFOLK, S. F. (1963). A comparison of the growth of chicks in the Gustafsson germ-free apparatus and in a conventional environment, with and without dietary supplements of penicillin. *Br. J. Nutr.*, **17**, 141.

COOK, G. M. W. & STODDART, R. W. (1973). *Surface carbohydrates of the eukaryotic cell.* Academic Press, London.

DE MAN, J. C., ROGOSA, M. & SHARPE, M. E. (1960). A medium for the cultivation of lactobacilli. *J. appl. Bact.*, **23**, 130.

DUBOS, R., SCHAEDLER, R. W., COSTELLO, R. & HOET, P. (1965). Indigenous, normal and autochthonous flora of the gastrointestinal tract. *J. exp. Med.*, **122**, 67.

ERLANDSEN, S. L. & CHASE, D. G. (1974). Morphological alterations in the microvillous border of villous epithelial cells produced by intestinal microorganisms. *Am. J. clin. Nutr.*, **27**, 1277.

FLETCHER, M. & FLOODGATE, G. D. (1973). An electron-microscopic demonstration of an acidic polysaccharide involved in the adhesion of a marine bacterium to solid surfaces. *J. gen. Microbiol.*, **74**, 325.

FRETER, R. (1969). Studies of the mechanism of action of intestinal antibody in experimental cholera. *Tex. Rep. Biol. Med.*, **27**, 299.

FULLER, R. (1973). Ecological studies on the lactobacillus flora associated with the crop epithelium of the fowl. *J. appl. Bact.*, **36**, 131.

FULLER, R. (1975). Nature of the determinant responsible for the adhesion of lactobacilli to chicken crop epithelial cells. *J. gen. Microbiol.*, **87**, 245.

FULLER, R. & TURVEY, A. (1971). Bacteria associated with the intestinal wall of the fowl (*Gallus domesticus*). *J. appl. Bact.*, **34**, 617.

GIBBONS, R. J. & VAN HOUTE, J. (1971). Selective bacterial adherence to oral epithelial surfaces and its role as an ecological determinant. *Infect. Immun.*, **3**, 567.

GIBBONS, R. J., VAN HOUTE, J. & LILJEMARK, W. F. (1972). Parameters that effect the adherence of *Streptococcus salivarius* to oral epithelial surfaces. *J. dent. Res.*, **51**, 424.i

HAMPTON, J. C. & ROSARIO, B. (1965). The attachment of microorganisms to epithelial cells in the distal ileum of the mouse. *Lab. Invest.*, **14**, 1464.

HOWARD, C. J. & GOURLAY, R. N. (1974). An electron-microscopic examination of certain bovine mycoplasmas stained with ruthenium red and the demonstration of a capsule on *Mycoplasma dispar*. *J. gen. Microbiol.*, **83**, 393.

JONES, G. W. & RUTTER, J. M. (1972). Role of the K88 antigen in the pathogenesis of neonatal diarrhoea caused by *Escherichia coli* in piglets. *Infect. Immun.*, **6,** 918.

LEAK, L. V. (1968). Fine structure of the mucilaginous sheath of *Anabaena* sp. *J. Ultrastruct. Res.*, **21,** 61.

LUFT, J. H. (1971). Ruthenium red and violet. II. Fine structural localization in animal tissues. *Anat. Rec.*, **171**, 369.

MOWRY, R. W. (1963). The special value of methods that colour both acidic and vicinal hydroxyl groups in the histochemical study of mucins. With revised directions for the colloidal iron stain, the use of Alcian blue G8X and their combinations with the periodic acid-Schiff reaction. *Ann. N.Y. Acad. Sci.*, **106,** 402.

PAPADIMITRIOU, J. M. (1966). Cell membrane changes during contact with some micro-organisms. *Nature, Lond.*, **212,** 631.

PATE, J. L. & ORDAL, E. J. (1967). The fine structure of *Chondrococcus columnaris* III. The surface layers of *Chondrococcus columnaris. J. Cell Biol.*, **35,** 37.

RAMBOURG, A. (1971). Morphological and histochemical aspects of glycoproteins at the surface of animal cells. *Int. Rev. Cytol.*, **31,** 57.

SAVAGE, D. C. (1972). Associations and physiological interactions of indigenous microorganisms and gastrointestinal epithelia. *Am. J. clin. Nutr.*, **25,** 1372.

SHEA, S. M. (1971). Lanthanum staining of the surface coat of cells. Its enhancement by the use of fixatives containing Alcian blue or cetylpyridinium chloride. *J. Cell Biol.*, **51,** 611.

SPRINGER, E. L. & ROTH, I. L. (1973). The ultrastructure of the capsules of *Diplococcus pneumoniae* and *Klebsiella pneumoniae* stained with ruthenium red. *J. gen. Microbiol.*, **74,** 21.

STALEY, T. E., JONES, E. W. & CORLEY, L. D. (1969). Attachment and penetration of *Escherichia coli* into intestinal epithelium of the ileum in newborn pigs. *Am. J. Path.*, **56,** 371.

WAGNER, R. C. & BARRNETT, R. J. (1974). The fine structure of prokaryotic-eukaryotic cell junctions. *J. Ultrastruct. Res.*, **48,** 404.

WETZEL, M. G., WETZEL, B. K. & SPICER, S. S. (1966). Ultrastructural localization of acid mucosubstances in the mouse colon with iron-containing stains. *J. Cell Biol.*, **30,** 299.

The Adhesion of Bacteria to Solid Surfaces

MADILYN FLETCHER AND G. D. FLOODGATE

Marine Science Laboratories, University College of North Wales, Menai Bridge, Gwynedd, Wales

Many bacteria which commonly live and grow in suspension in liquid media are also able to adhere to available surfaces and proliferate. These sessile bacteria may attach by means of pili, or, as with the stalked caulobacters, by holdfast structures, but many have no obvious attachment mechanism. These bacteria are assumed to possess extracellular adhesives, and many apparently produce large quantitites of extracellular polymers subsequent to attachment. In aquatic situations, such micro-colonies tend to develop primarily on the surfaces of dead plant and animal cells and inorganic materials (Corpe, 1970). This provides a surface film which can influence the successive attachment of macro-organisms, e.g. oyster larvae (Mitchell and Young, 1972). Bacterial films are undoubtedly important in many other habitats, particularly the human mouth, where they comprise dental plaque (McHugh, 1970), and the soil, where they may influence quality and texture (Whistler and Kirby, 1956).

The formation of microbial surface films in natural environments is a very complex phenomenon. It often involves the successional attachment of diverse organisms which may interact with one another (Corpe, 1972; Mitchell and Young, 1972). Environmental factors such as pH, temperature, nutrient concentration, may also affect film formation. Because of the complexity of film development, it is more easily studied in the laboratory under controlled conditions, so that the part played by bacteria can be determined.

A method is described below which limits the number of extraneous influences affecting a bacterium's initial attachment and the subsequent formation of a surface film. This method also allows determination of the importance of individual environmental factors. For example, in this study, the influence of Ca^{2+} and Mg^{2+} concentrations was investigated.

Materials and Methods

The organism

The marine bacterium used for this study was a marine *Pseudomonas* sp. (NCMB 2021), isolated from the Menai Strait, North Wales.

Culturing methods

Comparison of attached and free-living organisms. The cells were cultured in 200 ml of medium in 500-ml flasks containing pieces of MF-Millipore filter, to serve as an attachment surface. The medium comprised 0·1% (w/v) peptone, 0·1% (w/v) yeast extract in 80% (v/v) aged seawater in distilled water, pH 7·4. The cells were cultured for 22 h at 12–14° and were aerated throughout by bubbling. Both free-living cells and the Millipore filters with their attached populations were prepared for electron microscopy (see below).

Investigation into the effects of variations in Ca^{2+} and Mg^{2+} concentrations. The cells were cultured as above except that the medium comprised 0·2% (w/v) glucose in artificial seawater (Marshall, Stout and Mitchell, 1971), and incubation was extended to 3 days. The initial Ca^{2+} and Mg^{2+} concentrations of the seawater were 3×10^{-3}M and $3{\cdot}8 \times 10^{-2}$M, respectively.

After the incubation period, during which bacteria had attached to the Millipore filters and proliferated, the filters were transferred to media deficient in Ca^{2+}, Mg^{2+} or both ($Ca^{2+} = 3 \times 10^{-4}$M; $Mg^{2+} = 3{\cdot}8 \times 10^{-4}$M) and left for periods ranging from 5 to 90 min. The filters were then removed and prepared for electron microscopy, along with filters which were not treated with the cation-deficient media.

Electron microscopy

The fixation and embedding of free-living cells needed to parallel as close as possible the preparation of attached cells. Accordingly, the free-living cells from culture samples were artificially attached to filters, immediately after being combined with 2 equal volumes of 3·6% (v/v) glutaraldehyde fixative. The resulting suspension was filtered through a Millipore filter, and a portion of the cells attached to the filter and could be passed from one solution to the next by merely transferring the filter.

Fixation of free-living and attached cells was according to the ruthenium red method of Pate and Ordal (1967). The filters were subsequently

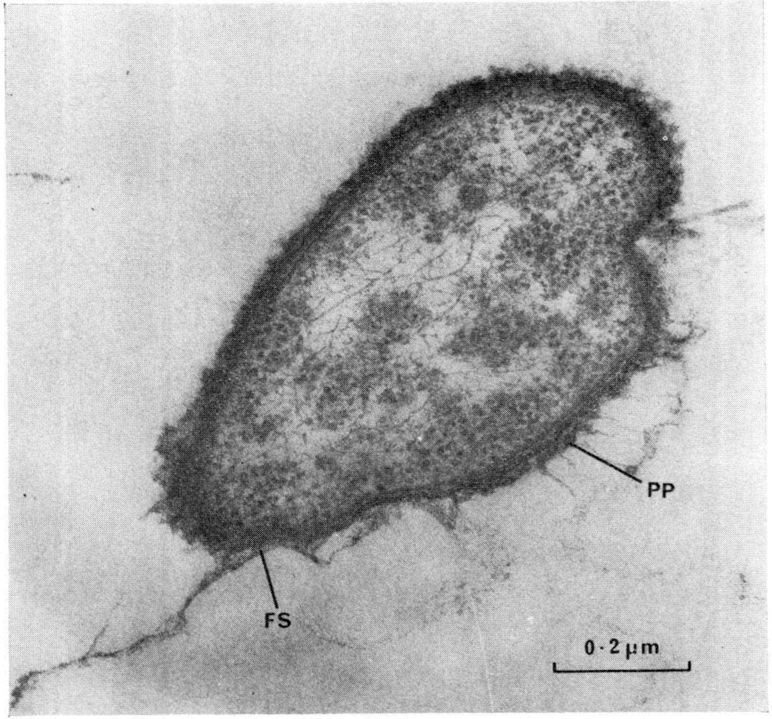

FIG. 1. A free-living bacterium, which was attached to the surface of a Millipore filter (FS) during fixation. Primary polysaccharide (PP) is present, indicating that the free-living cell is equipped with the adhesive so that attachment can occur when it comes into contact with a suitable surface.

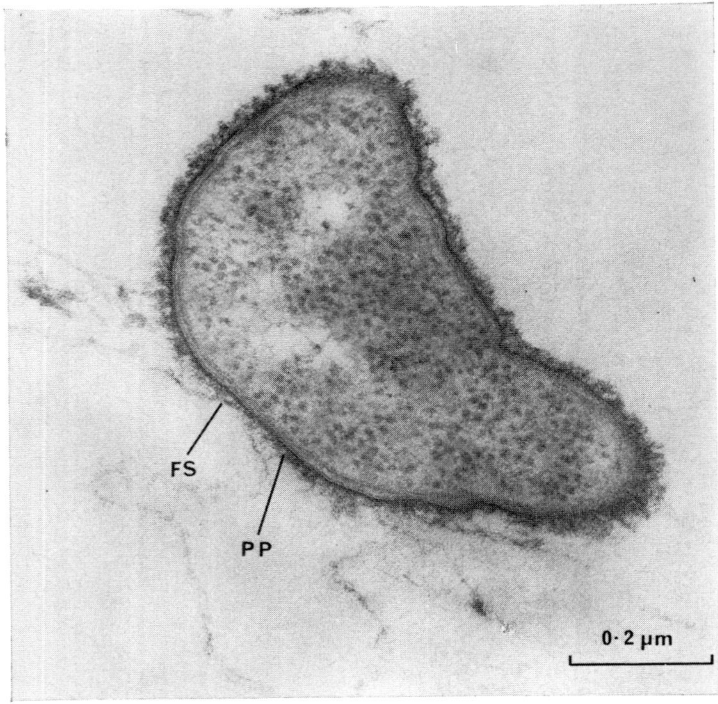

FIG. 2. A bacterium which has attached naturally to a Millipore filter. Primary polysaccharide (PP) bridges the gap between the cell and filter surfaces (FS).

dehydrated in ethanol and embedded in Araldite. Sections were cut on an LKB ultramicrotome and were poststained with uranyl acetate and lead citrate, or lead citrate only. An AEI EM6B electron microscope, operating at 60 kV, was used to examine the sections.

Results

Comparison of attached and free-living organisms

Ruthenium red staining demonstrated two types of extracellular polymer associated with the cells' attachment to Millipore filters (Fletcher and Floodgate, 1973). The nature of the ruthenium red staining (Luft, 1966, 1971) and supplementary experimental data (Fletcher, unpublished results) suggest that both polymers contain some acidic polysaccharide. One type of polymer, designated primary polysaccharide, formed a thin coat on the surface of both free-living and many attached cells (Figs 1–2) and was apparently responsible for initial attachment to the filters. The second type, secondary polysaccharide, was formed after adhesion and probably strengthened the cells' attachment to the filter surface (Fig. 3).

Investigation into the effects of variations in Ca^{2+} and Mg^{2+} concentrations

The morphology of bacterial films grown in the glucose-artificial seawater medium used in the Ca^{2+}-, Mg^{2+}-experiments differed from that of films which developed in peptone-yeast extract-aged seawater media (see above). Primary polysaccharide was retained, instead of giving way to secondary polysaccharide, and the bacteria rested in "pockets" in the intercellular matrix (Fig. 4).

When the bacterial films were transferred to the cation-deficient media, there was a severe disruption of secondary polysaccharide (Fig. 5). This effect was quite sudden, appearing within 5 min, and was not noticeably accentuated by leaving the filters in the deficient media for longer periods. Apparently Ca^{2+} and Mg^{2+} are important in the maintenance of the structure of the intercellular matrix, which is consistent with the presence of an acidic polysaccharide adhesive.

Discussion

Millipore filters appear to be an ideal attachment surface for the study of the fine structure of bacterial surface films. They are easily transferred from one solution to the next during fixation and dehydration, and they remain stable if organic solvents are avoided. The filters also have the advantage

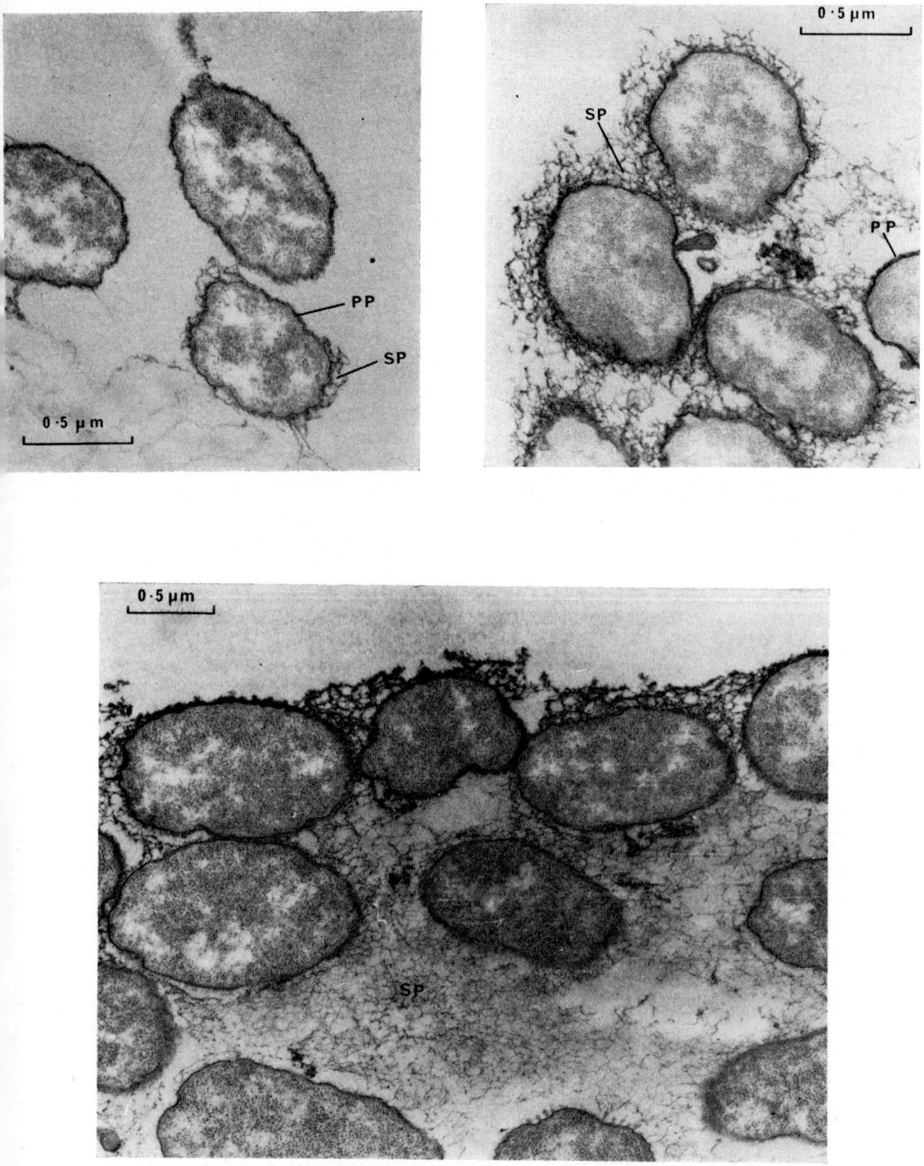

FIG. 3. Sections from bacterial films showing the bacteria and their associated primary (PP) and secondary (SP) polysaccharides. Top: The increase in the amount of secondary polysaccharide is apparently related to an increase in cell number. Bottom: Primary polysaccharide is eventually replaced by secondary polysaccharide which forms an intercellular matrix. The organism was grown in a peptone-yeast extract-aged seawater medium.

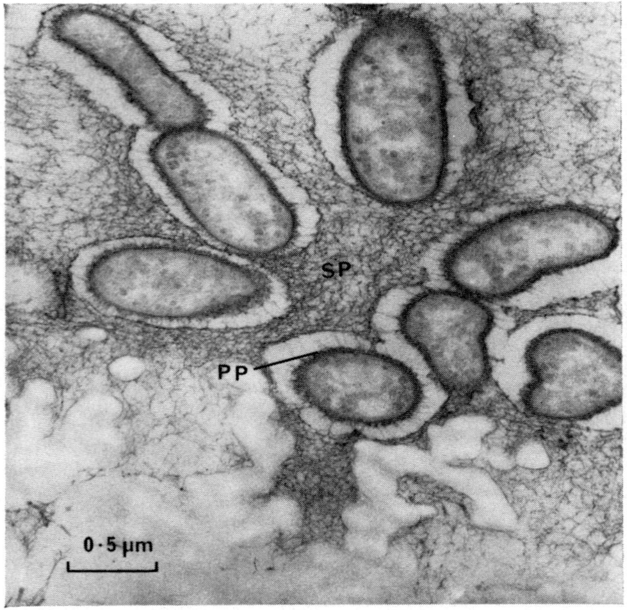

FIG. 4. A micro-colony from the glucose-artificial seawater medium, showing both primary (PP) and secondary (SP) polysaccharides.

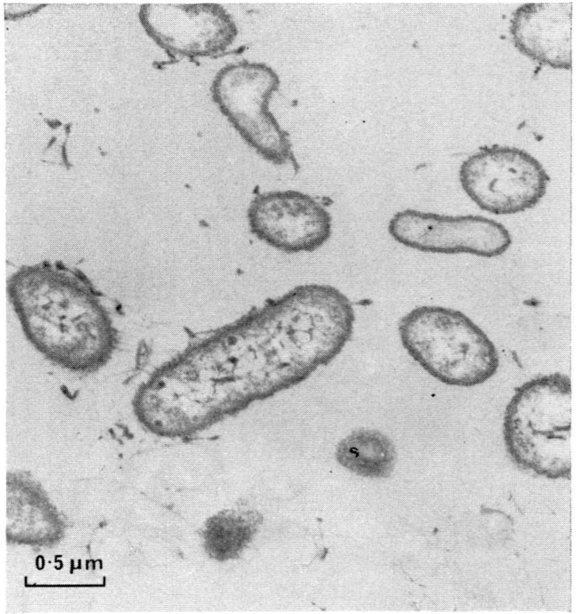

FIG. 5. A micro-colony which was transferred to a medium reduced in Ca^{2+} by a factor of 10 and reduced in Mg^{2+} by a factor of 100.

that they are easily sectioned using glass knives. Araldite embedding was used for this investigation, but some filters have been embedded in Spur, which has the advantage (or disadvantage, depending on the worker's needs) that the filter disappears when the Spur polymerizes. Thus for workers who require a convenient means of transporting cells, but who have no interest in the underlying surface, Millipore filters combined with Spur embedding may be very useful.

More important from a biological point of view, the filters provide an inert attachment surface for living organisms. They are non-toxic and in this study were apparently not degraded by the sessile cells. Their use simplifies the study of bacterial surface films and should easily be applied to investigations of other types of cell adhesion.

References

CORPE, W. A. (1970). Attachment of marine bacteria to solid surfaces. In *Adhesion in Biological Systems*. (Manley, R. S., ed.). Academic Press, New York and London, p. 73.

CORPE, W. A. (1972). Microfouling: The role of primary film forming bacteria. *Proc. 3rd International Congress on Marine Corrosion and Fouling* (Gaithersburg, Md., Oct. 2–6), p. 598.

FLETCHER, M. & FLOODGATE, G. D. (1973). An electron-microscopic demonstration of an acidic polysaccharide involved in the adhesion of a marine bacterium to solid surfaces. *J. gen. Microbiol.*, **74,** 325.

LUFT, J. H. (1966). Fine structure of capillary and endocapillary layer as revealed by ruthenium red. *Fed. Proc.*, **25,** 1773.

LUFT, J. H. (1971). Ruthenium red and violet. I. Chemistry, purification, methods of use for electron microscopy and mechanism of action. *Anat. Rec.*, **171,** 347.

MCHUGH, W. D., ed. (1970). *Dental Plaque*. E. & S. Livingstone, Ltd, Edinburgh and London.

MARSHALL, K. C., STOUT, R. & MITCHELL, R. (1971). Mechanism of the initial events in the sorption of marine bacteria to surfaces. *J. gen. Microbiol.*, **68, 337.**

MITCHELL, R. & YOUNG, L. (1972). The role of micro-organisms in marine fouling. Technical Report No. 3. U.S. Office of Naval Research Contract No. N00014–67–A–0298–0026 NR–306–025.

PATE, J. L. & ORDAL, E. J. (1967). The fine structure of *Chondrococcus columnaris*: III. The surface layers of *Chondrococcus columnaris*. *J. Cell Biol.*, **35,** 37.

WHISTLER, R. L. & KIRBY, K. W. (1956). Composition and behaviour of soil polysaccharides. *J. Am. Chem. Soc.*, **78,** 1755.

Some Aspects of the Cell Walls of *Vibrio* spp.

C. R. ROWLES*, R. PARTON†† AND M. H. JEYNES

Department of Bacteriology, Medical School, University of Birmingham, England

The behaviour of species of the genus *Vibrio*, when undergoing conversion to spheroplasts, has led us to believe that their cell wall structure differed from that usually found in Gram-negative bacteria, such as *Escherichia coli*.

Our investigations have revealed a number of interesting ultrastructural features which are reported below.

The Peptidoglycan Sacculus (Figs 1–4)

The peptidoglycan layer was not detected in ultrasections of normal *Vibrio* cells fixed by Kellenberger's method. If, however, cells were pre-fixed in glutaraldehyde, followed by the Kellenberger technique, a very thin, dense layer was visible in the region in which it is normally found in *E. coli*. The extremely thin nature of this layer suggested that an investigation of the isolated sacculus should be undertaken.

Sacculi were isolated from log phase cultures of several species of vibrios by means of 4% (w/v) sodium lauryl sulphate treatment and examined in the electron microscope. The sacculi presented a much more delicate appearance than that seen in the isolated peptidoglycan sacculus of *E. coli*, to the extent that good visualization was achieved only by a combination of uranyl acetate staining and metal shadowing techniques.

The identity of the peptidoglycan was confirmed by reacting the sacculi on the microscope grid with lysozyme (10–100 μg/ml at 37° for 30 min) which degrades it leaving granules of material which is presumably largely protein or peptide as these are digested by trypsin. After this treatment only the slight depressions in the support membrane, where the sacculi lay originally, are visible.

* Present address: 174 Landor Road, Whitnash, Leamington Spa.

†† Present address: Department of Microbiology, University of Glasgow, Alexander Stone Building, Garscube Estate, Bearsden, Glasgow.

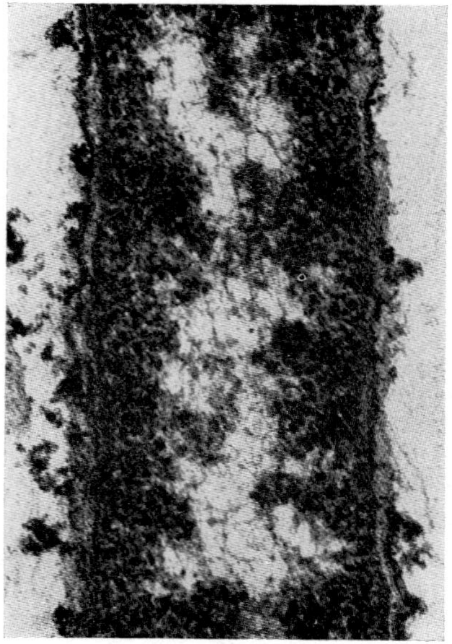

FIG. 1. *V. eltor*. Ultrasection, prefixed with glutaraldehyde, showing extremely thin peptidoglycan layer (×100 000).

FIG. 2. *V. eltor*. Isolated peptidoglycan sacculi (×30 000). Gold-palladium shadowed after uranyl acetate positive stain.

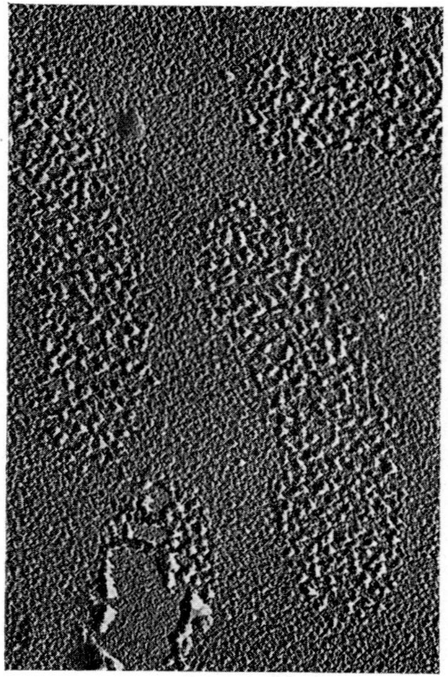

FIG. 3. *V. eltor*. Isolated sacculi after treatment with 100 μg/ml egg-white lysozyme at 37° for 30 min (× 30 000). Gold-palladium shadowed after uranyl acetate positive stain.

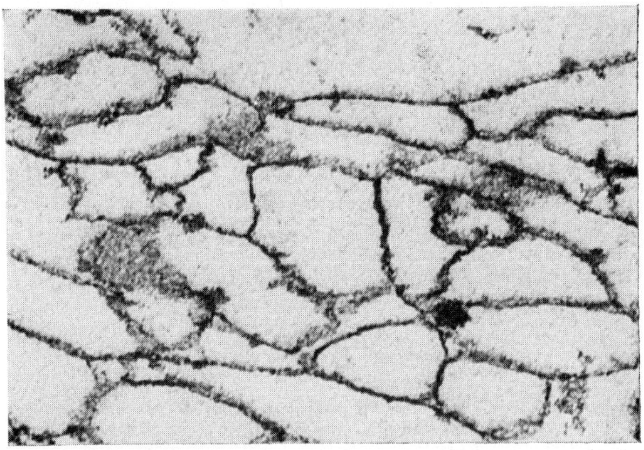

FIG. 4. *V. eltor*. Ultrasection of isolated sacculi (× 75 000). Uranyl acetate positive stain.

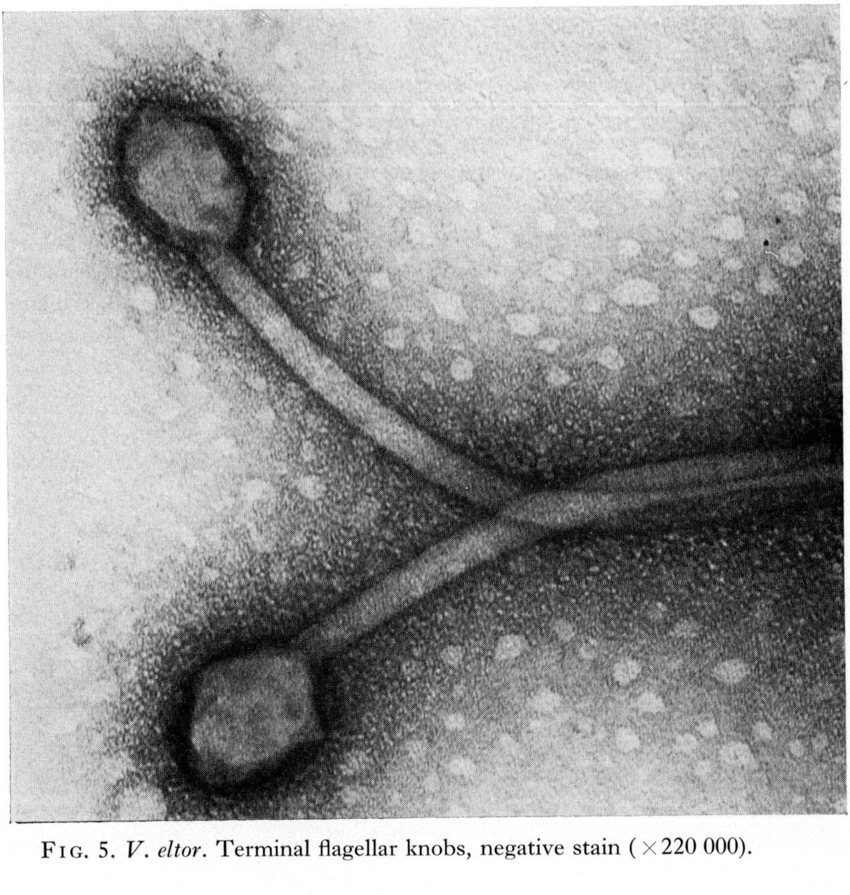

FIG. 5. *V. eltor*. Terminal flagellar knobs, negative stain (×220 000).

Ultrasections of the isolated sacculi confirm that they consist of a single dense layer which varies in thickness presumably due to the angle of sectioning.

This evidence would seem to support the suggestion that the peptidoglycan layer is not the only rigid structure in the cell wall of Gram-negative bacteria which is responsible for the determination and maintenance of cell shape (Henning and Schwarz, 1973).

Terminal Knobs (Fig. 5)

A stable feature of about 5% of the flagella of *V. cholerae* (biotype *eltor*) was the possession of terminal knobs. These were seen in negatively-stained and metal-shadowed preparations and appeared as rounded structures 40–100 nm in diameter at the distal end of the flagella of various lengths.

They usually appeared homologous with the flagella but occasionally a vesicular formation was observed. The flagella of *V. eltor* are sheathed and in some cases, where the sheath was discernible, the terminal knob appeared to be enclosed in this structure. Occasionally, the flagellum extended through the knob for a short distance and possessed a tapering tip. Tapering of the flagella on approach to the knobs was also observed.

Although these terminal structures were seen in a number of strains of *V. cholerae*, their occurrence was rare by comparison with *V. eltor*.

Similar structures have been seen on the sex pili (but never on common pili) of both *Salmonella* spp. and *E. coli* (Lawn, 1966; Meynell, Meynell and Datta, 1968). They appear as terminal knobs, 15–80 nm in diameter and usually spherical in shape although disc and cup shaped knobs were also observed. It was also noted that the pili often tapered as they approached the knobs.

It is suggested that the knobs might be composed of either cell wall material carried away as the new flagella emerge from the cell, or that they represent assemblies of less well organized flagellin sub-units. This would accord with the suggestion of Meynell and Lawn (1967) with regard to the pilus knobs and the findings of Lawn and Meynell (1970) that some knobs adsorbed pilus antibody.

It is a little difficult to reconcile the mode of growth of flagella by the extrusion of flagellin sub-units through the lumen of the flagella (Kerridge, 1973) with the position of the knobs. It might be expected that, with a mode of growth which is by means of extension at the end distal to the cell, the knob would be left behind and, therefore, would not be terminal.

The knobs observed in the present study cannot be confused with those, observed by other workers, on pili. The pili of *V. eltor* and *V. cholerae* are only 6–10 nm in diameter and are not restricted to the poles of the cell, Tweedy, Park and Hodgkiss, 1968; Adhikari and Chatterjee, 1969), whereas the flagella which possessed knobs 20–30 nm in diameter and always terminal. Additionally, pili were only observed rarely on our strains and did not possess terminal knobs.

Complex Cytoplasmic Membranes (Polar Organelles) (Fig. 6)

Ultrasections of *V. eltor* frequently showed structures which have been described as "complex cytoplasmic membranes" (Ritchie, Keeler and Bryner, 1966) or "Polar organelles" (Tauschel and Drews, 1969).

The appearance presented is that of a dense line some 10–12 nm inside the cytoplasmic membrane and linked to this by a series of fine dense lines 5–6 nm apart.

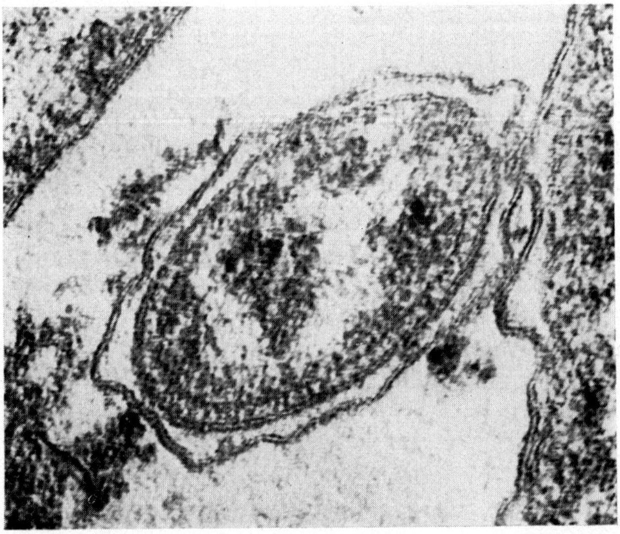

FIG. 6. *V. cholerae*. Ultrasection showing large area of complex cytoplasmic membrane (× 190 000).

When seen in transverse section they often extended around the entire circumference of the cell. In longitudinal sections, they were usually observed at or near the poles but more rarely appeared at areas positioned along the length of the cell.

Although these structures may be demonstrated in ultrasections of cells fixed by the normal osmium tetroxide/tryptone method, the results were often indistinct. Better results were obtained by either the use of 1%

(w/v) uranyl acetate in place of the tryptone in the fixative, or by the use of glutaraldehyde prefixation. The outer surface of the cytoplasmic membrane was often more densely stained in the region of the complex membrane and also appeared to be thickened.

Similar structures, termed "polar membranes" have been demonstrated at the ends of cells of *Spirillum serpens* and have been suggested as a possible basal apparatus for the flagella tuft (Murray and Birch-Anderson, 1963). Cohen-Bazine and Kunisawa (1963) described like structures which they termed "polar caps" in *Rhodospirillum rubrum*. They stated that the wall overlying these areas was unusually smooth and suggested that this might imply a greater rigidity than the remainder of the cell surface. Similar areas of complex cytoplasmic membrane were described in both *R. rubrum* and *R. molishianum* by Hickman and Frenkel (1956*a*; 1956*b*) although they were not always polar.

Ritchie, Keeler and Bryner (1966) found that in *V. fetus* areas of complex cytoplasmic membrane were more conspicuous at the poles but not restricted to such areas. In view of the identical appearance of transverse and longitudinal sections of such areas they suggested that their structure might consist of a layer of radially-orientated compartments which might have important biochemical functions related to energy production for flagella movement or cell wall synthesis.

Similar structures have been demonstrated in *V. proteus* (Rowles, 1972), *S. undula* and *V. metchnikovii* (Freer and Salton, 1971) and in the swarm cells of *Sphaerotilus natans* (Hoeniger, Tauschel and Stokes, 1973). They thus appear to be a regular anatomical feature in the family Spirillaceae and possibly occur in other bacteria as well. In view of the ordered structure observed, they may, like mesosomes, have important enzymic functions to perform.

References

ADHIKARI, P. C. & CHATTERJEE, S. N. (1966). Fimbriation and pellicle formation of *Vibrio El Tor*. *Ind. J. med. Res.*, **57**, 1897.

COHEN-BAZINE, G. & KUNISAWA, R. (1963). The fine structure of *Rhodospirillum rubrum*. *J. Cell Biol.*, **16**, 401.

FREER, J. H. & SALTON, M. R. J. (1971). In *Microbial Toxins*, Vol. IV, (Weinbaum, G., Kaolis, S. & Ajl, S. J., eds), p. 67. London: Academic Press.

HICKMAN, D. D. & FRENKEL, A. W. (1956*a*). Observations on the fine structure of *Rhodospirillum molischianum*. *J. Cell Biol.*, **25**, 261.

HICKMAN, D. D. & FRENKEL, A. W. (1965*b*). Observations on the structure of *Rhodospirillum rubrum*. *J. Cell Biol.*, **25**, 279.

HENNING, V. & SCHWARZ, V. (1973). In *Bacterial Membranes and Walls*, (Loretta Leive, ed.), p. 432. New York: Marcell Dekker Inc.

HOENIGER, J. F. M., TAUSCHEL, H.-D. & STOKES, J. L. (1973). The fine structure of *Sphaerotilus natans*. *Can. J. Microbiol.*, **19**, 309.

KERRIDGE, D. (1973). In *The Generation of Subcellular Structures*, pp. 151–152. Amsterdam and London: North-Holland.

LAWN, A. M. (1966). Morphological features of the pili associated with *E. coli* K12 carrying R factors or F factors. *J. gen. Microbiol.*, **45,** 377.

LAWN, A. M. & MEYNELL, E. (1970). Serotypes of sex pili. *J. Hyg. Camb.*, **68,** 683.

MEYNELL, G. G. & LAWN, A. M. (1967). Sex pili and common pili in the conjugational transfer of colicine factor Ib by *S. typhimurium. Genetical Res. Camb.*, **9,** 359.

MEYNELL, E., MEYNELL, G. G. & DATTA, N. (1968). Transmissible bacterial plasmids. *Bact. Rev.*, **32,** 55.

MURRAY, R. G. E. & BIRCH-ANDERSEN, A. (1963). Specialized structure in the region of the flagella tuft in *Spirillum serpens. Can. J. Microbiol.*, **9,** 393.

RITCHIE, A. E., KEELER, R. F. & BRYNER, J. H. (1966). Anatomical features of *Vibrio fetus*. An electron microscopic survey. *J. gen. Microbiol.*, **43,** 427.

ROWLES, C. R. (1972). *Bacterial spheroplasts, Thesis.* University of Birmingham, England.

TAUSCHEL, H.-D. & DREWS, G. (1969). Der Geisselapparat von *Rhodopseudomonas palustris.* I. Untersuchungen zur Feinstruktur des Polorganells. *Arch. Mikrobiol.*, **66,** 166.

TWEEDY, J. M., PARK, R. W. A. & HODGKISS, W. (1968). Evidence for the presence of fimbriae (pili) on vibrio species. *J. gen. Microbiol.*, **51,** 235.

The Structure of Clostridial Spores

P. D. WALKER, J. A. SHORT, GILLIAN ROPER

Wellcome Research Laboratories
Langley Court, Beckenham, Kent, England

AND

W. HODGKISS

Torry Research Station, Aberdeen, Scotland

The anaerobic spore-forming bacteria represent a very important economic group of organisms containing a number of animal and human pathogens. Organisms belonging to this genus produce a variety of diseases in man and animals which are associated with the production of potent lethal exotoxins.

A study of the structure of clostridial spores has applications in the following areas, (i) synthesis and assembly of spore components, (ii) the relationship between sporulation and toxin production and (iii) bacterial taxonomy. In the following article the use of electron microscope techniques to elucidate the spore structure of a number of species causing disease in man and animals is described.

Materials and Methods

Strains

The spores of several strains in each species were examined in order to check that the structures observed were representative. Cultures were obtained from the Wellcome Research Laboratories Culture Collection (CN), Torry Research Station Culture Collection (TRS), National Collection of Industrial Bacteria (NCIB), Leeds University Bacteriology Department Culture Collection (LUBC).

The following strains were selected for photography: *Clostridium bifermentans*, CN 1617, TRS 16PL TRS 288B; *Cl. sordellii*, CN 1734, TRS 35B, NCIB 6929; *Cl. botulinum* type A, CN 5008; *Cl. botulinum* type C, CN 4946; *Cl. botulinum* type E, NCIB 4248; *Cl. botulinum* type F, TRS 610B; *Cl. sporogenes*, LUBC 206; *Cl. welchii* type A, CN 3353.

Preparation of spores

Spores of *Cl. welchii* type A were prepared in the medium of Duncan and Strong (1968). Spores of all other species studied were prepared as described by Hodgkiss, Ordal and Cann (1967) and Walker, Thomson and Baillie (1967).

Methods

The following techniques were employed for examination of spores, ultrathin sectioning, negative staining, shadowing, freeze etching, immunochemical, histochemical and cytochemical staining. Reference to these techniques is made in another article in this publication (Short and Walker).

Results and Discussion

Ultrathin sectioning

Following classical Kellenberger fixation (Kellenberger, Ryter and Séchaud, 1958) and embedding in epoxy resins, ultrathin sectioning of clostridial species in various stages of sporulation and germination shows a pattern of events almost identical to similar stages seen in the genus *Bacillus*. (Bayen *et al.*, 1967; Fitz-James, 1962; Hashimoto, 1960; Hashimoto and Naylor, 1958; Hoeniger and Headley, 1968; Hoeniger, Stuart and Holt, 1968; Smith and Ellner, 1957; Takagi, Kawata and Yamamoto, 1960; Takagi *et al.*, 1960; Takagi, Nakamura and Ureda, 1965). Development of the forespore septum, cortical synthesis, sporecoat formation and maturation are illustrated (Figs 1–6). The laminated structure of the sporecoat can be seen both during formation and after completion of the coat (Figs 3 and 4). The appearance of the exosporium varies widely depending on the species and may appear as a single layer (Fig. 4) or, alternatively, as a complex structure composed of several laminated layers (Fig. 28b). In *Cl. welchii* spoke-like extensions of the sporecoat radiate towards the exosporium (Fig. 17).

The tubular appendages which have been described in various species of this genus can also be seen in thin section where they are seen as extensions of the sporecoat (Figs 5 and 24). The crystalline structure associated with sporulation in some food poisoning strains of *Cl. welchii* (Duncan, King and Frieben, 1973) can also be demonstrated (Fig. 7).

Abbreviations (Figs 1–38)

A = Axial filament
FM = Forespore membrane
OM = Outer forespore membrane
IM = Inner forespore membrane
SC = Sporecoat
CO = Cortex
E = Exosporium
TU = Tubule
Cy = Crystalline inclusion
S = Spokes
CW = Cell wall

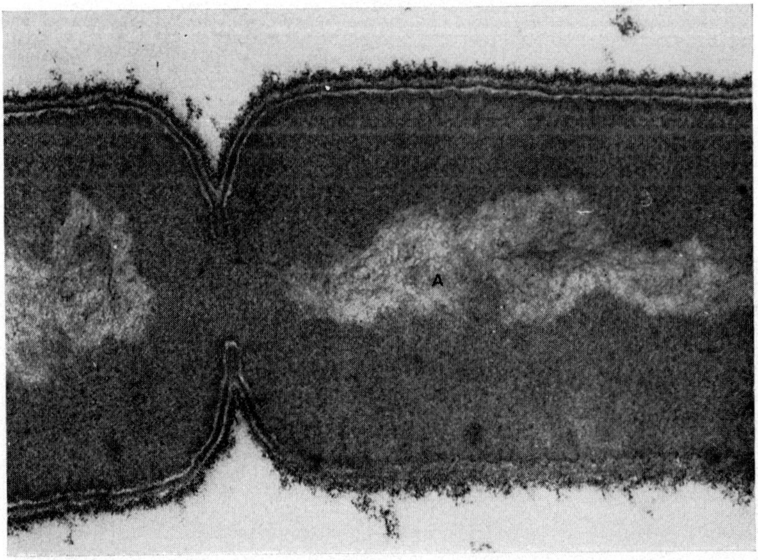

FIG. 1. Ultrathin section of vegetative cell of *Cl. sporogenes* LUBC 206 showing axial filament (× 122 000).

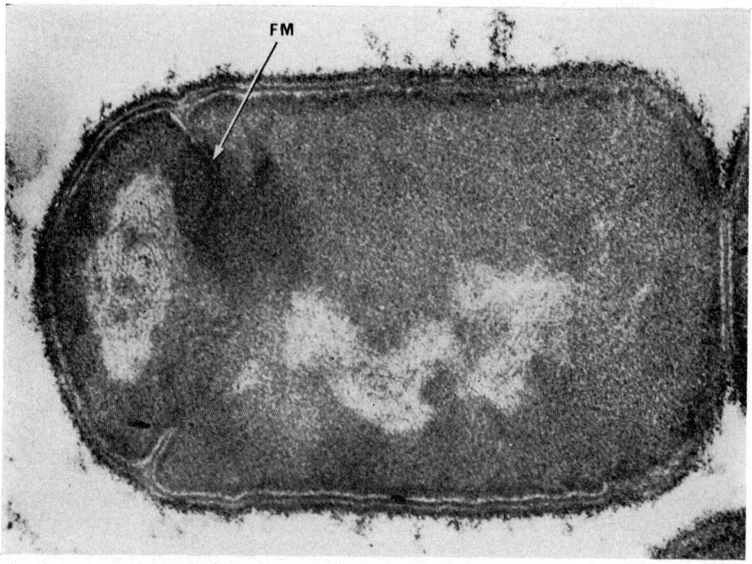

FIG. 2. Ultrathin section of sporulating cell of *Cl. sporogenes* LUBC 206 showing fore-spore septum development (× 164 000).

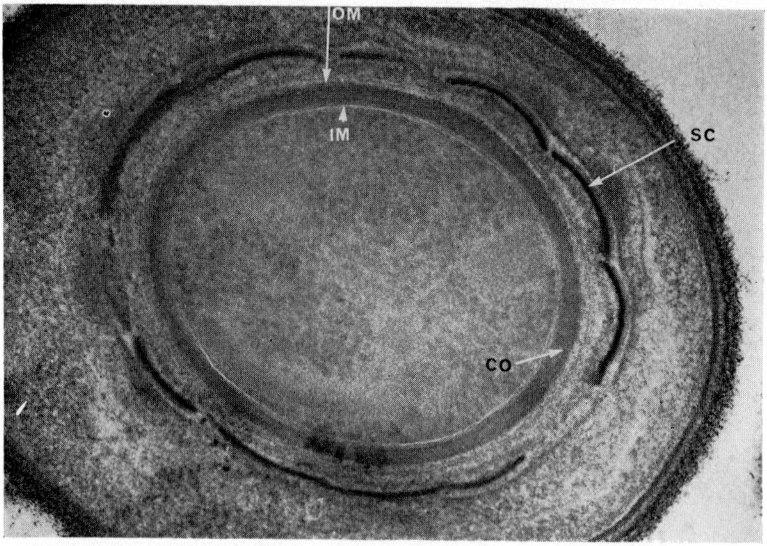

FIG. 3. Ultrathin section of sporulating cell of *Cl. botulinum* type C, CN4946 showing developing cortex and sporecoat (× 110 000).

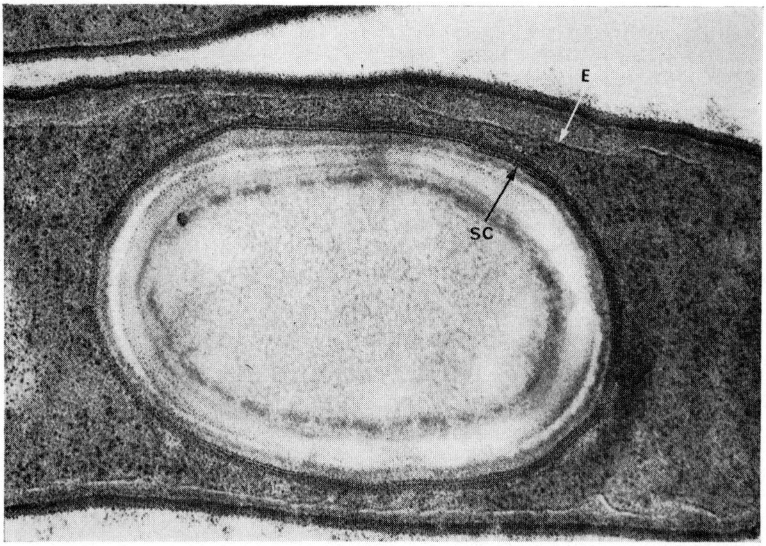

FIG. 4. Ultrathin section of sporulating cell of *Cl. bifermentans* CN 1617 showing mature spore (× 100 000).

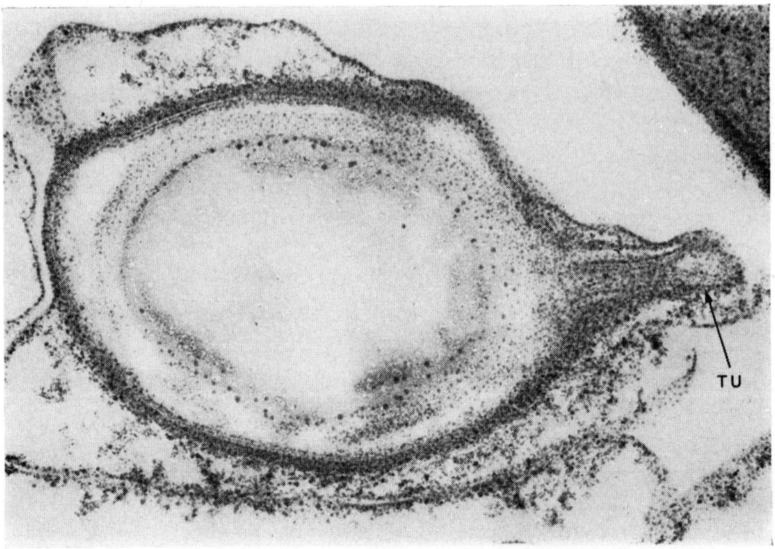

FIG. 5. Ultrathin section of mature spore of *Cl. bifermentans* CN 1617 showing tubular appendage (× 123 000).

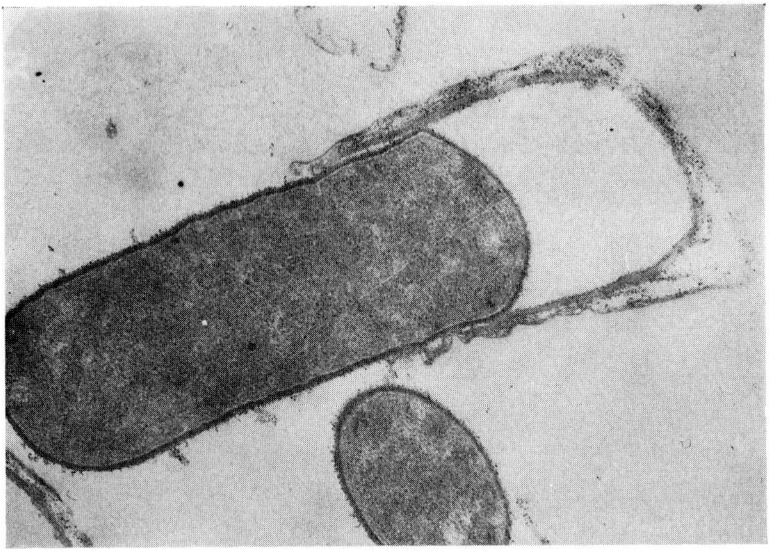

FIG. 6. Ultrathin section of germinating spore of *Cl. bifermentans* CN 1617 (× 51 000).

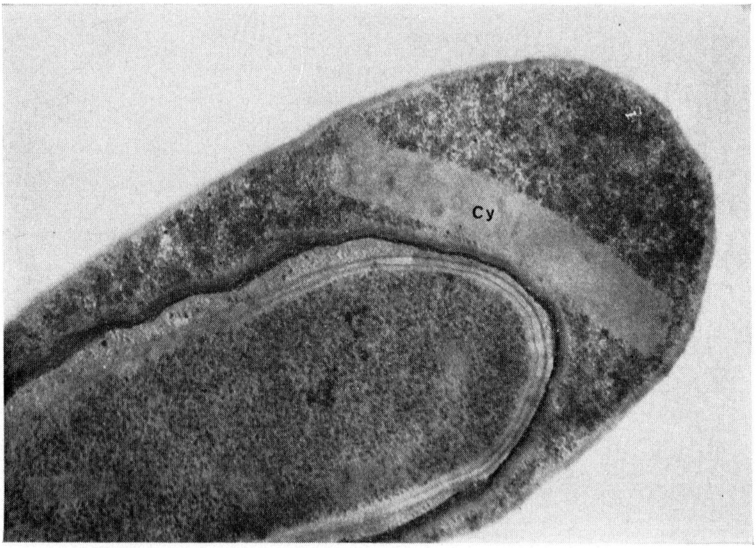

FIG. 7. Ultrathin section of sporulating cell of *Cl. welchii* NCTC 10239 showing spore and associated crystalline inclusion body (× 66 000).

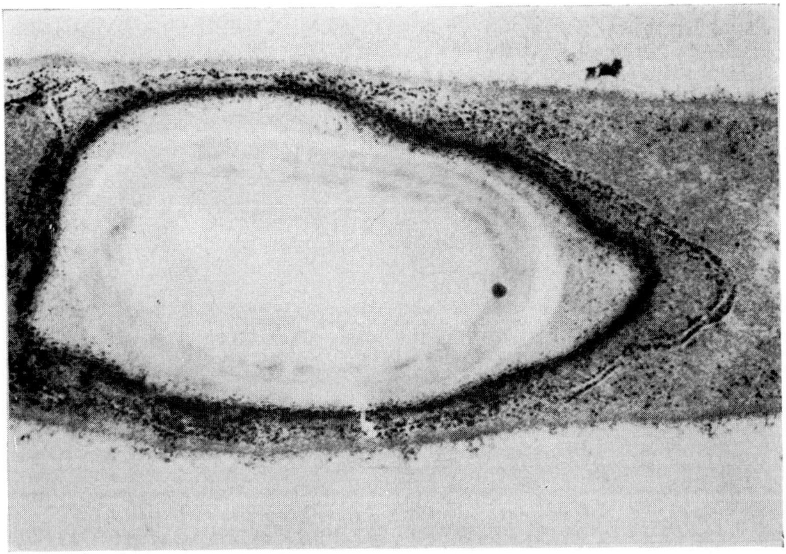

FIG. 8. Ultrathin section of sporulating cell of *Cl. bifermentans* CN 1617 after incubation in thiolacetic acid in the presence of lead. Deposits of lead sulphide are seen around the developing sporecoat and exosporium. (× 61 000).

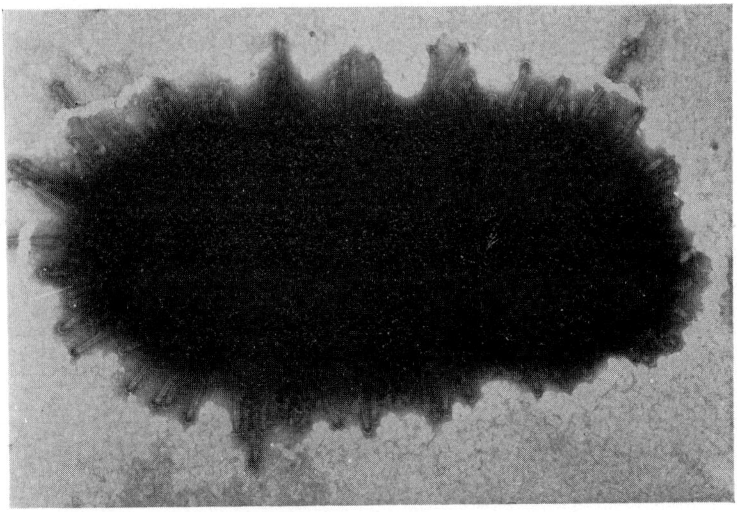

FIG. 9. Negatively-stained preparation of spore of *Cl. botulinum* type E, NCIB 4248 showing numerous tubular appendages (× 40 000).

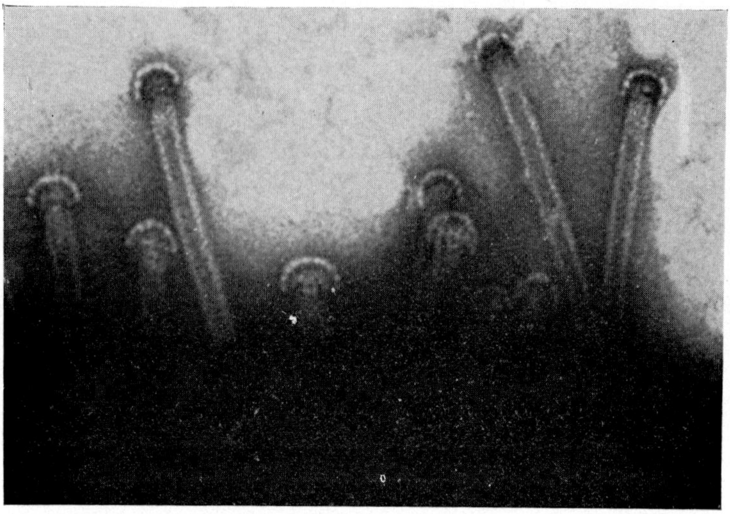

FIG. 10. Negatively-stained preparation of appendages of *Cl. botulinum* type E, NCIB 4248 showing ultrastructural detail (× 240 000).

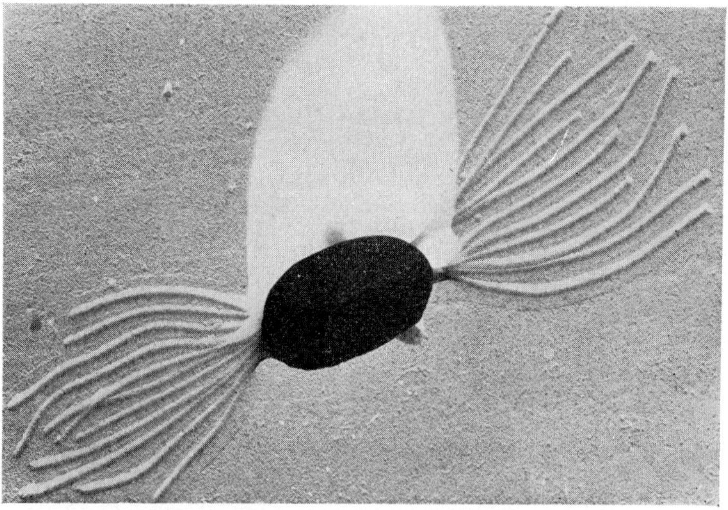

FIG. 11. Shadowed preparation of spore of *Cl. bifermentans* TRS 16PL showing tubular appendages (× 20 000).

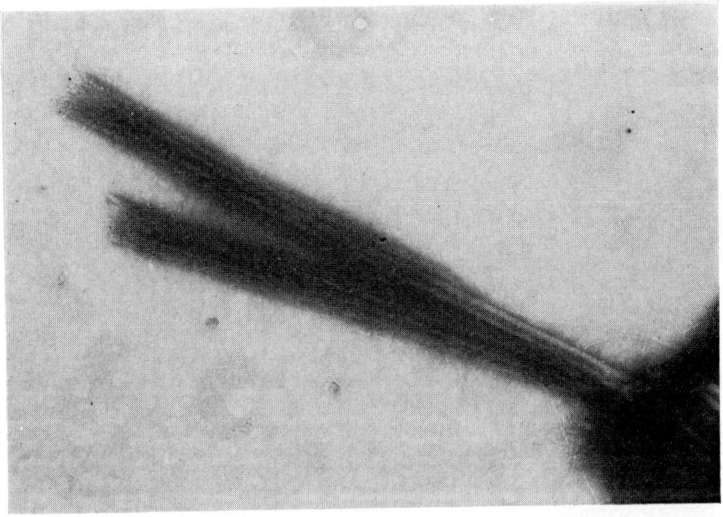

FIG. 12. Negatively-stained preparation of *Cl. bifermentans* TRS 288B spore appendages showing the tubular nature of the appendage and the micro-fibrillar sheath of the distal portion of the tubular wall (× 60 000).

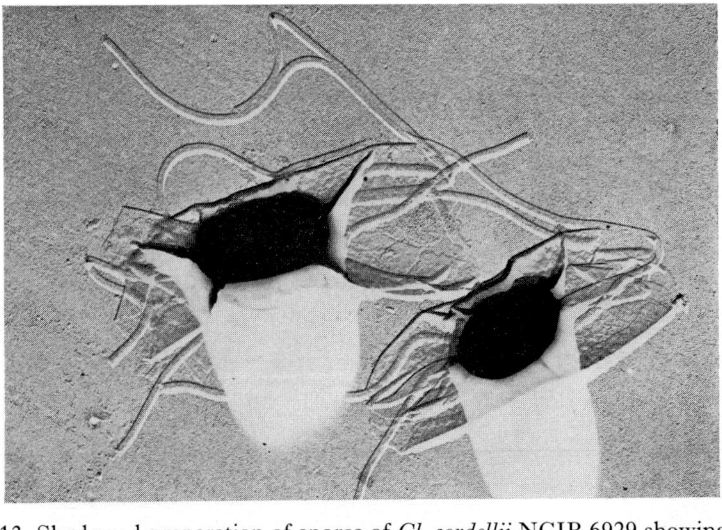

FIG. 13. Shadowed preparation of spores of *Cl. sordellii* NCIB 6929 showing open end coiled appendages penetrating through the exosporium surrounding the spore (× 12 000).

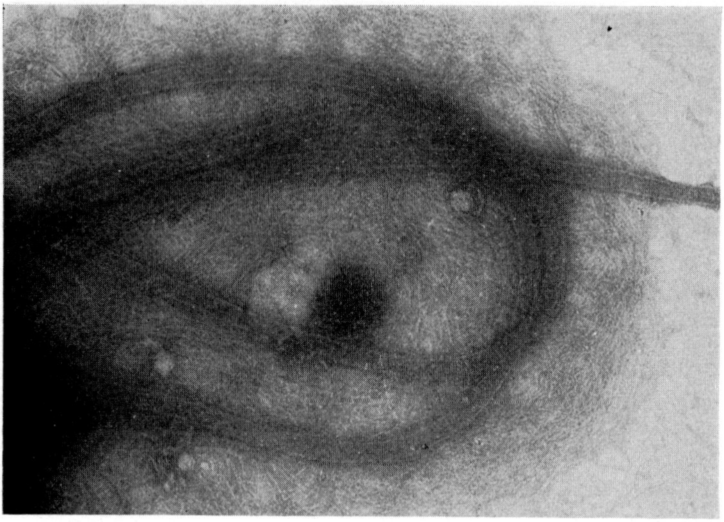

FIG. 14. Negatively-stained preparation of appendages of *Cl. sordellii* TRS 35B spores coiled within the micro-fibrillar exosporium (× 40 000).

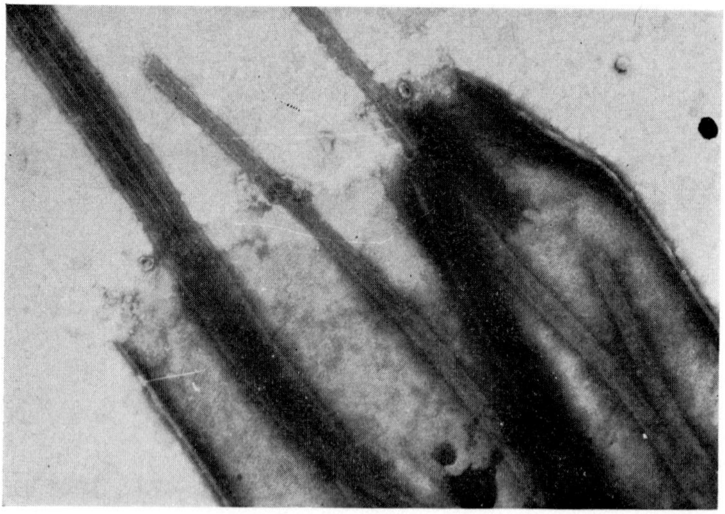

FIG. 15. Negatively-stained preparation of appendages of *Cl. sordellii* NCIB 6929 spores showing the tubular nature of the appendages and homogenous exosporium (× 40 000).

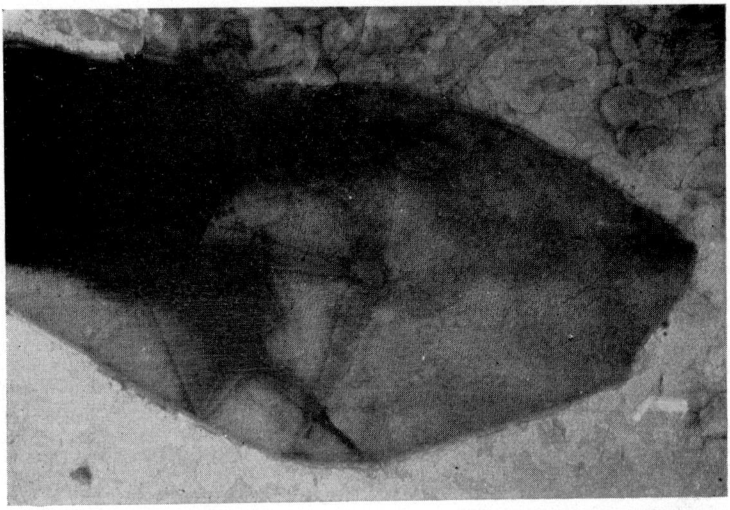

FIG. 16. Negatively-stained preparation of exosporium of *Cl. botulinum* type F, TRS 610B showing hexagonal pattern structure (× 50 000).

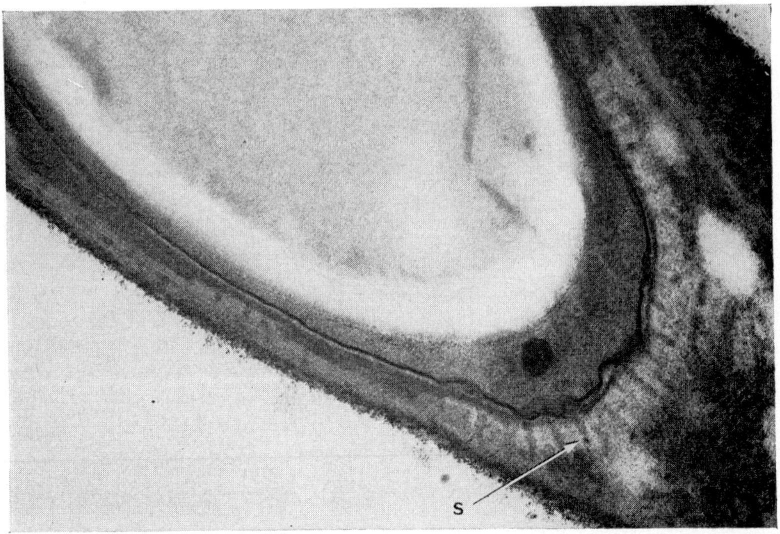

FIG. 17. Ultrathin section of sporulating cell of *Cl. welchii* CN 3353 showing spoke-like projections from the sporecoat (× 102 500).

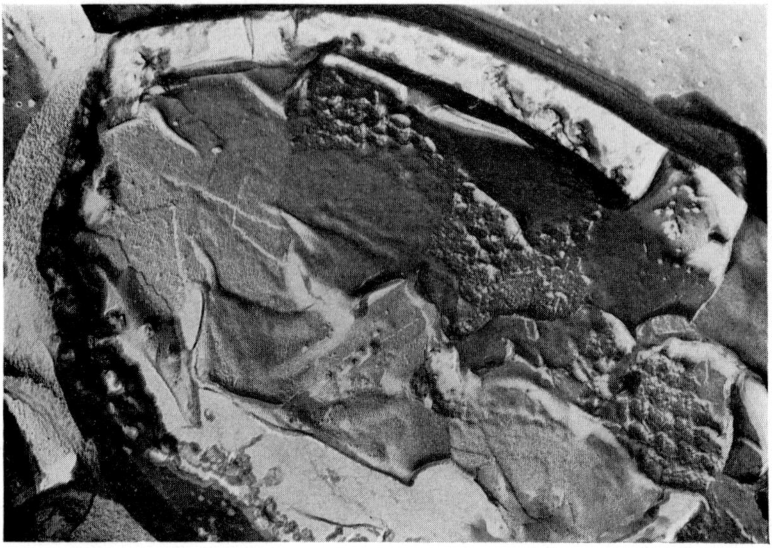

FIG. 18. Freeze-etched preparation of spore of *Cl. welchii* CN 3353 showing outer covering of the exosporium. A hexagonal pattern can be discerned on the outer surface and where this has been stripped off the spoke-like projections emanating from the sporecoat can be seen (× 61 000).

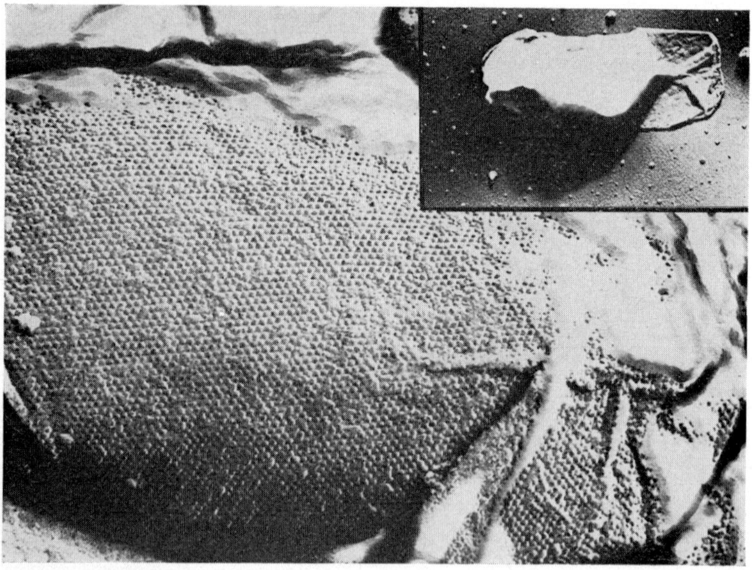

FIG. 19. Freeze-etched preparation of spore of *Cl. bifermentans* CN 1617. Note the hexagonally arranged globular particles on the surface approximately 9 nm in diameter (× 123 000).
Insert: Shadowed preparation of *Cl. bifermentans* showing exosporium (× 20 000).

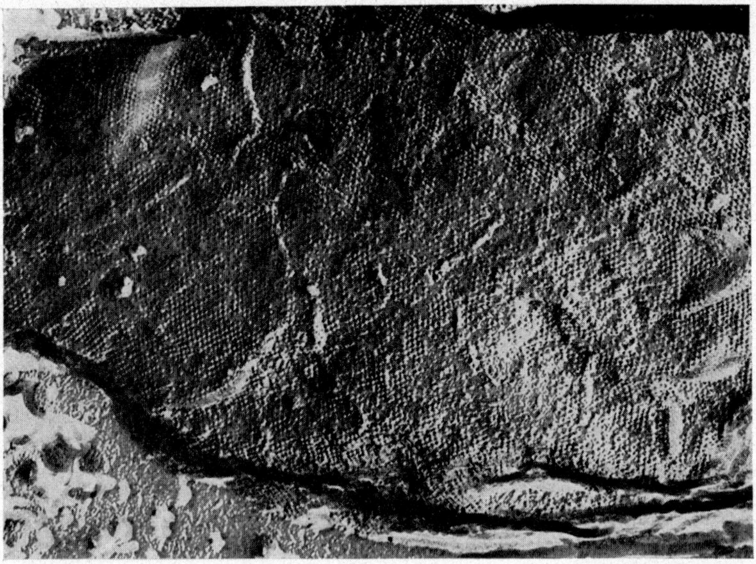

FIG. 20. Freeze-etched preparation of *Cl. bifermentans* CN 1617 showing spherical particles arranged on a smooth basement membrane (× 91 500).

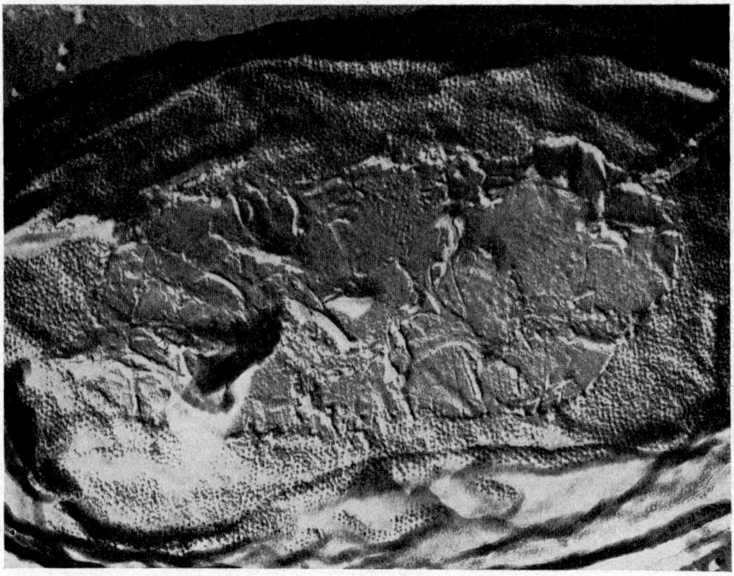

FIG. 21. Freeze-etched preparation of spore of *Cl. bifermentans* CN 1617 showing deeper layers of the exosporium (× 91 500).

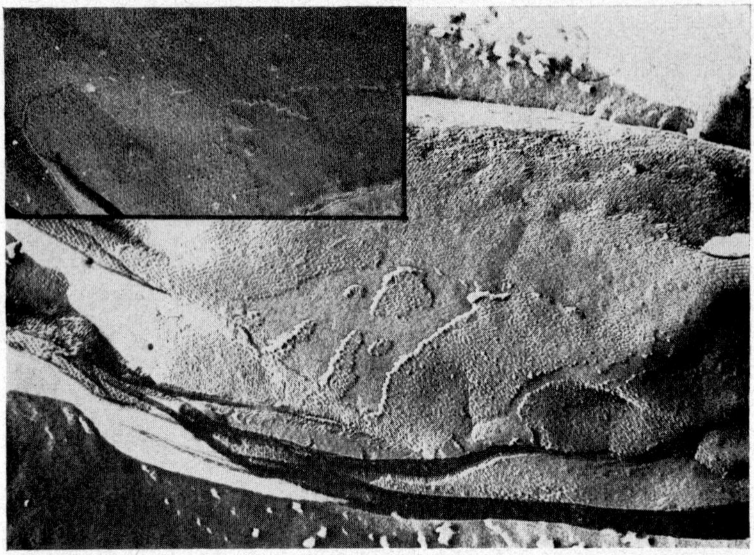

FIG. 22. Freeze-etched preparation of spore of *Cl. sordellii* CN 1734 showing the surface layer on which a fine hexagonal pattern can be discerned (× 80 000). *Insert:* (× 80 000).

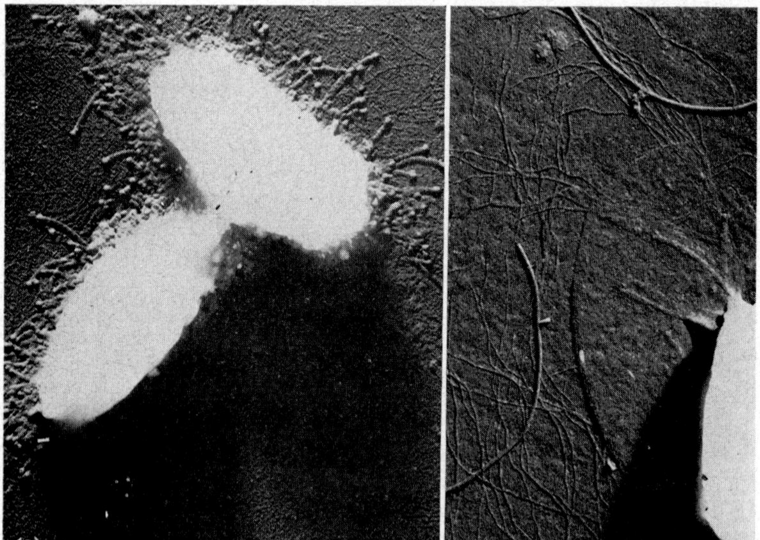

FIG. 23 (a and b). Shadowed preparations of spores of *Cl. botulinum* type E, NCIB 4248 showing tubular appendages emanating from the surface and thin exosporium. ((a) × 25 000, (b) × 46 000.)

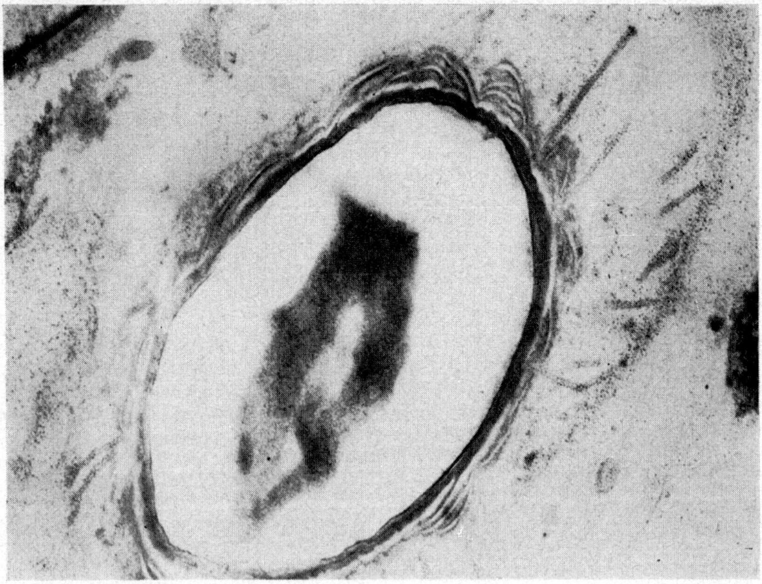

FIG. 24. Ultrathin section of spore of *Cl. botulinum* type E, NCIB 4248 showing tubular appendages arising from the sporecoat and enveloped within the thin exosporium (× 82 000).

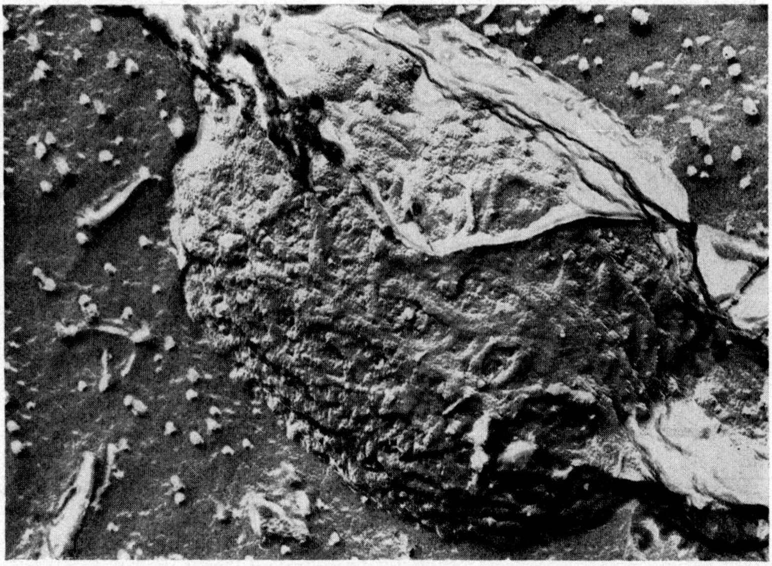

FIG. 25. Freeze-etched preparation of spore of *Cl. botulinum* type E, NCIB 4248 showing rough surface of the exosporium with underlying appendages (× 50 000).

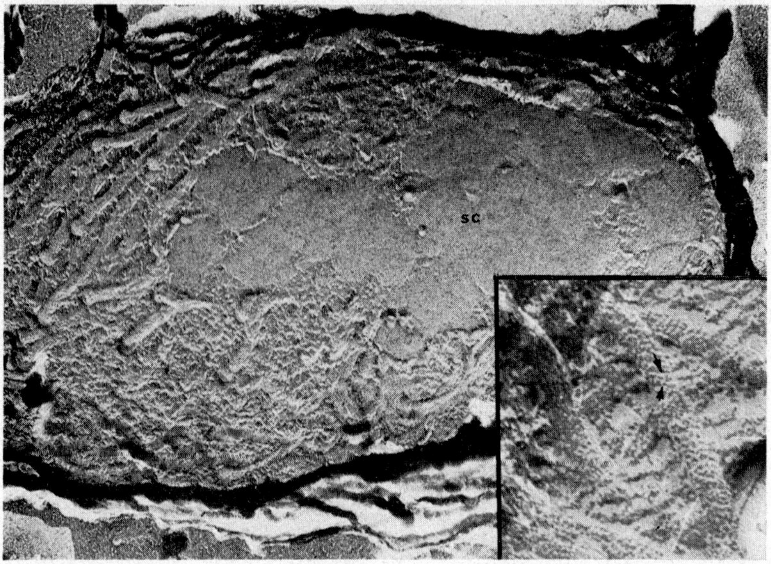

FIG. 26. Freeze-etched preparation of spore of *Cl. botulinum* type E, NCIB 4248 showing tubular appendages and underlying sporecoat ($\times$ 72 259).
Insert: Details of tubular appendages composed of spherical particles arranged in a helix ($\times$ 120 250).

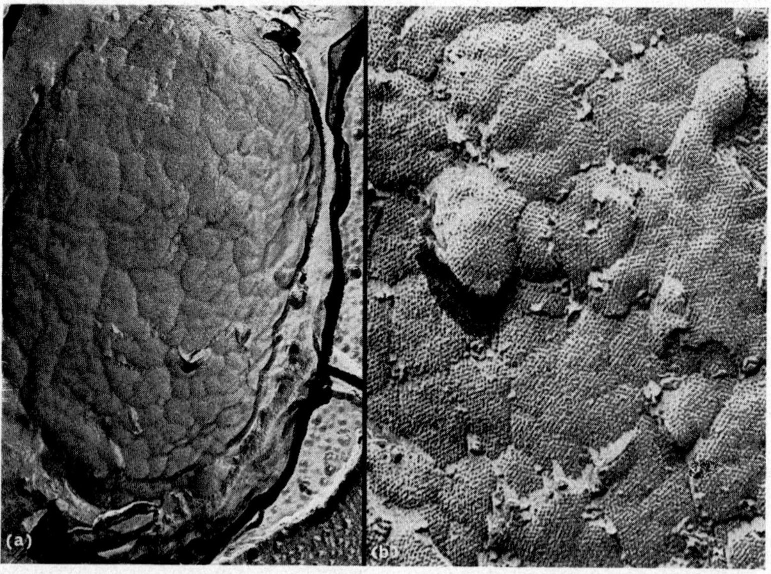

FIG. 27. Freeze-etched preparation of spores of *Cl. botulinum* type E, NCIB 4248 showing details of the sporecoat. The sporecoat is composed of tightly knit plates composed of particles approximately 3 nm in diameter arranged in a hexagonal pattern. ((a) $\times$ 61 000, (b) $\times$ 152 500.)

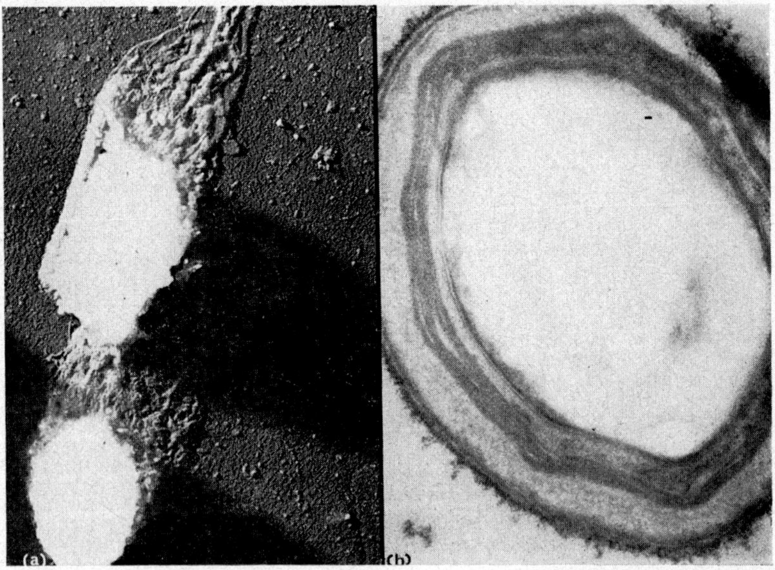

FIG. 28 (a). Shadowed preparation of spore of *Cl. botulinum* type A, CN 5008 showing thick voluminous exosporium (× 40 000). (b) Ultrathin section of sporulating cell of *Cl. botulinum* type A CN5008 showing laminated exosporium (× 73 350).

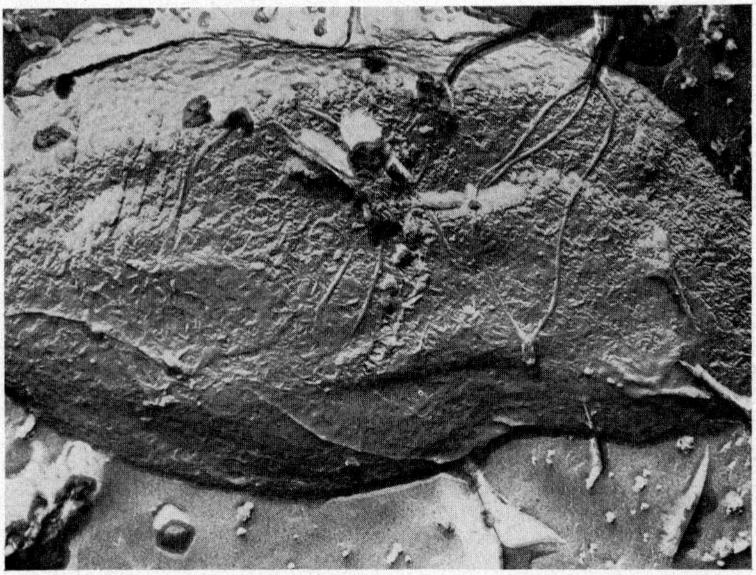

FIG. 29. Freeze-etched preparation of spore of *Cl. botulinum* type A, CN 5008 showing smooth surface covered with fibres (× 61 000).

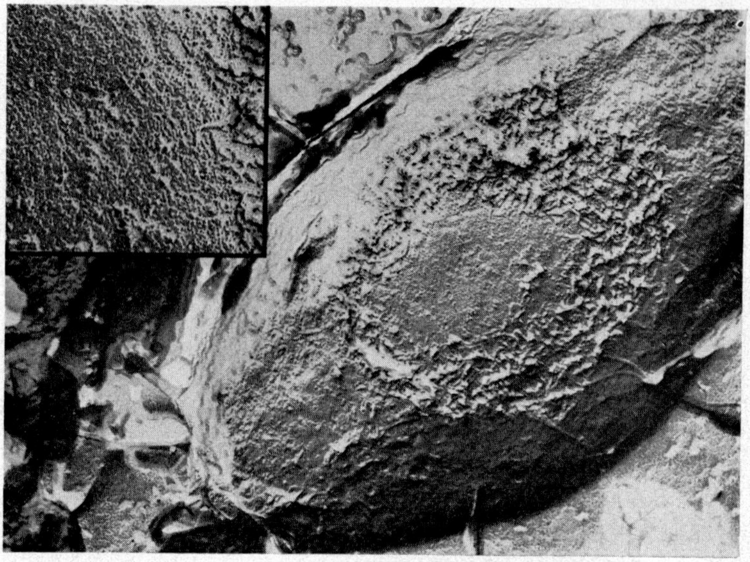

FIG. 30. Freeze-etched preparation of spore of *Cl. botulinum* type A, CN 5008 showing portion of exosporium removed and underlying layers showing hexagonal pattern (× 82 000).
Insert: (× 110 000).

FIG. 31. Freeze-etched preparation of *Cl. botulinum* type A, CN 5008 showing layers of the exosporium (× 102 500).

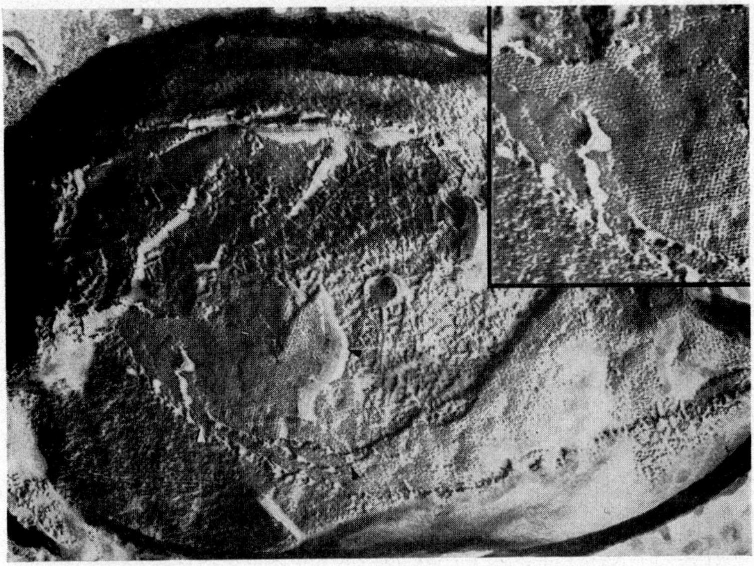

FIG. 32. Freeze-etched preparation of spore of *Cl. sporogenes* LUBC 206 showing details of exosporial layers and hexagonal pattern. (× 91 500). *Insert:* (× 152 500).

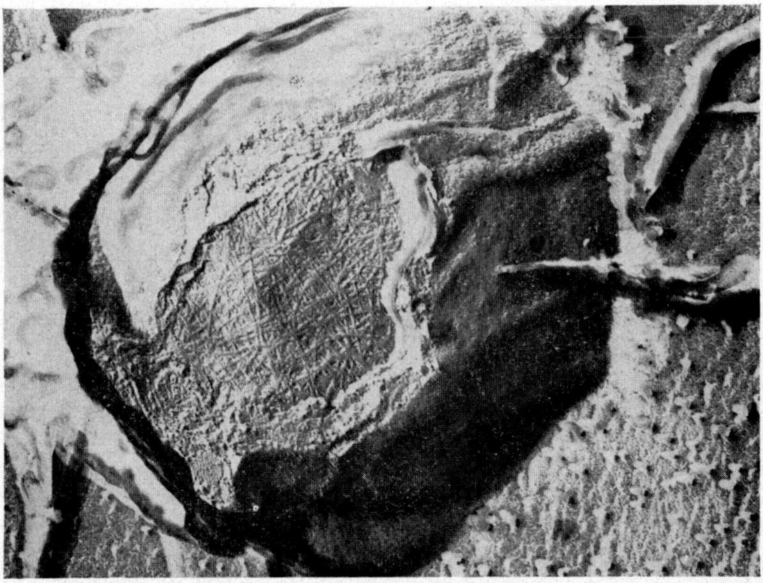

FIG. 33. Freeze-etched preparation of spore of *Cl. sporogenes* LUBC 206 showing exosporium removed and underlying sporecoat composed of randomly arranged fibres (× 72 250).

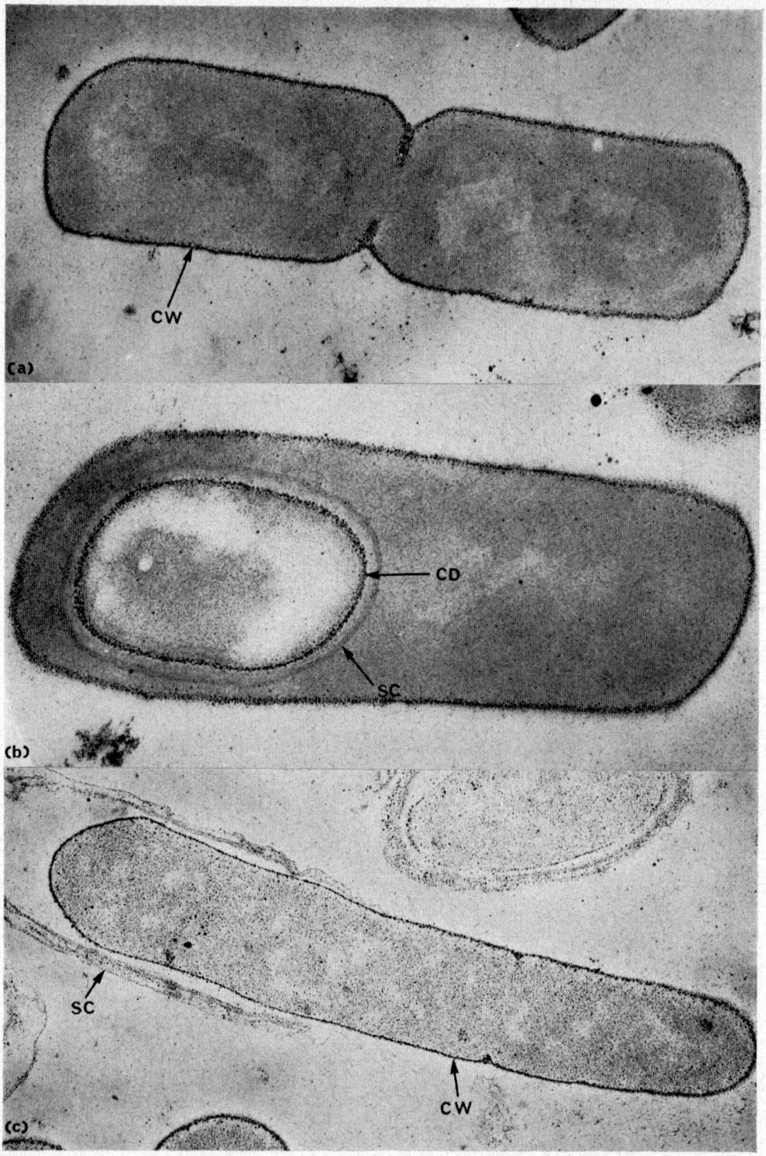

FIG. 34. (a) Ultrathin section of young vegetative cell of *Cl. bifermentans* CN 1617 after oxidation with periodic acid and treatment with silver methanamine staining technique. Deposits of silver are seen along the cell wall and developing cross walls. (× 61 000). (b) Sporulating cell of *Cl. bifermentans* CN 1617 similarly stained. Deposits of silver are seen along the cell wall and developing cortex (× 82 000). (c) Germinating spore of *Cl. bifermentans* CN 1617 showing outgrowth similarly stained. Deposits of silver are seen along the outgrowing cell wall but not the sporecoat (× 46 000).

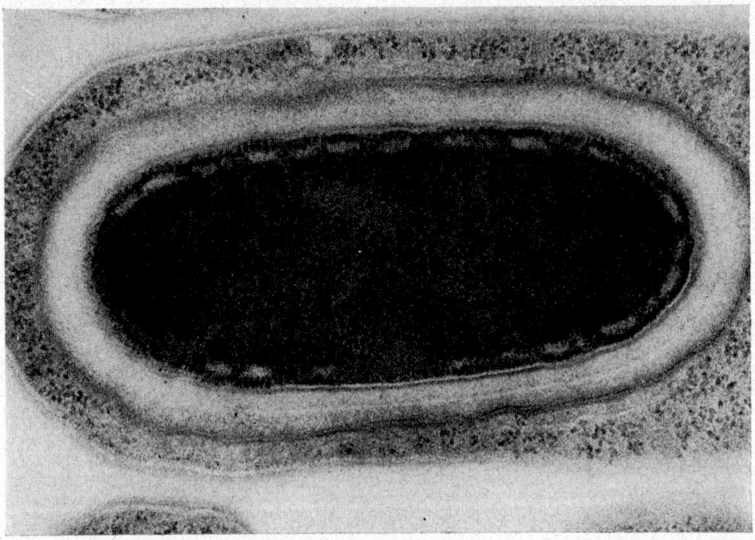

FIG. 35. Ultrathin section of sporulating cell of *Cl. bifermentans* CN 1617 fixed in glutaraldehyde and embedded in glycol methacrylate. Poststained with lead citrate. Note ribosomes in cytoplasm and developing spore and good preservation of the sporecoat (× 92 000).

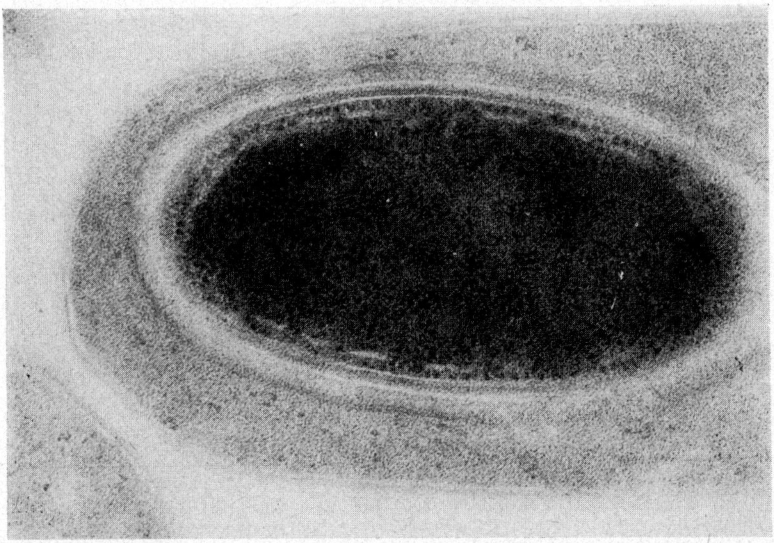

FIG. 36. Similar section to Fig. 35 after treatment with ribonuclease. Note removal of ribosomes from the vegetative cell but not the spore (× 92 000).

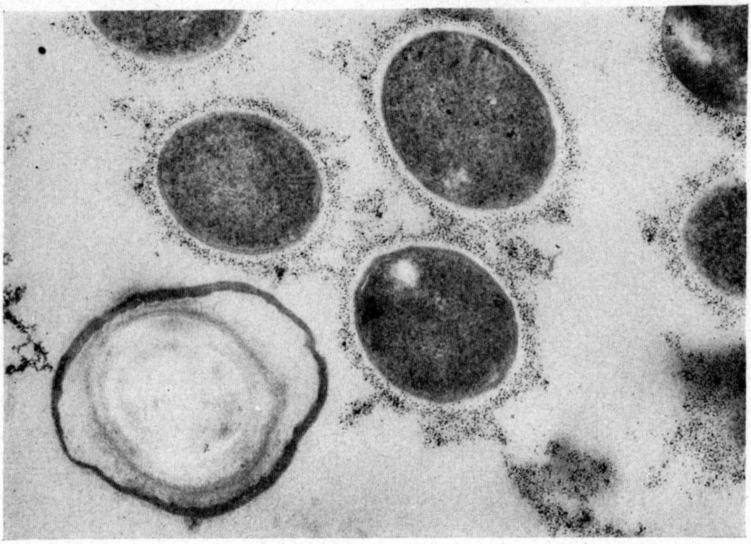

FIG. 37. Ultrathin section of mixture of spores and vegetative cells of *Cl. sordellii* CN 1734 after treatment with vegetative ferritin labelled antibody. Note deposits of ferritin around the vegetative cells but not the spore (× 47 600).

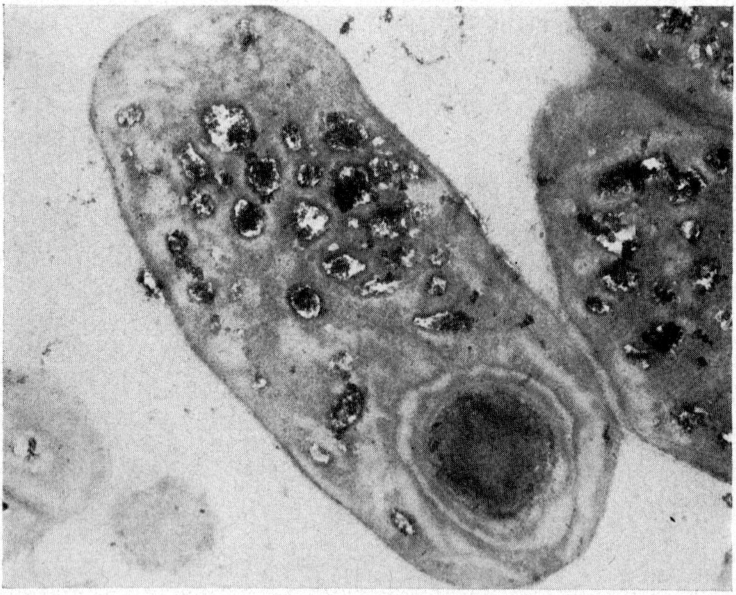

FIG. 38. Ultrathin section of *Cl. welchii* CN 3353 fixed in glutaraldehyde and embedded in glycol methacrylate after staining with unlabelled antibody and peroxidase anti-peroxidase (PAP) technique. Note deposits in the vacuoles present in the cytoplasm of the vegetative cell (× 18 750).

Negative staining and shadowing

Negative staining and shadowing can be used with advantage to demonstrate the structure of the exosporium and tubular appendages which have been described in various species. (Hodgkiss and Ordal, 1966; Hodgkiss, Ordal and Cann, 1966; Hodgkiss, Ordal and Cann, 1967; Krasil'nikov, Duda and Sokolov, 1964; Pope, Yolton and Rode, 1967; Rode, Crawford and Glenn Williams, 1967; Rode and Smith, 1971; Walker and Batty, 1967; Yolton, *et al.* 1968). A number of examples are illustrated.

Negative staining of *Cl. botulinum* type E spores shows numerous stalk-like appendages slightly swollen at the ends surrounding the spores (Figs 9 and 23a). The appendages are tubular, varying in length but have a uniform diameter of approximately 20 nm (Fig. 10). The lumen of each tubule terminates approximately 40 nm from the distal end of the tube which appears to be solid or to have a plug of material closing it and is surmounted by a hemispherical cap.

In some strains of *Cl. bifermentans* a further type of appendage is seen. This is approximately 2 μm in length, the distal portion being thicker than the proximal (Fig. 11). Negative staining again demonstrates the tubular nature and the distinctive fine structure of these appendages. The distal portion is thickened by micro-fibrils approximately 50 nm in length and 2·5–4·0 nm in diameter. These are arranged in a parallel array and inserted into the tubule wall at an angle of 40° to the long axis of the tubule (Fig. 12).

A third type of appendage is seen in some strains of *Cl. sordellii* and consists of open ended tubules (Fig. 13) commonly coiled within the exosporium (Fig. 14 and 15). Studies using ultrathin sectioning suggest that the tubules are coiled within the exosporium during formation of the spore but uncoil and break through the exosporium on lysis to the vegetative cell and release of the spore.

The structure of the exosporium as seen in negative staining shows considerable variation from species to species and even within species, but appears to fall into three main groups. Either the exosporium is thick and homogeneous showing no discernable ultrastructural detail, e.g. *Cl. sporogenes*, or alternatively, may appear as a thin membranous structure composed of fine fibrils, e.g. *Cl. botulinum* type E. In some species, e.g. *Cl. botulinum* type F negative staining demonstrates a hexagonal pattern of holes with a centre to centre spacing of approximately 10 nm (Fig. 16).

Freeze etching

The complexity of the exosporium in various clostridial species is particularly revealed by this technique.

In contrast to the lack of detail in shadowed preparations the exosporium of *Cl. bifermentans* is seen to be composed of globular particles approximately 9 nm in diameter which are arranged on a smooth basement membrane (Figs 19 and 20). Removal of this layer reveals a further series of layers which in some replicas show a fine hexagonal pattern of granules approximately 3 nm in diameter (Fig. 21). Similarly constituted layers were seen in other species investigated. Unlike the closely related *Cl. bifermentans* the exosporium of *Cl. sordellii* reveals only a series of layers showing the hexagonal pattern (Fig. 22). In *Cl. botulinum* type E, the thin fragile exosporium seen in thin sections to envelope the hair-like projections emanating from the surface of the spore (Fig. 24) is revealed by freeze etching as an essentially rough layer and the hair-like projections can easily be seen underneath this thin outer covering (Fig. 25). Removal of the exosporium reveals the hair-like projections which can be seen to be composed of small spherical units arranged in a helix manner (Fig. 26). The sporecoat is composed of a series of tightly knit plates (Fig. 27a) which have a hexagonal pattern on their surface similar to that seen in some of the layers of the exosporium (Fig. 27b).

In *Cl. botulinum* type A the thick voluminous exosporium recognized in both shadowed preparations and in thin section is seen to be composed of a number of laminated layers (Figs 28a and 28b). The outer surface of the spore appears to be comparatively smooth although a number of hair-like projections can be seen (Fig. 29). Fracturing of the outer layer reveals an underlying rough layer which, in turn, covers underlying layers showing a faint hexagonal pattern (Fig. 30).

Spores of the closely related *Cl. sporogenes* reveal an almost identical structure to *Cl. botulinum* type A. The layers of the exosporium and its hexagonal pattern can clearly be seen (Figs 31 and 32). The sporecoat of this organism appears to be composed of randomly arranged fibres covering a smooth layer (Figs 32 and 33).

The surface of the exosporium of spores of *Cl. perfringens* type A shows a smooth layer on which a fine hexagonal pattern can be discerned (Fig. 18). Where the exosporium has been fractured the spoke-like projections emanating from the sporecoat seen in thin section (Fig. 17) can be seen protruding in the form of hexagonal plates.

Histochemical and cytochemical staining

Three methods are illustrated.

Using the silver methanamine staining technique the chemical similarity between the cortex and cell wall of sporulating organism can be demonstrated visually (Figs 34a and 34b). The method detects compounds containing 1:2 glycol groups, e.g. *N*-acetylglucosomine, which are part of structural polysaccharides such as mucopeptide (Walker and Short, 1968; 1969). The sporecoat which is protein in nature does not stain (Figs 34b and 34c).

Following fixation in glutaraldehyde and embedding in methacrylates the appearance of organisms in ultrathin sections differs from that observed with classical Kellenberger fixation (Granboulan and Leduc, 1967). There is an absence of unit membranes and nuclear preservation is poor. On the other hand, unlike Kellenberger fixation, chemical components are preserved in the section in a manner which makes them still susceptible to enzyme digestion. Sections of sporulating bacilli fixed and embedded in this way when floated onto solutions of ribonuclease result in removal of ribosomes from the vegetative cells and developing spores. As maturation of the spore proceeds, however, ribosomes can no longer be removed from the spore although vegetative ribosomes are still susceptible to digestion (Figs 35 and 36). It appears that during maturation of the spore the ribosomes undergo changes, possibly due to chelation with substances such as dipicolinic acid which mask the enzyme sites (Walker, 1969; Walker, Thomson and Short, 1971).

Using the technique of metal "capture" it is possible to demonstrate esterases during sporulation of clostridial species (Walker *et al.*, 1967). Incubation of washed, sporulating cells in thiolacetic acid in the presence of lead sulphide at the sites of enzyme activity. In young vegetative cells deposits are found on the cytoplasmic membrane and during sporulation around the developing spore membranes (Fig. 8).

Immunochemical staining

The labelling of antibodies with ferritin and enzymes has proved an effective marker for use with the electron microscope. Use of ferritin labelled antibodies to demonstrate antigens on the surface of spores and vegetative cells of clostridial species is illustrated in Fig. 37 (Walker *et al.*, 1967). Unfortunately, due to the non-specific adherence of ferritin to commonly employed embedding media this marker can only effectively be used with micro-organisms using the preembedding staining technique. For this reason, the use of enzyme labelled and other antibody techniques have been introduced. In Fig. 38 the use of one of these alternatives, the soluble peroxidase anti-peroxidase complex (PAP) in conjunction with unlabelled antibodies (Sternberger *et al.*, 1970), is illustrated. Suitably fixed and embedded sections of sporulating cells of *Cl. welchii* have been stained with unlabelled antibody to purified enterotoxin followed by treatment with the PAP complex and incubation in the appropriate peroxidase substrate to mark the antibody. It can be seen that deposits are present in vacuoles in the cytoplasm associated with the developing spore (Walker, Short and Roper, 1975).

General Discussion

By combining information obtained from different techniques a comprehensive picture of the synthesis and assembly of spore components can be built up. A feature of the majority of the pathogenic species studied is the presence of an exosporial membrane surrounding the spore. In the majority of cases this is revealed, particularly by freeze etching, as a structure of considerable complexity. Apart from *Cl. botulinum* type E, the exosporium of most species is composed of several layers of hexagonally arranged granules with variations in the structure of the outer layer. This varies from the globular particles observed with *Cl. bifermentans*, to the relatively smooth surface of *Cl. botulinum* type A. Fractures of the spore, while revealing much of the complexity of the exosporium, show little detail of the sporecoat. The presence of the exosporium appears to make it difficult to obtain fractures revealing this structure. The one exception is *Cl. botulinum* type E and it is perhaps significant that in this organism the exosporium is thin and relatively simple in nature. In this species the coat is seen to be composed of a series of tightly knit plates composed of granules arranged in a hexagonal pattern similar to that seen in some layers of the exosporium. Many workers prefer to regard the exosporium as a modified outer sporecoat and the similarity in structure adds support to this. The only other fractures revealing sporecoat structure appear in *Cl. sporogenes*, where the coat consists of randomly arranged fibres arranged on a basement membrane.

The tubular appendages seen in various species clearly originate from the sporecoat. The function of these appendages which reveal considerable structural complexity is unknown. Many of the strains investigated by Hodgkiss *et al.* (1967), on which these appendages can be demonstrated originate from marine environments or from fish products and the appendages may play some part in attachment to surfaces in the natural state. A serious attempt to relate types of appendages to the ecology of the organisms might well be rewarding. Similarly, several attempts have been made to assess the presence or absence of appendages as an aid to classification (Hodgkiss *et al.*, 1967; Rode, *et al.*, 1971). Clearly the presence of different types of appendage on different strains within the same species makes their use in general classification doubtful (Hodgkiss *et al.*, 1967). However, it could well be that grouping of organisms within species on the basis of appendages might prove fruitful in providing morphological information to gain more insight into a more meaningful classification of bacteria. That surface structure may well indicate deeper fundamental differences is illustrated by the observations of Rode (1968) who showed that different germination patterns of spores within a species could be correlated with differences in surface structure as revealed by the carbon replica technique.

Finally, the relationship of sporulation to toxin production needs to be further examined. In *Bacillus thuringiensis* a crystalline protein toxin is formed during sporulation and assembled on the developing exosporium (Somerville and James, 1970; Somerville, 1971). Recent work by Short *et al.* (1974) using the electron microscope and the PAP unlabelled antibody technique has demonstrated the antigenic similarity between this toxic material and layers of the sporecoat and exosporium. This confirms previous immunological and biochemical observations (Delafield, Somerville and Rittenberg, 1968; Somerville, Delafield and Rittenberg, 1968, 1970; Lecadet, Chevrier and Dedonder, 1972). A similar crystalline inclusion has been demonstrated in a number of food poisoning strains of *Cl. welchii* and it is possible that this represents a structure analogous to that found in *B. thuringiensis*. On the other hand, an enterotoxin producing strain of *Cl. welchii* studied by Walker *et al.* (1975) which did not produce crystals showed location of the enterotoxin in vacuoles developing in the cytoplasm at the time of sporulation. The evidence suggested that the enterotoxin was a spore-related rather than a spore-dependent substance. Whether other clostridial toxins are similarly related is not known.

The more recent developments in the area of immunological staining of sections of bacteria should enable many of these and other problems associated with the sites of synthesis and assembly of spore components to be elucidated. Already in *B. cereus* it has been shown that exosporial protein is synthesized in the cytoplasm of the mother cell and transported and assembled around the developing spore (Walker *et al.*, 1975; Short and Walker, 1975). Further studies using antisera to other purified spore components should provide further information on these aspects of spore structure.

References

BAYEN, H., FREHEL, C., RYTER, A. & SEBALD, M. (1967). Etude cytologique de la sporulation chez *Clostridium histolyticum*. Souche sporogene et mutants de sporulation. *Annls Inst. Pasteur Paris*, **113**, 163.

DELAFIELD, F. P., SOMERVILLE, H. J. & RITTENBERG, S. C. (1968). Immunological homology between crystal and spore protein of *Bacillus thuringiensis*. *J. Bact.*, **96**, 713.

DUNCAN, C. L., KING, G. J. & FRIEBEN, W. R. (1973). A paracystalline inclusion formed during sporulation of enterotoxin producing strains of *Clostridium perfringens* type A. *J. Bact.*, **114,** 845.

DUNCAN, C. L. & STRONG, D. H. (1968). Improved medium for sporulation of *Clostridium perfringens. Appl. Microbiol.*, **16,** 82.

FITZ-JAMES, P. C. (1962). Morphology of spore development in *Clostridium pectinovorum. J. Bact.*, **84,** 104.

GRANBOULAN, P. & LEDUC, E. H. (1967). Ultrastructural cytochemistry of *Bacillus subtilis. J. ultrastruct. Res.*, **20,** 111.

HASHIMOTO, T. & NAYLOR, H. B. (1958). Studies of the fine structure of microorganisms. II. Electron microscopic studies on sporulation of *Clostridium sporogenes. J. Bact.*, **75,** 647.

HASHIMOTO, T. (1960). Studies on the cytological basis of spore resistance and the origin of the first spore coat. *Tokushima J. exp. Med.*, **7,** 36.

HODGKISS, W. & ORDAL, Z. J. (1966). Morphology of the spore of some strains of *Clostridium botulinum* type E. *J. Bact.*, **91,** 2031.

HODGKISS, W., ORDAL, Z. J. & CANN, D. C. (1966). The comparative morphology of the spores of *Clostridium botulinum* type E and the spores of the "OS MUTANT". *Can. J. Microbiol.*, **12,** 1283.

HODGKISS, W., ORDAL, Z. J. & CANN, D. C. (1967). The morphology and ultrastructure of the spore and exosporium of some *Clostridium* species. *J. gen. Microbiol.*, **47,** 213.

HOENIGER, J. F. M. & HEADLEY, C. L. (1968). Cytology of spore germination in *Clostridium pectinovorum. J. Bact.*, **96,** 1835.

HOENIGER, J. F. M., STUART, P. F. & HOLT, S. C. (1968). Cytology of spore formation in *Clostridium perfringens. J. Bact.*, **96,** 1818.

KELLENBERGER, E., RYTER, A. & SÉCHAUD, J. (1958). Electron microscope study of DNA-containing plasms. II. Vegetative and mature phage DNA as compared with normal bacterial nucleoids in different physiological states. *J. biophys. biochem. Cytol.*, **4,** 671.

KRASIL'NIKOV, N. A., DUDA, V. I. & SOKOLOV, A. A. (1964). Protrusions on the surface of spores of anaerobic bacteria of the genus *Clostridium. Mikrobiologiya*, **33,** 454.

LECADET, M. M., CHEVRIER, G. & DEDONDER, R. (1972). Analysis of a protein fraction in the sporecoats of *Bacillus thuringiensis. Eurpn. J. Biochem.*, **25,** 349.

POPE, L., YOLTON, D. P. & RODE, L. J. (1967) Appendages of *Clostridium bifermentans* spores. *J. Bact.*, **94,** 1206.

RODE, L. J. (1968). Correlation between spore structure and spore properties in *Bacillus megaterium. J. Bact.*, **95,** 1979.

RODE, L. J., CRAWFORD, D. & GLENN-WILLIAMS, M. (1967). Clostridium spores with ribbon like appendages. *J. Bact.*, **93,** 1160.

RODE, L. J. & SMITH, L. D. S. (1971). Taxonomic implications of spore fine structure in *Clostridium bifermentans. J. Bact.*, **105,** 349.

RODE, L. J., POPE, L., FILIP, C. & SMITH, L. DS. (1971). Spore appendages and taxonomy of *Clostridium sordellii. J. Bact.*, **108,** 1384.

SHORT, J. A. & WALKER, P. D. (1975). The location of bacterial antigens on sections of *Bacillus cereus* by use of the soluble peroxidase anti-peroxidase complex and unlabelled antibody. *J. gen. Microbiol*, **89,** 93.

SHORT, J. A. & WALKER, P. D. (1975). Techniques for transmission electron microscopy. (This publication.)

SHORT, J. A., WALKER, P. D., THOMSON, R. O. T. & SOMERVILLE, H. J. (1974). The fine structure of *Bacillus finitimus* and *Bacillus thuringiensis* spores with special reference to the location of crystal antigen. *J. gen. Microbiol.*, **84**, 261.

SOMERVILLE, H. J. (1971). Formation of the parasporal inclusion of *Bacillus thuringiensis*. *Eurpn. J. Biochem.*, **18**, 226.

SOMERVILLE, H. J., DELAFIELD, F. P. & RITTENBERG, S. C. (1968). Biochemical homology between crystal and spore protein of *Bacillus thuringiensis*. *J. Bact.*, **96**, 721.

SOMERVILLE, H. J., DELAFIELD, F. P. & RITTENBERG, S. C. (1970). Urea-mercaptoethanol-soluble protein from spores of *Bacillus thuringiensis* and other species. *J. Bact.*, **101**,, 551.

SOMERVILLE, H. J. & JAMES, C. R. (1970). Association of the crystalline inclusion of *Bacillus thuringiensis* with the exosporium. *J. Bact.*, **102**, 580.

SMITH, A. G. & ELLNER, P. D. (1957). Cytological observations on the sporulation process of *Clostridium perfringens*. *J. Bact.*, **73**, 1.

STERNBERGER, L. A., HARDY, P. D., CUCULIS, J. J. & MEYER, H. G. (1970). The unlabelled antibody enzyme method of immunohistochemistry. Preparation and properties of soluble antigen-antibody complex (horseradish peroxidase-anti-horseradish peroxidase) and its use in identification of spirochetes. *J. Histochem. Cytochem.*, **18**, 315.

TAKAGI, A., KAWATA, T. & YAMAMOTO, S. (1960). Electron microscope studies on ultrathin sections of spores of the *Clostridium* group, with special reference to the sporulation and germination process. *J. Bact.*, **80**, 37.

TAKAGI, A., KAWATA, T., YAMAMOTO, S., KUBO, T. & OKITA, S. (1960). Electron microscopic studies on ultrathin sections of spores of *Clostridium tetani* and *Clostridium histolyticum* with special reference to sporulation and spore germination process. *Jap. J. Microbiol.*, **4**, 137.

TAKAGI, A., NAKAMURA, K. & UREDA, M. (1965). Electron microscope studies of the intracytoplasmic membrane system in *Clostridium tetani* and *Clostridium botulinum*. *Jap. J. Microbiol.*, **9**, 131.

WALKER, P. D. (1969). The location of chemical components on ultrathin sections of *Bacillus cereus* embedded in glycol methacrylate. *J. appl. Bact.*, **32**, 463.

WALKER, P. D. & BATTY, I. (1967). Spore morphology with special reference to the identity of the "O.S." Variants of *Clostridium botulinum* type E. *J. Path. Bact.*, **93**, 340.

WALKER, P. D., THOMSON, R. O. T. & SHORT, J. A. (1971). Location of chemical components on ultrathin sections of bacteria. In *Spore Research* (Barker, A. N., Gould, G. W. and Wolf, J., eds), p. 201. New York and London: Academic Press.

WALKER, P. D. & SHORT, J. A. (1968). The location of mucopolysaccharides on ultrathin sections of bacteria by the silver methanamine staining technique. *J. gen. Microbiol.*, **52**, 467.

WALKER, P. D. & SHORT, J. A. (1969). Location of bacterial polysaccharide during various phases of growth. *J. Bact.*, **98**, 1342.

WALKER, P. D., SHORT, J. A. & ROPER, G. (1975). Location of antigens on ultrathin sections of spore-forming bacteria. In *Spores VI* (Halvorsen, H. O., Hanson R. and Campbell, L. L., eds) American Society for Microbiology, p. 572.

WALKER, P. D., THOMSON, R. O. T. & BAILLIE, A. (1967). Fine structure of clostridia with special reference to the location of antigens and enzymes. *J. appl. Bact.*, **30**, 444.

YOLTON, D. P., POPE, L., GLENN WILLIAMS, M. & RODE, L. J. (1968). Further electron microscope characterization of spore appendages of *Clostridium bifermentans*. *J. Bact.*, **95**, 231.

Germination and Outgrowth of Bacillus Spores

G. J. DRING, G. W. GOULD, SUSAN C. DICKENS AND J. M. STUBBS

Unilever Research Laboratory, Colworth House, Sharnbrook, Bedford, England

When incubated in a suitable medium, bacterial endospores lose their extreme resistance and dormancy and grow to form new vegetative cells through changes termed germination and outgrowth. The sequence of physical and chemical changes that accompany germination take place in temporal sequence and include sensitization of the spores to heat and radiation, release of calcium and dipicolinic acid, darkening under the phase contrast microscope, hydration and swelling of the spore, loss of resistance to staining, release of low molecular weight fragments of solubilized peptidoglycan and a fall in the extinction of spore suspensions (Levinson and Hyatt, 1966; Hashimoto, Frieben and Conti, 1969; Dring and Gould, 1971). Respiration and other metabolic activities increase at this time. Studies with *Bacillus megaterium* spores show that during this period *de novo* protein synthesis is supported by endogenous proteolysis of dormant spore protein; amino acids liberated being initially excreted and then reabsorbed (Setlow, 1974, 1975; Setlow and Primus, 1975). Provided that suitable nutrients are then present, outgrowth will occur. During outgrowth, transport of ions (Eisenstadt and Silver, 1972) and nutrients commences, followed by the synthesis of new protein, RNA, DNA, enzymes and vegetative cell wall polymers as the new vegetative cell is formed (Strange and Hunter, 1969).

These physical and chemical changes are accompanied by a number of well-defined cytological changes which are illustrated in the following electron micrographs. First, the structure of the dormant spore is considered, and second, the major cytological changes that accompany germination and outgrowth.

Methods

Germination and outgrowth

The methods for initiating and monitoring germination and outgrowth were as described by Gould (1971). Samples for electron microscopy were removed from cultures by centrifugation prior to resuspension in solutions of fixatives (Hamilton and Stubbs, 1967).

Fixation

The central regions of ungerminated spores are not readily penetrated by glutaraldehyde-based fixatives (Rode, Lewis and Foster, 1962). Ungerminated spores were therefore fixed by resuspension in freshly made and filtered potassium permanganate solution (2%, w/v) and incubation at 20° for 90 min prior to washing by centrifuging in water at least 10 times.

Germinating spores and outgrowing forms were fixed either with potassium permanganate as above or with glutaraldehyde-osmium tetroxide. Pellets were resuspended in glutaraldehyde (5%, w/v) in sodium phosphate buffer (100 mM; pH 7·3) and, following incubation at 4° for 90 min were centrifuged, washed once in buffer and resuspended in osmium tetroxide (1%, w/v) at 4° for 2 h prior to washing with water.

Fixed pellets were embedded in Epon 812. Following sectioning, glutaraldehyde-osmium tetroxide-fixed specimens were stained with lead citrate (1·5%, w/v; pH 12·0).

Freeze etching

Spores were suspended in 20% glycerine and then freeze-etched at −100°. The preparations were shadowed with carbon–platinum, and then cleaned with chromic acid for examination in the electron microscope.

The Dormant Spore

The diagram on p. 149 is a representation of the major structures in a bacterial endospore and gives the terms commonly used to describe them. The spore consists of a small central quantity of cytoplasm (the core or

FIGS 1–6. These figures illustrate the basic similarity of the structure of all bacterial endospores, as seen in ultrathin sections. In particular all have a central core surrounded by a plasma membrane and a wide cortex (see Fig. 1). The main variations lie in the differing complexity of the outer coat layers. An exosporium may or may not be present.

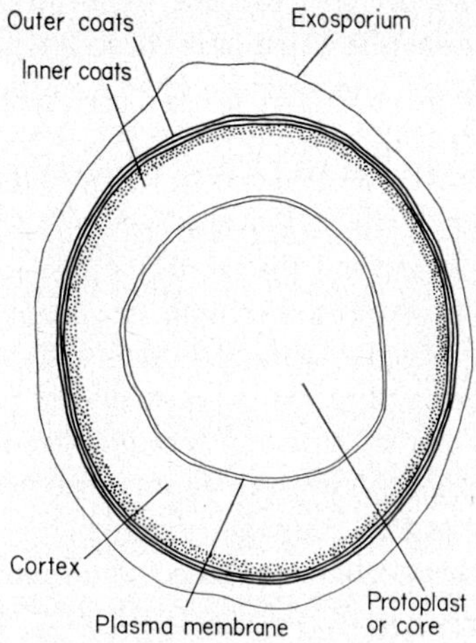

DIAGRAM. Major structures of a typical bacterial endospore.

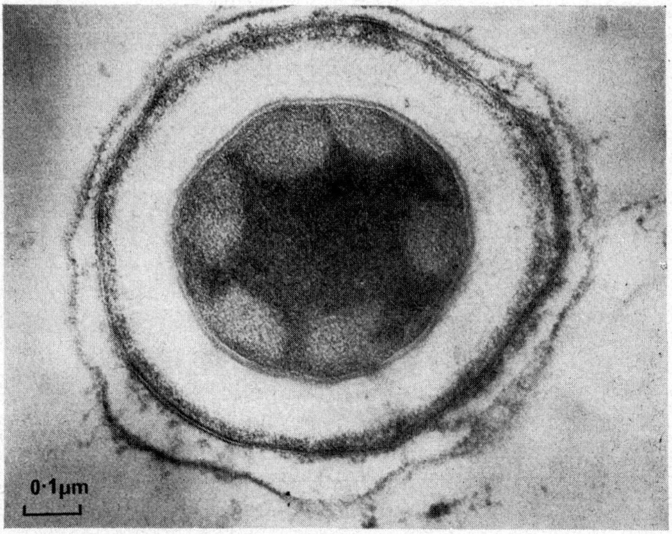

FIG. 1. *B. cereus*. Relatively simple coat layer in terms of the number of discernible layers present. Note presence of the surrounding loose and irregularly-shaped exosporium.

protoplast) enclosed within the plasma membrane and surrounded by thick specialized layers peculiar to spores; first, a wide electron-translucent zone known as the cortex and outside this, a number of laminated coat layers and, in some species only, a loose outer structure, the exosporium.

The major structures can all be seen in the sections of spores of *Bacillus* species in Figs 1, 2 and 3. These plates also illustrate that the major ultra-structural differences between spores of different species reside in the complexity of the outer coat layers and the presence or absence of an exosporium. For comparison, the structure of spores of *Clostridium* species (see *Cl. sporogenes*, Fig. 4) is basically similar to that of *Bacillus* species, and the endospores of two *Thermoactinomyces* species shown in Figs 5 and 6 are also remarkably similar to those of some *Bacillus* spores; e.g. *B. subtilis* (Fig. 2) and *B. coagulans* (Fig. 3).

The complexity of sporecoat structure is illustrated by the enlarged portion of the sporecoat of *B. polymyxa* shown in Fig. 7 and by the freeze-etch micrograph of the outer coat surface of a *B. coagulans* spore shown in Fig. 8. The freeze-etched spore of *B. cereus* in Fig. 9 has fractured to reveal the smooth outer surface of the cortex, the thickness of the spore-coat and the thin outer exosporium.

Germination and Outgrowth

Bacterial endospores may be divided into two groups according to the apparent mode of emergence of the new cell from the sporecoat during outgrowth as seen in the light microscope. Spores of one group appear to dissolve most of the coat; spores of the other group appear to split the coat with little indication of dissolution. The coat dissolvers include most of the large-celled species (e.g. *B. cereus*, *B. megaterium*) whereas the coat splitters include most of the small-celled species (e.g. *B. subtilis*, *B. coagulans*, *B. polymyxa*).

Figure 10 shows a germinating spore of *B. cereus* (a dissolver) in which lysis of petidoglycan in the cortex region is well advanced (compare with Fig. 1), allowing expansion of the central protoplast. In Fig. 11 the lytic process is about complete and the cortex has almost disappeared. Emergence of the new vegetative cell from the remaining sporecoats and exosporium is shown in Fig. 12. At this stage the new vegetative cell-wall is well defined. Eventually residues of the coats and exosporia are discarded and remain as debris in the outgrowing culture (Fig. 13).

Figures 14 and 15 illustrate germination and outgrowth of *B. polymyxa*, a coat splitter. The spore swells on germination and the cortex becomes more stainable and electron opaque (Fig. 14) but does not rapidly disappear as it does during the germination of coat dissolvers. Emergence of the

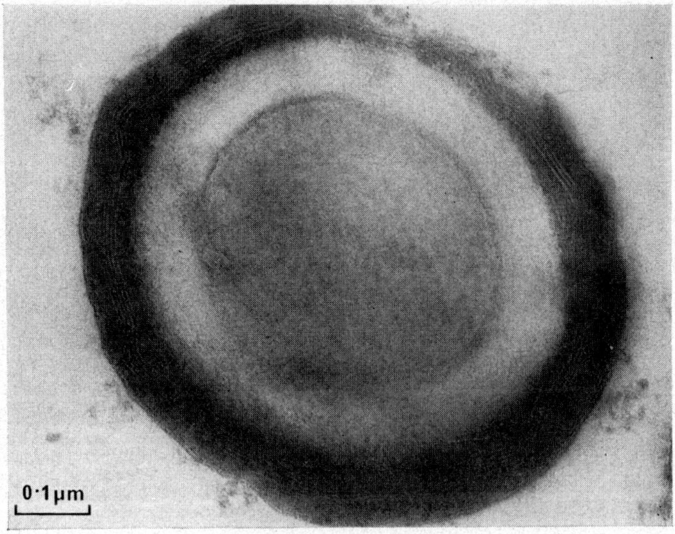

FIG. 2. *B. subtilis* and FIG. 3. *B. coagulous*. Coat layers of differing complexity.

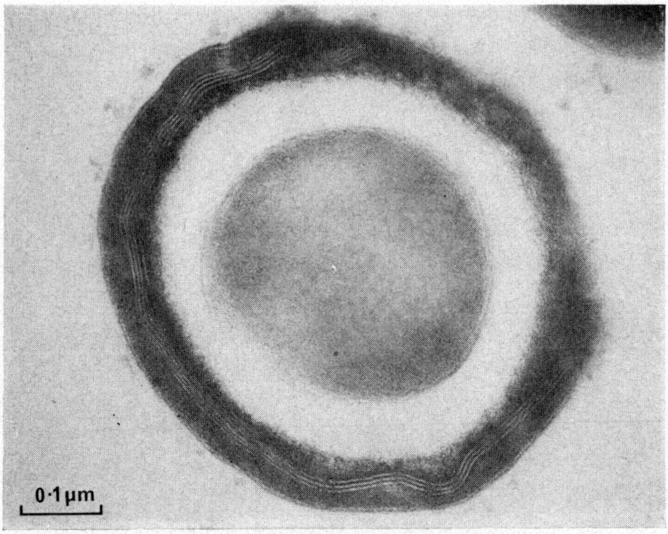

FIG. 3

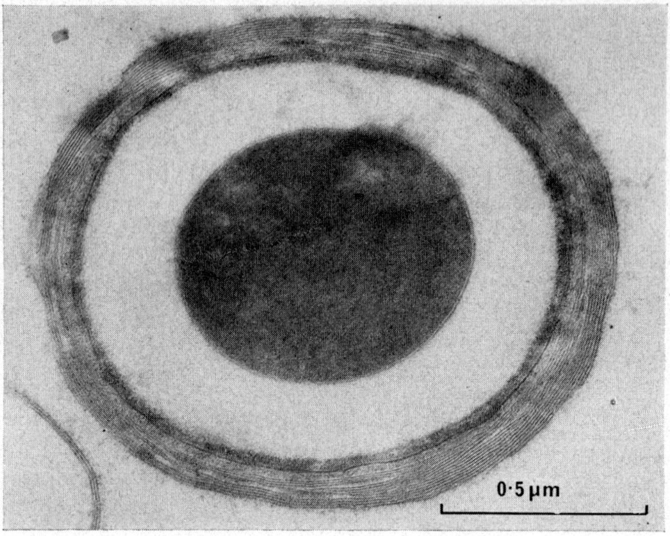

FIG. 4. *Cl. sporogenes*. Extremely multilaminated coat layers (Gould and Dring, 1974).

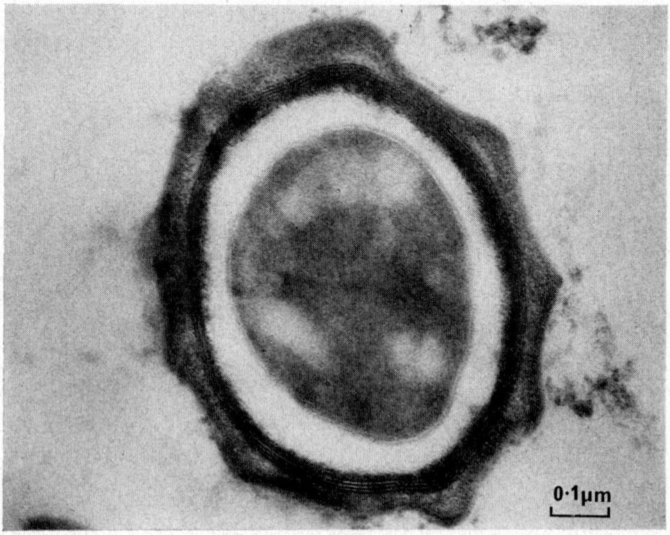

FIG. 5. *Thermoactinomyces vulgaris*. Note the multilayered inner coat region together with a surface-sculptured outer coat.

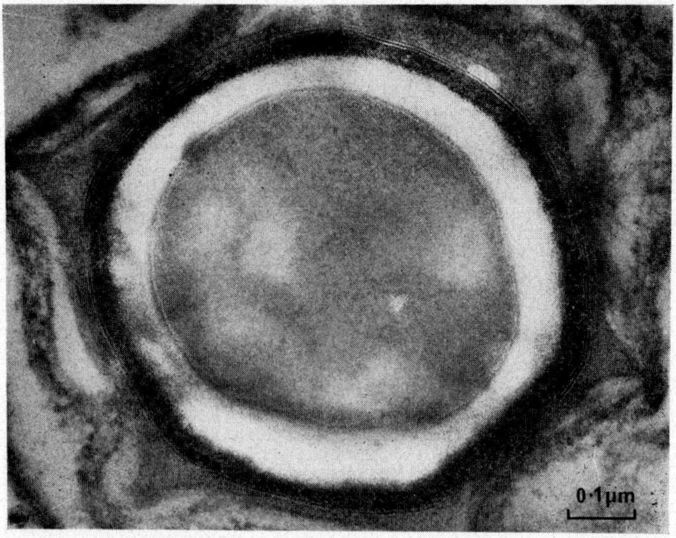

FIG. 6. *Actinobifida dichotomica*. Section of a spore within a vegetative hypha. Note marked sculpturing of the outer coat layer.

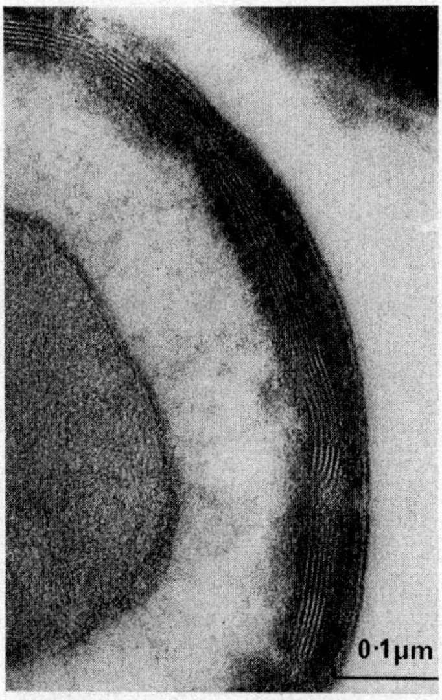

FIG. 7. *B. polymyxa*. Complex multilaminated coat layer.

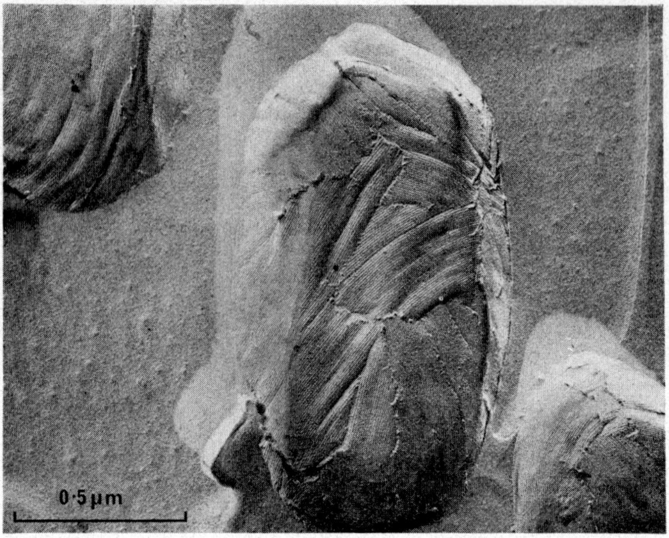

FIG. 8. *B. coagulans*. Freeze-etched spore showing "bandaged" appearance of the outer sporecoat layer (Gould, Stubbs and King, 1970).

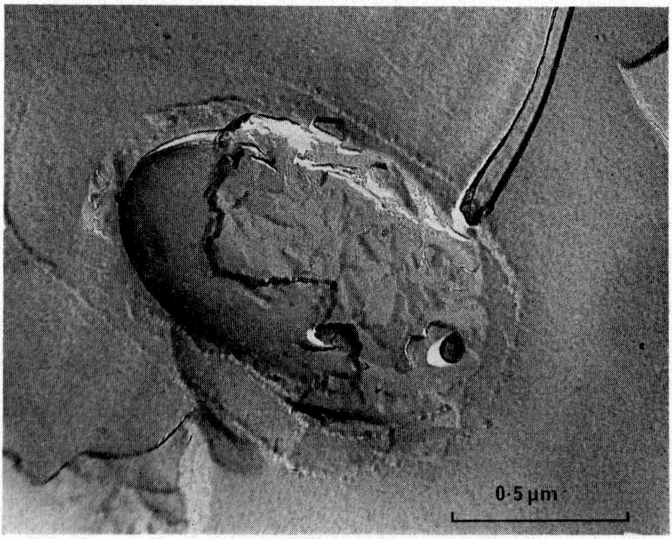

FIG. 9. *B. cereus*. Freeze-etched spore showing exosporium, depth of sporecoats and smooth outer surface of the cortex.

Figs. 10–15

These plates illustrate the gross cytological changes taking place during germination of *B. cereus* and *B. polymyxa* spores.

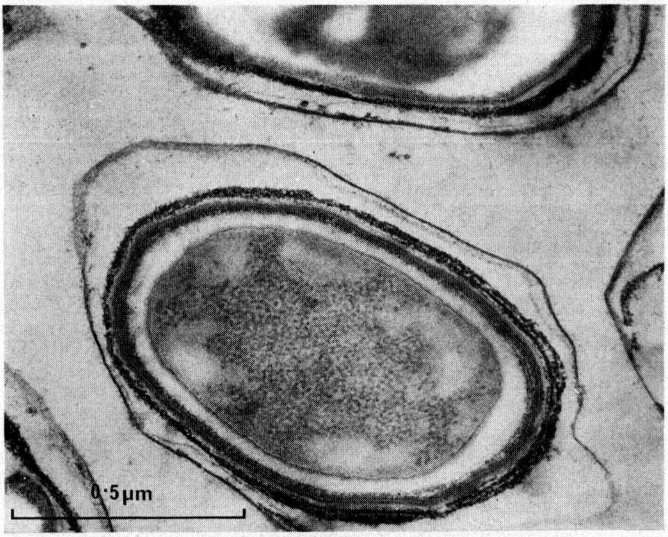

FIG. 10. *B. cereus* (coat dissolver). Early phase of germination. The core is beginning to rehydrate and swell whilst the cortex becomes thinner as a result of peptidoglycan hydrolysis.

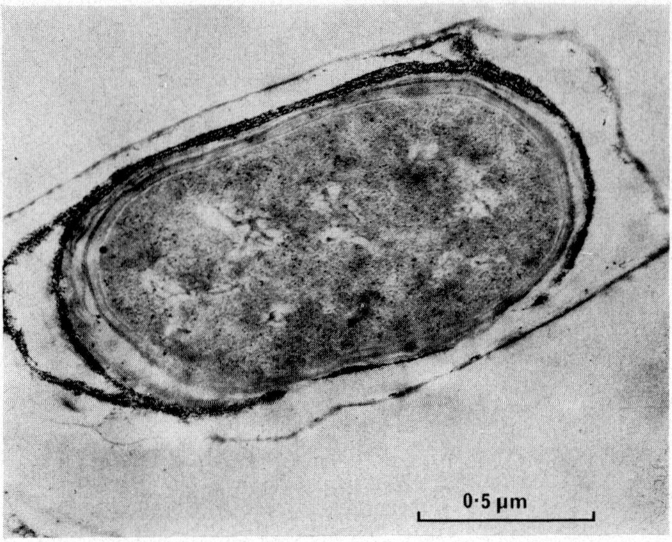

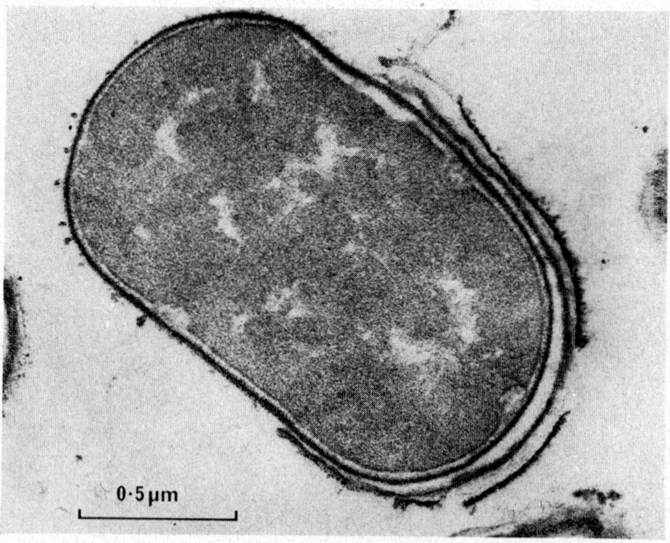

FIGS 11 AND 12. *B. cereus* (coat dissolver). Little cortex now remains. The coats are beginning to split and synthesis of new vegetative cell wall is proceeding (Hamilton and Stubbs, 1967).

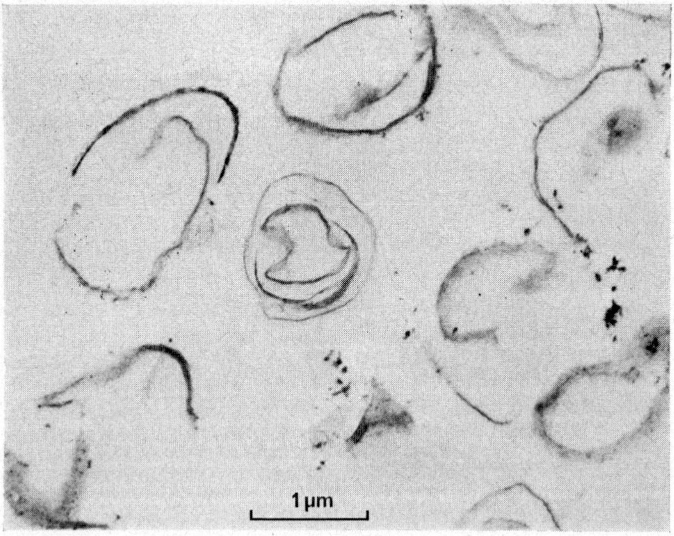

FIG. 13. *B. cereus* (coat dissolver). Remnants of sporecoats, devoid of residual cortex, and exosporia following outgrowth.

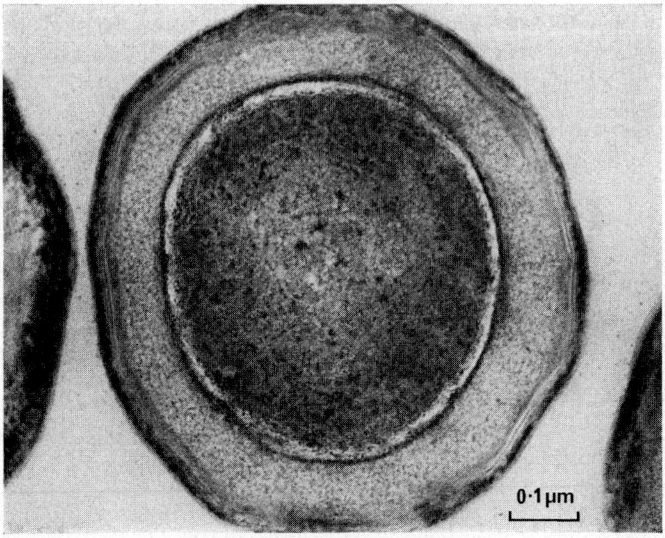

FIG. 14. *B. polymyxa* (coat splitter). Early germination showing swollen core but with cortex largely retained. Glutaraldehyde/osmium tetroxide fixation was used to show structure in the cortex.

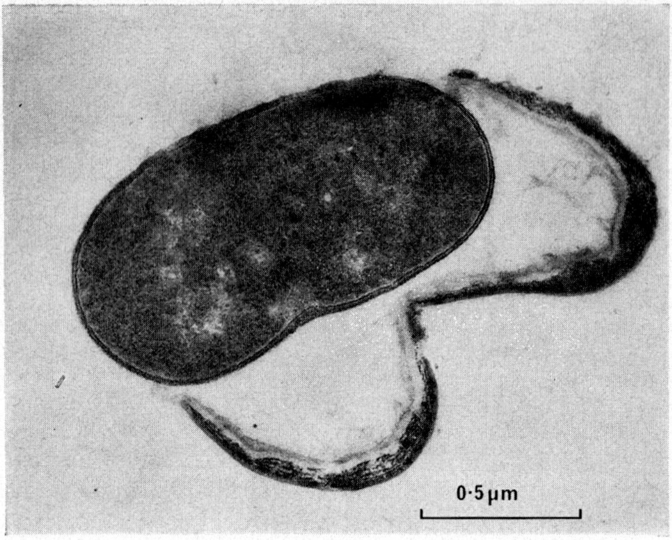

FIG. 15. *B. polymyxa* (coat splitter). Emergence, elongation and commencement of new vegetative cell division together with shedding of sporecoats which still retain some cortical material (Hamilton and Stubbs, 1967).

new vegetative cell surrounded by well-defined, newly-synthesized cell-wall is shown in Fig. 15. The outgrowing cell has split the coat, which still retains recognizable fragments of undissolved cortex on its inner surface.

References

DRING, G. J. & GOULD, G. W. (1971). Sequence of events during rapid germination of spores of *Bacillus cereus*. *J. gen. Microbiol.*, **65,** 101.

EISENSTADT, E. & SILVER, S. (1972). Restoration of cation transport during germination. In *Spores V*, (Halvorson, H. O., Hanson, R. and Campbell, L. L., eds), 443. Washington, D.C.: American Society for Microbiology.

GOULD, G. W. (1971). Methods for studying bacterial spores. In *Methods in Microbiology*, (Norris, J. R. and Ribbons, D. W., eds), Vol. 6A, p. 327. London and New York: Academic Press.

GOULD, G. W. & DRING, G. J. (1974). Mechanisms of spore heat resistance. In *Adv. Microbial Physiol.*, (Rose, A. H. and Tempest, D. W., eds), Vol. 11, 137. London: Academic Press.

GOULD, G. W., STUBBS, J. M. & KING, W. L. (1970). Structure and composition of resistant layers in bacterial spore coats. *J. gen. Microbiol.*, **60,** 347.

HAMILTON, W. A. & STUBBS, J. M. (1967). Comparison of the germination and outgrowth of spores of *Bacillus cereus* and *Bacillus polymyxa*. *J. gen. Microbiol.*, **47,** 121.

HASHIMOTO, T., FRIEBEN, W. C. & CONTI, S. F. (1969). Microgermination of *Bacillus cereus* spores. *J. Bact.*, **100,** 1385.

LEVINSON, H. S. & HYATT, M. T. (1966). Sequence of events during *Bacillus megaterium* spore germination. *J. Bact.*, **91,** 1811.

RODE, L. J., LEWIS, C. W. & FOSTER, J. W. (1962). Electron microscopy of spores of *Bacillus megaterium* with special reference to the effects of fixation and thin sectioning. *J. Cell Biol.*, **13,** 423.

SETLOW, P. (1974). Identification of several unique low molecular weight basic proteins in dormant spores of *Bacillus megaterium* and their degradation during spore germination. *Biochem. Biophys. Res. Commun.*, **61,** 1110.

SETLOW, P. (1975). Protein metabolism during germination of *Bacillus megaterium* spores. II. Degradation of pre-existing and newly synthesised protein. *J. biol. Chem.*, **250,** 631.

SETLOW, P. & PRIMUS, G. (1975). Protein metabolism during germination of *Bacillus megaterium* spores. I. Protein synthesis and amino acid metabolism. *J. biol. Chem.*, **250,** 623.

STRANGE, R. E. & HUNTER, J. R. (1969). Outgrowth and synthesis of macromolecules. In *The Bacterial Spore*, (Gould, G. W. and Hurst, A., eds), 445. London: Academic Press.

The Analysis of Mesosomes in *Bacillus* Species by Sectioning and Negative Staining

PETER J. HIGHTON

Department of Molecular Biology, Kings Buildings, Mayfield Road, Edinburgh, Scotland

Mesosomes were first seen by Chapman and Hillier (1953), in *Bacillus cereus*. With the preparative procedures then available the mesosomes appeared as small, ill-defined structures at the edge of the cytoplasm. They were given the name mesosomes by Fitz-James (1960), and can now be defined as an arrangement of membranes contained within an invagination of the plasma membrane. They appear to be a common feature of Gram-positive bacteria, but have been seen much less often in Gram-negative bacteria, and under the standard conditions used we have found only ill-defined structures in *Escherichia coli* B/r. However, more visible membranous structures have been seen in Gram-negative cells under some conditions (for details see Hoffman *et al*., 1973; Weigand *et al*., 1973). The following discussion will be concerned entirely with Gram-positive bacteria.

The Structure of Mesosomes

Standard fixation, dehydration and embedding procedure

Fixation (*Ryter and Kellenberger*, 1958; *Kellenberger, Ryter and Séchaud* 1958; *Glauert*, 1965).

(1) Prefixation: Remove culture from shaker and immediately add 1/10 volume of fixative (1% (w/v) OsO_4 in veronal acetate buffer).
(2) Fixation: Spin down in bench centrifuge without delay, and immediately resuspend in 1 ml fixative + 0·1 ml tryptone (1% (w/v) Bacto-tryptone, 0·5% (w/v) NaCl in H_2O). Leave overnight.
(3) Add 4 ml veronal acetate buffer and spin down.
(4) Mix pellet with 1 drop of agar (2% (w/v) agar in veronal acetate buffer) at 60°.
(5) Transfer to a glass slide at room temperature with a Pasteur pipette and cut up into small pieces.

(6) Postfixation: Place in uranyl acetate (0·5% (w/v) in veronal acetate buffer) for 2 h. (Essential to preserve detailed structure of mesosomes).

Dehydration (*Glauert*, 1965)

30% (v/v) acetone 15 min
50% (v/v) acetone 15 min
75% (v/v) acetone 15 min
90% (v/v) acetone 30 min
100% acetone 30 min (Na_2SO_4 added to remove H_2O)
100% acetone 30 min

Embedding (*Glauert, Rogers and Glauert*, 1956; *Glauert and Glauert*, 1958).

(1) 1 volume Araldite* + 1 volume 100% acetone—4 to 6 h at 48°.
(2) Araldite—24 h at 48°.
(3) Araldite + 1/40 volume accelerator (DMP 30)—24 h at room temperature.
(4) Araldite + 1/40 volume accelerator in capsules—60° until hard (several days).

Poststain (*Reynolds*, 1963). Cut sections on to 10% (v/v) acetone, and pick up on a carbon coated grid. Without drying, add a drop of lead citrate, Wash off with a few drops of H_2O and drain with filter paper.

Comments. Process at room temperature, except where indicated, and postfix, dehydrate and embed in small glass bottles. At each change suck off solutions, or pour off if viscous.

In most experiments KCN was added to give the same molarity as the OsO_4. This was originally done in a study of the localization of penicillinase by precipitation of lead penicilloate. Since lead penicilloate reacted with OsO_4, KCN was added as it has been shown to stabilize some OsO_4 reaction products in aqueous solution (Highton *et al.*, 1968). Only a few samples have been prepared without KCN, and these have not been sufficiently studied to be able to say whether they are less well preserved.

The appearance after standard fixation and sectioning

Sections of mesosomes in exponentially growing *B. licheniformis* 749C prepared by the above technique (Highton, 1970*a*) contained arrays of pairs of dense lines, like the "unit membrane" structure of Robertson

* Araldite—Resin (CY212): hardener (DDSA 964): plasticizer (dibutyl phthalate) = 10:10:1.

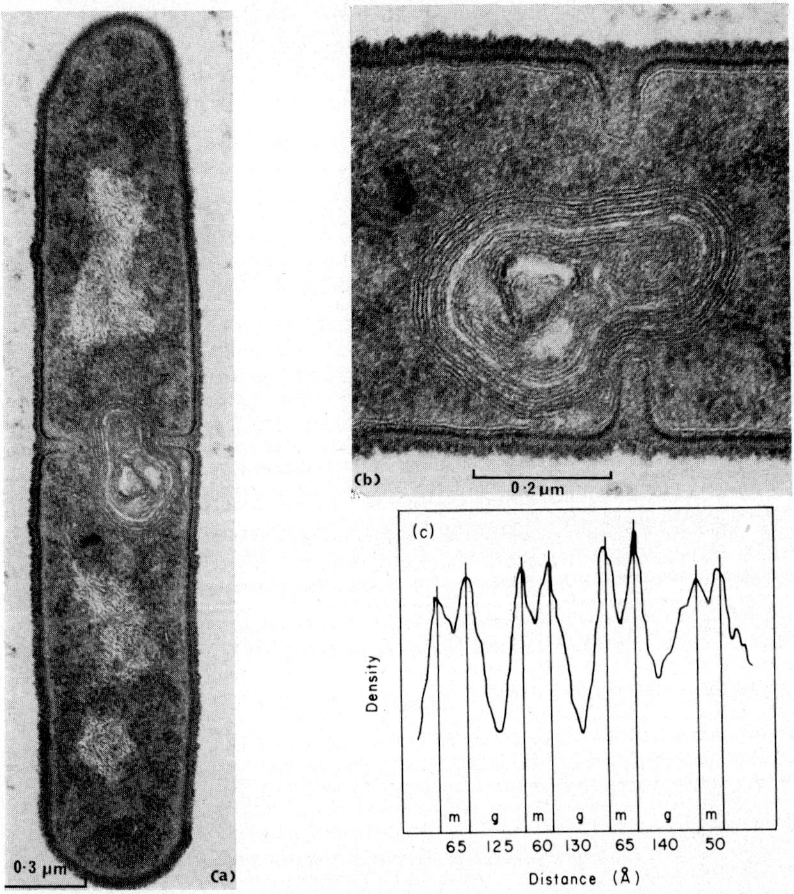

FIG. 1. (a) *B. licheniformis* 749C harvested in exponential phase. (b) Mesosome in (a) enlarged. (c) Densitometer trace across mesosomal membranes (m), separated by gaps (g).

(1959), (Fig. 1). Serial sections suggested that the membranes extended over distances of more than 50 nm perpendicular to the plane of the sections. The whole structure could thus be a system of curved sheets of membrane (lamellae), appearing as two parallel lines where cut perpendicularly, and structureless regions where cut obliquely. In a few mesosomes small circular or elliptical components (vesicles) were seen apparently bounded by only one dense line. The results of this study, and a previous one (Highton, 1969) were consistent with the existence of only one mesosome per bacterium.

The same result was obtained for *B. subtilis* 172, and models of complete bacteria were constructed from serial sections. However, lamellar structures were relatively uncommon, and the mesosomes were largely vesicular (Highton, 1970*b*). On the other hand, *B. cereus* 569H/24 appeared to have several lamellar mesosomes per bacterium (Highton and Hobbs, 1972).

No difference was found between the mesosomes of the different penicillinase mutants of *B. cereus* and *B. licheniformis* studied (Table 1).

TABLE 1. Bacteria studied

Species	Strain	Penicillinase production
B. licheniformis	749	Inducible
	749C	Magnoconstitutive
	749C3	Magnoconstitutive
	749C3/P22	Microconstitutive
	749C/72	Microconstitutive
	749C/72/IIIg	Microconstitutive
B. cereus	569	Inducible
	569H	Constitutive
	569H/24	Microconstitutive
B. subtilis	168	—
	172	—
	W23	—

Changes produced by variations in processing

If OsO_4 prefixation was omitted, and the bacteria were resuspended in 1M sucrose in veronal acetate buffer for 30 min at 20° before fixation, the plasma membrane of *B. lichenformis* 749C was withdrawn from the wall by plasmolysis, and the mesosome was fragmented and distributed around the bacterium as vesicles (Fig. 2a). Similar but less extensive fragmentation occurred in this species and in *B. subtilis* 172, if the culture was simply allowed to stand for 30 min, without shaking, at 20° before fixation.

By resuspending in 5 mM $Pb(NO_3)_2$ in veronal acetate buffer for 30 min at 0° after harvesting and omitting the OsO_4 prefixation, it was possible to unfold the mesosomal membranes of *B. licheniformis* 749C so that they lay between the plasma membrane and the wall (Fig. 2b). The $Pb(NO_3)_2$ reacted with structures throughout the bacterium so that no poststaining with lead citrate was necessary. Additional dense deposits, not seen after

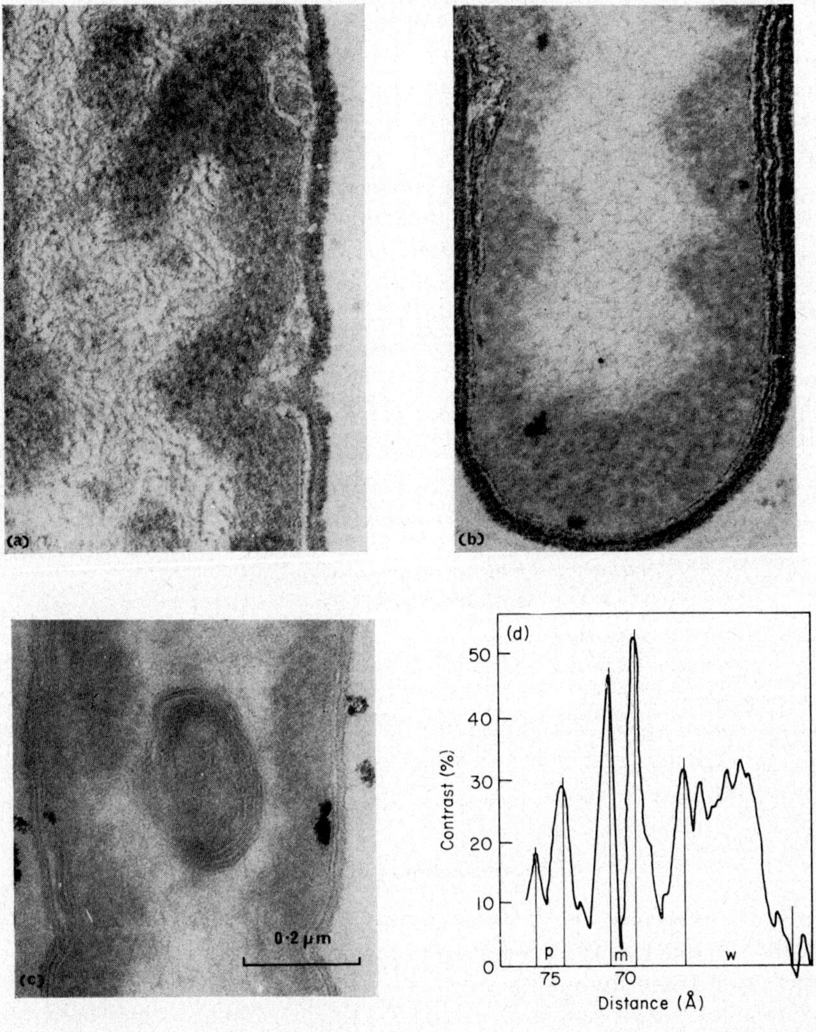

FIG. 2. Mesosomes of *B. licheniformis* 749C. (a) Suspended in 1 M sucrose for 30 min at 0° before OsO_4 fixation. (b) Suspended in 5 mM $Pb(NO_3)_2$ for 30 min at 0° before OsO_4 fixation and not poststained. (c) Suspended in 5 mM $Pb(NO_3)_2$ for 10 min at 20° before fixation in 2·5% glutaraldehyde and not poststained. (d) Densitometer trace across plasma membrane (p), extruded mesosomal membrane (m) and wall (w).

the standard procedure, occurred in the cytoplasm, generally abutting on the plasma membrane. Similar deposits were also seen in *B. subtilis* 168 by reacting with $Pb(NO_3)_2$ before fixation. The density profile in Fig. 2d shows that the mesosomal membranes had greater density than the plasma membrane after reaction with $Pb(NO_3)_2$, and such traces also showed that the separation of the two dense layers of the mesosomal membranes was often increased when unfolded.

Granboulan and Leduc (1967) obtained good preservation of structure in *B. subtilis* 168 by fixing with glutaraldehyde and embedding in glycol-methacrylate (GMA), but our attempts at glutaraldehyde fixation of *B. licheniformis* 749C followed by Araldite embedding were most unsuccessful. Fixation with OsO_4 after glutaraldehyde still did not preserve the mesosomes. However, glutaraldehyde fixation after reaction with $Pb(NO_3)_2$ at 20° did partly preserve the unit-membrane structure of mesosomes (Fig. 2c) and produced dense deposits in the cytoplasm, as with OsO_4 fixation.

Interpretation

(1) In *B. licheniformis* 749C the appearance of the mesosomes is strongly dependent on the fixation process.
(2) Vesicles are produced by the breakdown of lamellar structures during processing—possibly by separation of the two layers of the unit-membrane.
(3) Variations in the appearance of different species under the same conditions may reflect differences in sensitivity to destructive effects during processing.

More detailed analysis of sections

In some mesosomes of *B. licheniformis* 749C and related strains, and in *B. subtilis* 168 and W23, membranes apparently composed of three dense lines have been found (Fig. 3). If stained between OsO_4 prefixation and fixation, by suspension in 5 mM $Pb(NO_3)_2$ in veronal acetate buffer for 10 min at 20°, the stain apparently penetrated between the unit-membranes and these "triple" structures. It possibly reacted with material which was between, and connected to, the surface of the membranes, and which was poorly preserved by the standard procedure (Highton, 1970*a*). This supports the interpretation of the "triples" as structural units. Possibly they are composed of two unit-membranes close together, and are conceivably a step in membrane synthesis. Lead was also deposited in the gap normally seen between the plasma membrane and the wall, but it did not penetrate the plasma membrane as in the absence of prefixation (Fig. 2), so that there were no dense deposits, and poststaining with lead citrate was needed to show up the ribosomes and DNA.

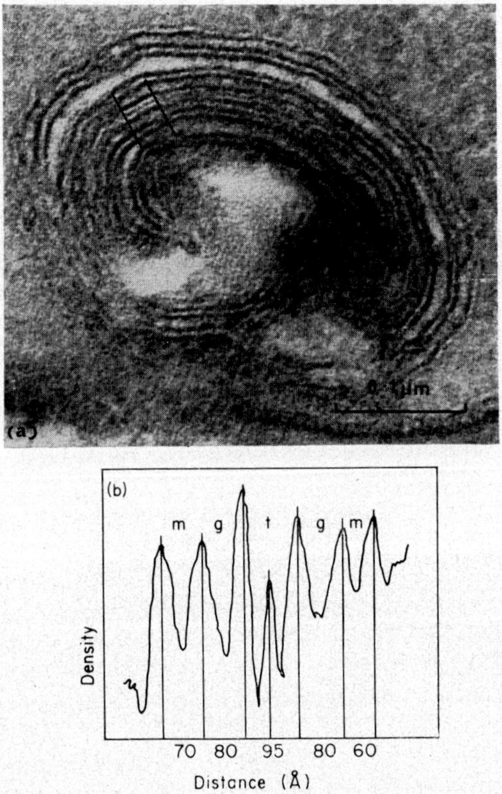

FIG. 3. (a) Mesosome of *B. licheniformis* 749C showing "triple" membrane. (b) Densitometer trace across "triple" (t), and unit membranes (m) separated by gaps (g), between lines marked in (a)

Negative staining

Initial attempts to look at mesosomes in *B. licheniformis* 749C, by putting a drop of culture on to a carbon film mounted on an electron microscope grid, adding a drop of stain, and drying on filter paper, produced cells with apparently many invaginations of the plasma membrane into which stain had penetrated. From the results described previously it was concluded that the mesosomes had been disrupted and the fragments distributed around the cell, between the plasma membrane and the wall.

Attempts were made to preserve the mesosomes by fixation, and by the addition of bovine serum albumin (BSA) to make the stain dry as a more uniform layer over the supporting film. Prefixation with OsO_4 as described above, followed by resuspension in veronal acetate buffer and addition of 0·1% (w/v) BSA before staining, produced cells with intact mesosomes (Fig. 4) (Highton, Lloyd and Whitfield, 1973).

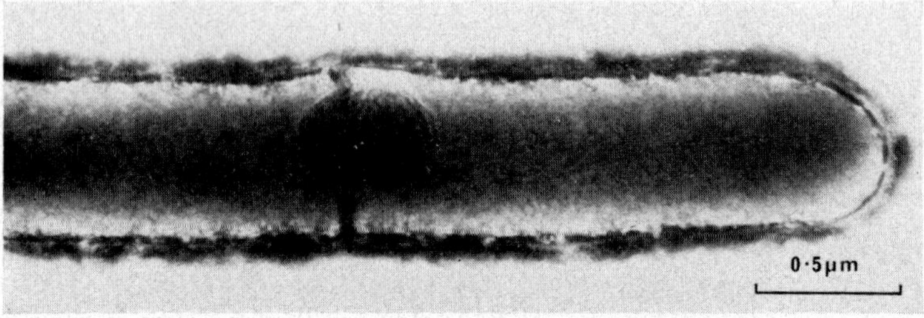

FIG. 4. *B. licheniformis* 749C prefixed in OsO_4 and suspended in 0·1% (w/v) bovine serum albumin before negative staining with sodium phosphotungstate (pH 7·0).

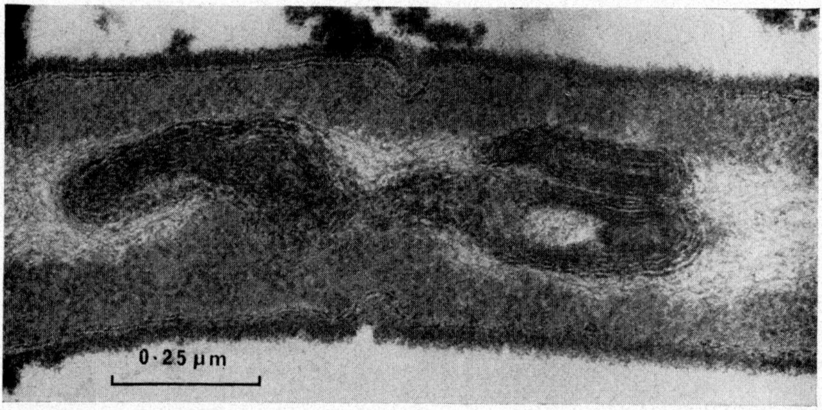

FIG. 5. Dividing mesosome in *B. licheniformis* 749C. Suspended in 5 mM $Pb(NO_3)_2$ for 10 min at 20° between OsO_4 prefixation and fixation.

Three stains, 2% (w/v) sodium phosphotungstate (pH 7·0), 2% (w/v) sodium silicotungstate (pH 7·2) and 2% (w/v) ammonium molybdate (pH 7·2) were used, and all gave the same result. Uranyl acetate did not penetrate the cell wall but showed up the flagellar attachment sites.

Similar results were obtained with *B. licheniformis* 749 and *B. subtilis*

168 and W23. It has also been possible to use grids made hydrophilic by coating with cytochrome *c* (Harford and Beer, 1972), or by ionic bombardment (Highton and Whitfield, 1974), instead of by adding BSA.

Mesosomes in the Cell Cycle

Exponential growth

In a cell in which there is always one, and only one, mesosome in exponential growth, the mesosome must double and divide with the cell. Figure 5 shows a section of a mesosome of *B. licheniformis* 749C thought to be in the process of division. The mesosome must also "move", as the cell grows, from one pole of the cell, where it is left after division, to the centre of the cell, ready for the next division. Attempts were made to demonstrate this process in *B. licheniformis* 749C (Highton, 1969) and *B. subtilis* 172 (Highton, 1970*b*), by plotting mesosome position against half-cell length. The analysis was complicated by a cell-length heterogeneity, giving a range of more than two, but the results were consistent with the model. A study of germinating spores of *B. licheniformis* 749C showed that mesosomes appeared immediately, along with the other vegetative cell components, and that the cell-length heterogeneity was established in the first few generations (Garland and Highton, 1973).

Sporulation

Marked changes in mesosome structure were found in *B. cereus* 569H/24 prior to sporulation (Highton, 1972). Beyond exponential phase, the cell mass doubled 3 to 4 times at a steadily decreasing rate. The mesosomes increased in density and their structure became ill-defined. The outer layer of the plasma membrane also increased in density, and the DNA appeared much coarser (compare Fig. 6, a and b).

Membrane density was produced by the reaction with OsO_4, and the binding of UO_2^{2+} and Pb^{2+} ions, and changes in density are thought to result from changes in macromolecular composition of the membranes. The loss of detail in the mesosomes may simply have resulted from the inability of the fixatives to preserve membranes of altered composition. Further, the coarsening of the DNA may have resulted from reduced permeability of the plasma membrane to UO_2^{2+} ions, as a similar result was produced in *E. coli* by omitting uranyl acetate postfixation (Ryter and Kellenberger, 1958), and in *B. licheniformis* 749C by the same omission in our own experiments.

The same changes were also demonstrated in *B. cereus* 569 and 569H,

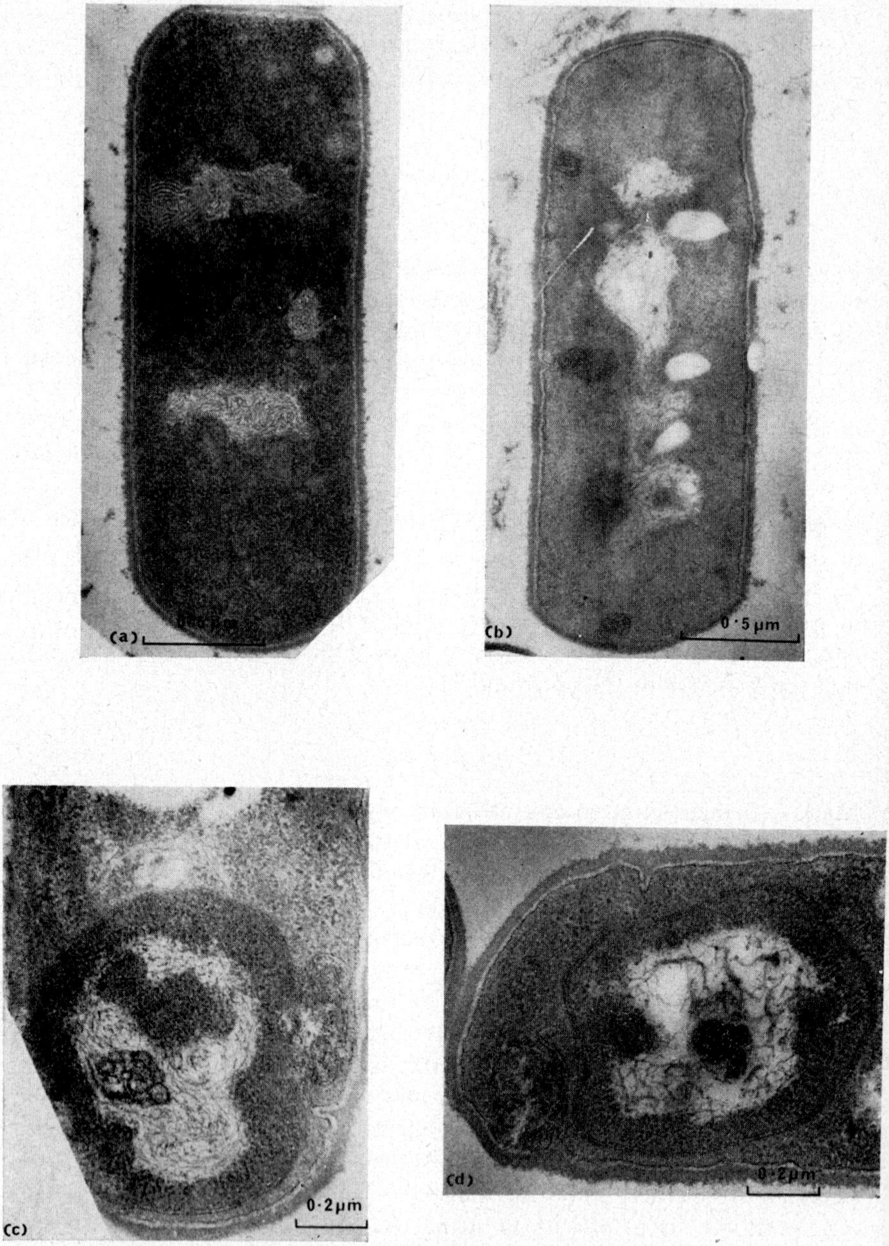

FIG. 6. *B. cereus* 569H/24. (a) Harvested in exponential phase. (b) Beyond exponential phase. (c) Developing forespore and associated mesosomes. (d) Forespore completely surrounded by double membrane.

but were not found in *B. licheniformis* 749 or *B. subtilis* 168 and W23, which all grew less beyond exponential phase than did *B. cereus*. However, there may have been macromolecular changes associated with sporulation but not revealed by the stains employed.

Most developing forespores in *B. cereus* had coarse DNA as in the rest of the cell. In some, however, it was fine and fibrillar, as in exponentially growing cells, and these showed mesosomes embedded in the DNA, presumably within an invagination of the inner forespore membrane (Fig. 6c). However, it was not possible to determine whether these mesosomes were actually incorporated into the spore, as structural preservation was poor once the forespore became surrounded by its two layers of membrane (Fig. 6d). Other mesosomes, in invaginations of the plasma membrane, were also generally present close to developing forespores (Fig. 6 c and d).

On germination mesosomes appeared immediately along with other vegetative cell components (Fig. 7), as in *B. licheniformis* 749C.

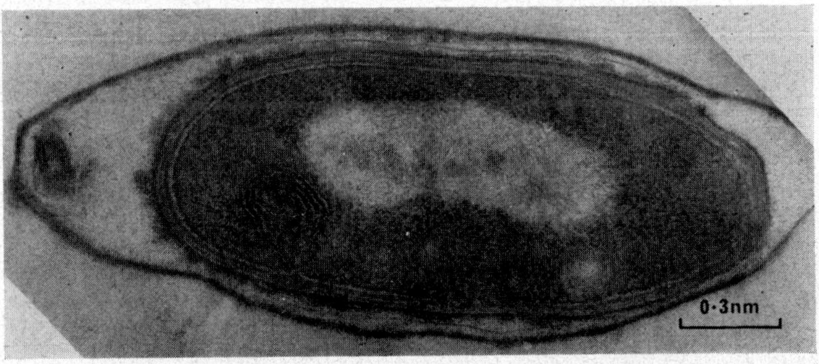

FIG. 7. Mesosome in germinated spore of *B. cereus* 569 (Courtesy of Dr J. M. Garland).

FIG. 8. *B. licheniformis* 749C/72, 75 min after addition of penicillin to culture at a concentration of 1 unit/ml.

The Effects of Penicillin

The addition of penicillin to cultures of *B. licheniformis* 749C/72 (penicillinase microconstitutive), at a concentration of 1 unit/ml, stopped growth, i.e. optical density increase, in 75 min. Marked changes in wall and cross-wall structure, and in DNA distribution occurred, and mesosomes disappeared (compare Figs 1a and 8). Growth was reinitiated by addition of penicillinase and mesosomes reappeared (Highton and Hobbs, 1971).

A similar cessation and reinitiation of growth, with loss and reappearance of mesosomes, was also effected with the penicillinase microconstitutive strain *B. cereus* 569/24 (Highton and Hobbs, 1972). However, since the wall structure had clearly been changed by exposure to penicillin, the interpretation of changes in internal structure must be done cautiously, as they may simply result from changed permeability to fixatives.

Acknowledgements

I am grateful to D. G. Hobbs, J. Lloyd and M. Whitfield for assistance with this work.

References

Chapman, G. B. & Hillier, J. (1953). Electron microscopy of ultra-thin sections of bacteria. 1. Cellular division in *Bacillus cereus*. *J. Bact.*, **66,** 362.

Fitz-James, P. C. (1960). Participation of the cytoplasmic membrane in the growth and spore formation of bacilli. *J. biophys. biochem. Cytol.*, **8,** 507.

Garland, J. M. & Highton, P. J. (1973). Mesosomes in early outgrowth of spores of *Bacillus licheniformis* 749c. *J. Ultrastruct. Res.*, **42,** 457.

Glauert, A. M. (1965). The fixation and embedding of biological specimens. In *Techniques for Electron Microscopy*, (Kay, D. H., ed.), 166. Oxford: Blackwell.

Glauert, A. M. & Glauert, R. H. (1958). Araldite as an embedding medium for electron microscopy. *J. biophys. biochem, Cytol.*, **4,** 191.

Glauert, A. M., Rogers, G. E. & Glauert, R. H. (1956). A new embedding medium for electron microscopy. *Nature, Lond.*, **178**, 803.

Granboulan, P. & Leduc, E. H. (1967). Ultrastructural cytochemistry of *Bacillus subtilis*. *J. Ultrastruct. Res.*, **20,** 111.

Harford, A. G. & Beer, M. (1972). Electron-microscopic localization of the binding of *Escherichia coli* RNA polymerase to T7 DNA *in vitro*. *J. mol. Biol.*, **69,** 179.

Highton, P. J. (1969). An electron microscopic study of cell growth and mesosomal structure of *Bacillus licheniformis*. *J. Ultrastruct. Res.*, **26,** 130.

Highton, P. J. (1970*a*). An electron microscopic study of the structure of mesosomal membranes in *Bacillus licheniformis*. *J. Ultrastruct. Res.*, **31,** 247.

Highton, P. J. (1970*b*). An electron microscopic study of mesosomes in *Bacillus subtilis*. *J. Ultrastruct. Res.*, **31,** 260.

HIGHTON, P. J. (1972). Changes in the structure of mesosomes and cell membrane of *Bacillus cereus* during sporulation. In *Spores V* (Halvorson, H. O., Hanson, R. and Campbell, L. L., eds), 13. Washington, D.C.: American Society for Microbiology.

HIGHTON, P. J. & HOBBS, D. G. (1971). Penicillin and cell wall synthesis: a study of *Bacillus licheniformis* by electron microscopy. *J. Bact.*, **106,** 646.

HIGHTON, P. J. & HOBBS, D. G. (1972). Penicillin and cell wall synthesis: a study of *Bacillus cereus* by electron microscopy. *J. Bact.*, **109,** 1181.

HIGHTON, P. J., LLOYD, J. & WHITFIELD, M. (1973). Mesosomes of *Bacillus* species, seen by negative staining. *J. gen. Microbiol.*, **78,** 375.

HIGHTON, P. J., MURR, B. L., SHAFA, F. & BEER, M. (1968). Electron microscopic study of base sequence in nucleic acids. VIII. Specific conversion of thymine into anionic osmate esters. *Biochemistry*, **7,** 825.

HIGHTON, P. J. & WHITFIELD, M. (1974). The control of the configuration of nucleic acid molecules deposited for electron microscopy, by ionic bombardment of carbon films. *J. Microscopy*, **100,** 299.

HOFFMANN, H., GEFTIC, S. G., HEYMANN, H. & ADAIR, F. W. (1973). Mesosomes in *Pseudomonas aeruginosa*. *J. Bact.*, **114,** 434.

KELLENBERGER, E., RYTER, A. & SECHAUD, J. (1958). Electron microscope study of DNA-containing plasms. II. Vegetative and mature phage DNA as compared with normal bacterial nucleoids in different physiological states. *J. biophys. biochem. Cytol.*, **4,** 671.

REYNOLDS, E. S. (1963). The use of lead citrate at high pH as an electron-opaque stain in electron microscopy. *J. Cell Biol.*, **17,** 208.

ROBERTSON, J. D. (1959). The ultrastructure of cell membranes and their derivatives. *Biochem. Soc. Symp.* (Cambridge, Engl.), **16,** 3.

RYTER, A. & KELLENBERGER, E. (1958). Étude au microscope électronique de plasmas contenant de l'acide désoxyribonucliéque. *Z. Naturforsch.* 13b, 597.

WEIGAND, R. A., HOLT, S. C., SHIVELY, J. M., DECKER, G. L. & GREENAWALT, J. W. (1973). Ultrastructural properties of the extra membranes of *Escherichia coli* OIIIa as revealed by freeze-fracturing and negative-staining techniques. *J. Bact.*, **113,** 433.

The Ultrastructure of *Proteus mirabilis* Swarmers

JUDITH P. ARMITAGE AND D. G. SMITH

Department of Botany and Microbiology, University College London, Gower Street, London, England

Proteus mirabilis and *P. vulgaris* undergo cyclical morphological changes on solid nutrient media. Macroscopically this gives rise to concentric rings of growth over the surface of the medium (Fig. 1), a phenomenon which has caused problems in diagnostic laboratories due to overgrowth of other colonies. Microscopically the normal, sparsely-flagellate, short rods

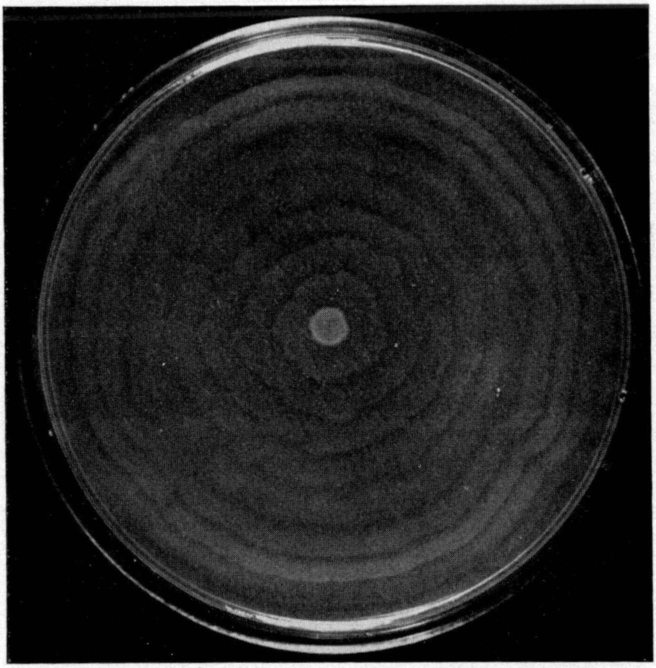

FIG. 1. Swarming colony of *P. mirabilis* on nutrient agar, centrally inoculated and incubated for 24h at 30°.

differentiate at the edge of the colony after about 5h growth, into multinucleate, highly-flagellate, non-septate cells up to 80μm long. These long cells are able to move through the surface moisture away from the initial colony, presumably under a negative chemotactic influence. The cells move out a few mm and then divide into normal short rods, growing and dividing for several generations (about 2½h) after which the elongation and swarming process is repeated at the edge of the ring. The swarming process has been reviewed by Smith (1972).

Electron microscopy has revealed no ultrastructural differences between the long, non-dividing cells and the short cells, however experiments have indicated that there may be a change in the envelope composition during the swarming process (Armitage, Rowbury and Smith, 1975).

Long, swarming cells have been isolated from actively swarming colonies and compared with non-swarming, agar-grown cells and exponential, broth-grown cells. We have shown that protein synthesis is required for division of the swarming cells into short rods, but DNA synthesis is not (Armitage, Rowbury and Smith, 1974).

There is also a difference in sensitivity of the long cells to several antibacterial agents (rifampicin, actinomycin D, sodium deoxycholate) whose activity is known to be dependent on the permeability of the cell envelope (Armitage, Rowbury and Smith, 1975).

Media and Growth

In all the experiments described here *P. mirabilis* was grown on Oxoid nutrient Broth No. 2. For the solid swarming medium 1.75% agar (Difco) was added, and for the solid swarm-inhibiting medium 1% activated charcoal was added to the swarming medium (Alwen and Smith, 1967).

Swarming cells were grown on 250 ml agar in 12×12 in. plates (Shandon) and inoculated at 8 sites with a 24h broth-grown culture. The non-swarming cells were streaked onto the charcoal medium and the plates and broth were incubated at 30° (the optimal swarming-temperature) for 18h before harvesting.

Isolation of the Cells

Swarming cells were isolated while actively in the third swarm, by cutting the inner, non-swarming region out of the agar and washing the remaining long cells off the agar into the experimental medium. This prodecure yielded a broth suspension of long cells and it was shown by Coulter Counter sizing (Coulter Electronics Ltd) that the majority were longer than the broth-grown and non-swarming, agar-grown cells (Fig. 2). Non-swarming cells were washed off the charcoal agar with the experimental medium and all the cells were washed if appropriate.

Electron Microscopy

Negative staining

Aqueous suspensions of cells were mixed with approximately equal volumes of 2% potassium phosphotungstate (pH 7) on a formvar coated electron microscope grid. The mixture was dried down with filter paper and the dried grid examined in a Siemens Elmiskop 1 electron microscope (Fig 3).

Heavy-metal shadowing

Cell suspensions were dried down on formvar coated grids and shadowed at an angle of 22° using gold–palladium (Fig. 4).

Sectioning

Suspensions of the three cell types in Kellenberger buffer (pH 6) were pre-fixed with osmium tetroxide and then fixed according to the method of

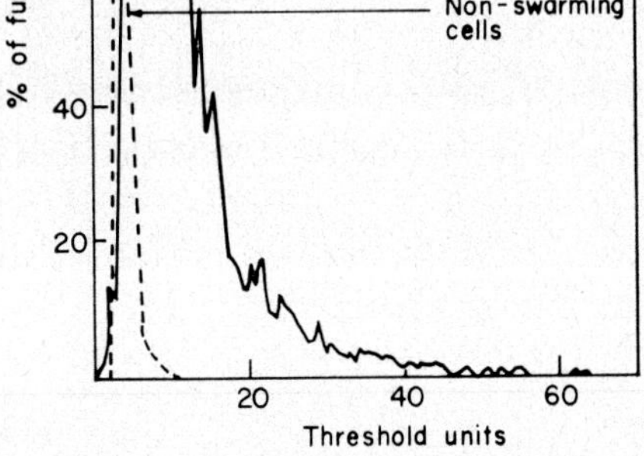

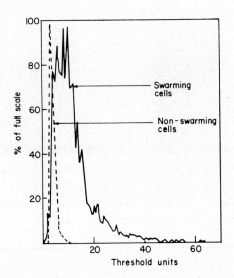

FIG. 2. Cell size distribution of preparations of swarming and non-swarming *P. mirabilis*. The Coulter Counter measurements are proportional to cell length since the cell diameter is virtually constant.

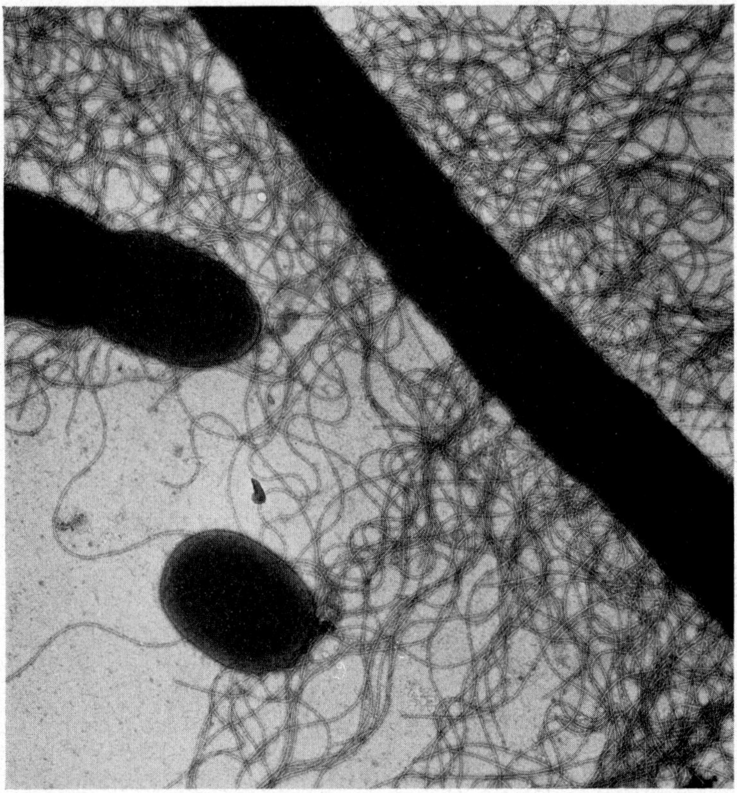

FIG. 3. Negatively-stained preparation of *P. mirabilis* showing part of a heavily flagellated swarming cell and a short, sparsely-flagellated, non-swarming cell (bottom left) ($\times$ 28 800)

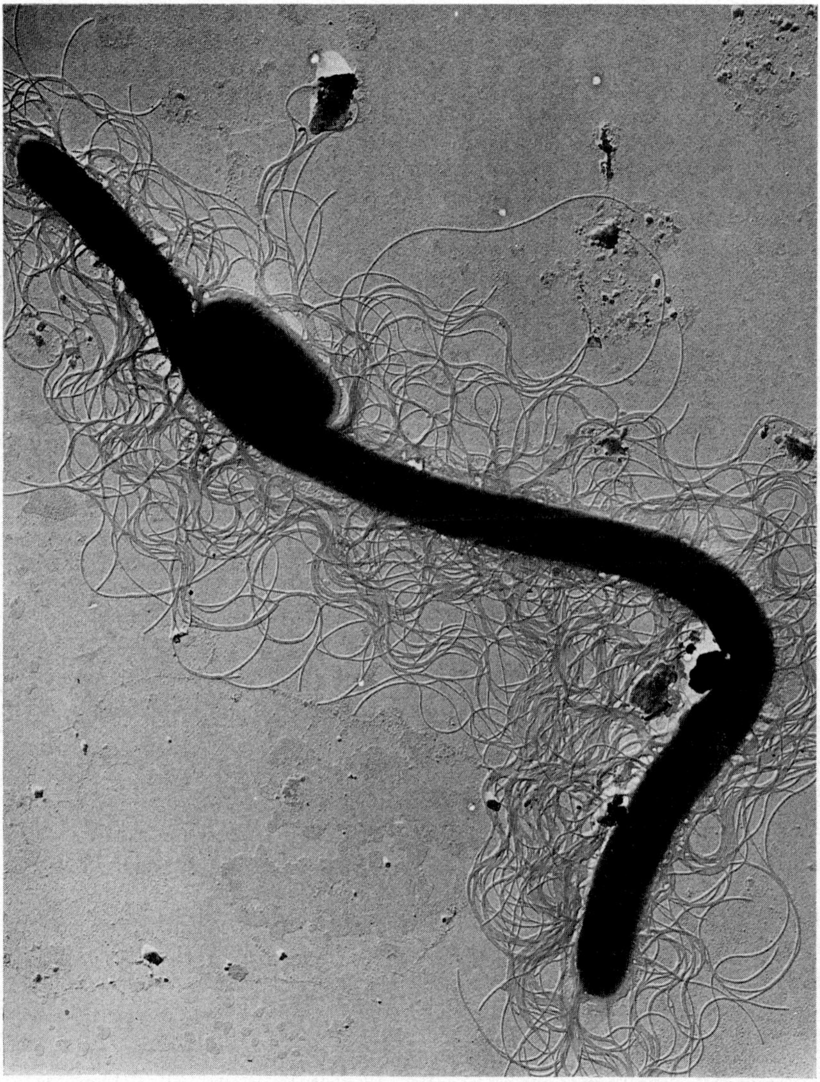

FIG. 4. Heavy-metal, shadowed preparation of *P. mirabilis* showing a medium sized swarming cell ($\times$ 12 1500).

Kellenberger, Ryter and Séchaud (1958). Pellets of cells were dehydrated in a series of tertiary butyl alcohols before embedding in Araldite.

Thin sections were cut on an L.K.B. ultramicrotome and stained for 20 min in ethanolic uranyl acetate and examined in the electron microscope (Fig. 5).

Ultrastructure of Swarming Cells

The ultrastructural details of the long, swarming cells of *P. mirabilis* vary little from those of normal, Gram-negative bacteria. The cells have the typical Gram-negative envelope profile of outer membrane, mucopeptide and inner cytoplasmic membrane. The nuclear material is compact and in discrete areas of the cell as in the short cells and the cytoplasm of the short and long cells shows a similar ribosomal density. However the multiple nuclear bodies in the swarming cells are not separated by septa, this has been shown not only with the electron microscope but also by plasmolysis (Hoeniger, 1966). The distance separating the nuclear bodies is similar to that of segregated nuclear bodies in the normally dividing cell (about 1.5 μm), and division takes place between the nuclear bodies at the end of the swarming period. The number of flagella per μm^3 increases from about 3.2 to 150 during the development of the swarming cells, indicating a large increase in the synthesis of the protein flagellin (Hoeniger, 1965*a*).

The flagella of the long cells appear to be identical to those of the short cells being about 200 mm in diameter and originating from a basal hook, although there are reports that the flagella of the swarming cells may be longer than those of the short cells (Hoeniger, 1965*a*). The conformation of the flagellin is sensitive to pH; as the pH is lowered there is a shortening of the flagella wavelength, and motility ceases (Hoeniger, 1965*b*).

The Effect of Penicillin on *Proteus mirabilis* Swarmers

Long, swarming cells were isolated and suspended in nutrient broth containing 15 i.u. penicillin G/ml, and the samples were fixed at intervals for thin sectioning.

Under normal conditions long cells suspended in nutrient broth divide very quickly into short cells, taking about 2h to complete the process. However, when suspended in penicillin-containing broth no divisions took place, but elongation continued. The cells increased in length but otherwise appeared normal for about 45 min when an increase in the diameter of the cells became obvious, this was followed by the appearance of several obvious bulges along the length of the swarm cell (Fig. 6). After a further 1 to 2h these bulges were extremely large but still surrounded by a complete

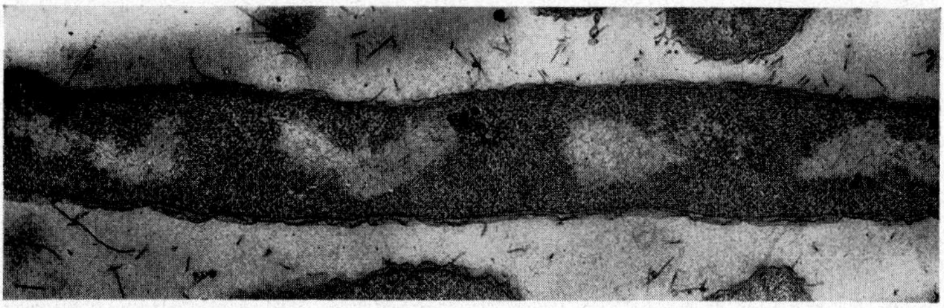

FIG. 5. Thin section of a swarming cell of *P. mirabilis* showing regularly spaced nucleoids and the absence of septa. Random sections of flagella are visible in the preparation (× 28 800).

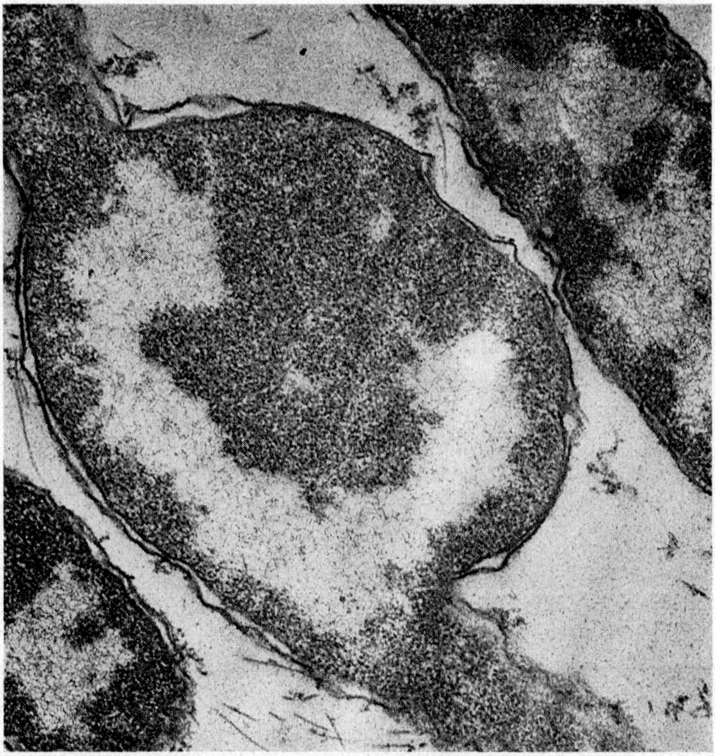

FIG. 6. A bulging swarmer of *P. mirabilis* produced after 120 min treatment with *N* 15 pencilin i.e./ml. The cell envelope still appears to be intact (× 46 800).

cell envelope, this is possibly because of the continued activity of the transpeptidases (Martin *et al.*, 1974), allowing the mucopeptide to maintain a high degree of integrity although weakened by the action of the penicillin. The cells finally formed spheroplasts and burst about an hour later (Fig. 7).

After approximately 1h exposure to penicillin the nuclear material in the cells lost its discrete appearance and became dispersed throughout the cell (Fig. 8). This may be due to the attachment of the nucleoid to the cell envelope, and penicillin, by stopping septation, also inhibits segregation, but as some intercalation of cell-wall material still takes place, the nuclear material is spread through the cell as the elongation continues.

The penicillin-treated *P. mirabilis* cells seem to be unlike natural long cells. Not only is there a dispersion of the nuclear material throughout the cell, but there is also an increase in sensitivity to lyzozyme in the penicillin cells (Fleck, Martin and Mock, 1974).

Discussion

The swarming cells of *Proteus* spp. have caused interest for many years, without knowledge of the phenomenon advancing much beyond the observational stage. The reason for this is the difficulty in obtaining any biochemical information *in situ*, as the cells must be removed from the agar at the beginning of the swarm to do any but the most gross studies. It is not even known whether the initial event in swarming is flagella increase or the inhibition of division, or whether the two events happen simultaneously. At the beginning of the swarm it is possible to see short rods covered with flagella, but it is possible that the event causing the inhibition of division has already taken place, but does not become evident for at least one division period. There have also been difficulties in isolating and identifying the chemical component which may be involved in the negative chemotaxis, although there is good evidence for its existence (Hughes, 1957; Grabow, 1972).

Despite these difficulties, we suggest that the following sequence of events may take place during the swarming phenomenon:

(1) The short rods inoculated onto the nutrient agar surface multiply normally for about 5h.
(2) During this period the cells produce, from certain nutrients, a diffusible compound(s).
(3) At a critical concentration this compound(s) inhibits cell division and stimulates flagella synthesis. This is most obvious in the rapidly growing cells at the edge of the colony.
(4) The long, highly-flagellate cells are able to move down a concentration gradient of a negatively-chemotactic agent, which may or may

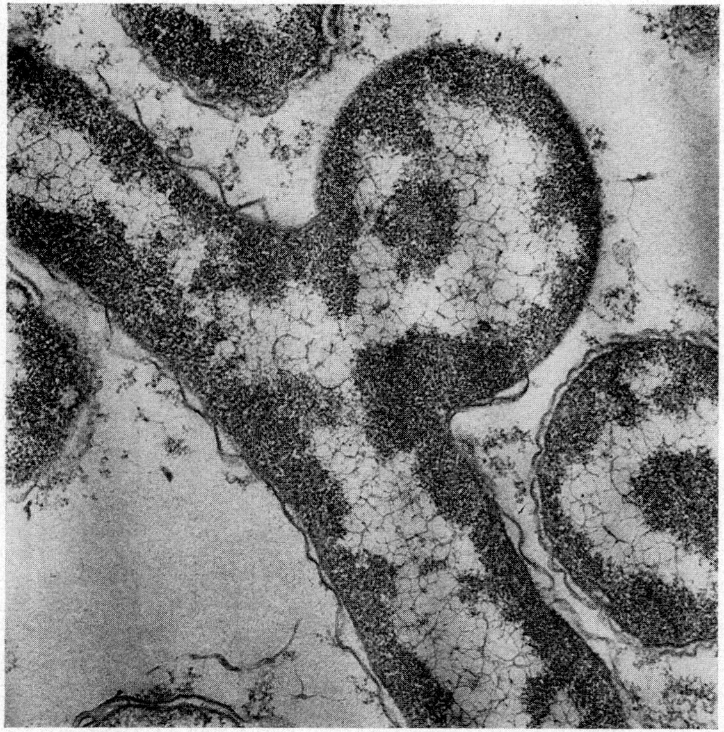

FIG. 7. Spheroplast formation in a swarmer of *P. mirabilis* after 180 min treatment with 15 penicillin i.u./ml. The outer membrane of the cell envelope is now ruptuted (× 42 300).

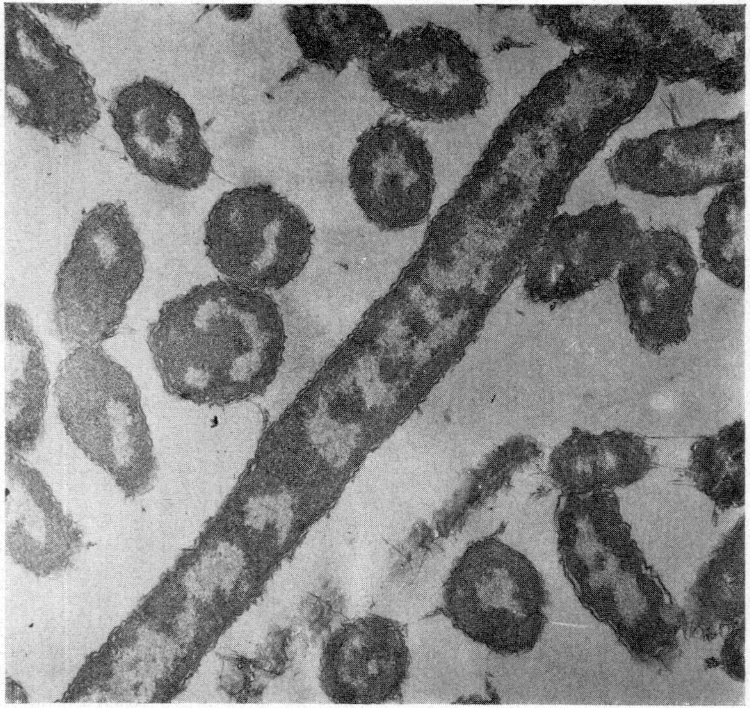

FIG. 8. Swarming cell of *P. mirabilis* after 20 min. treatment with 15 penicillin i.u./ml showing marked dispersion of the nucleoids (× 20 790).

not be the same as the inhibitor of division and/or stimulator of flagella synthesis.

(5) The cells move until the chemotactic agent becomes undetectable and then divide into normal short cells until the inhibitor and negative-chemotactic agent again build up to the critical concentration.

As can be seen from the above description swarming is the result of three events which could be independent or related; the increase in flagella synthesis, the inhibition of division and the movement from the initial colony.

We postulate that there is a change in the cell envelope composition leading to an inability of the cell to form septa. The evidence for this idea comes from the increase in the permeability of the long cells, indicating an envelope change, and a need for protein synthesis before septation. This change in the envelope composition may also trigger an imbalance in the flagellin synthesis. The envelope structure of *P. mirabilis* is certainly unusual when compared to other members of the Enterobacteriaceae, in that it is insensitive to both EDTA and lysozyme (Gray and Wilkinson, 1965) and to some extent penicillin.

References

ALWEN, J. & SMITH, D. G. (1967). A medium to suppress swarming in *Proteus*. *J. appl. Bact.*, **30,** 389.

ARMITAGE, J. P., ROWBURY, R. J. & SMITH, D. G. (1974). The effects of chloramphenicol, nalidixic acid and penicillin on the growth and division of swarming cells of *Proteus mirabilis*. *J. med. Microbiol.*, **7,** 459.

ARMITAGE, J. P., ROWBURY, R. J. & SMITH, D. G. (1975). Indirect evidence for cell wall and membrane differences between filamentous swarming cells and short non-swarming cells of *Proteus mirabilis* *J. gen. Microbiol.*, **89**, 199.

FLECK, J., MARTIN, J. P., & MOCK, M. (1974). Action of lysozyme on penicillin-induced filaments of *Proteus vulgaris*. *J. Bact.*, **120,** 929.

GRABOW, W. O. K. (1972). Growth-inhibiting metabolites of *Proteus mirabilis*. *J. med. microbiol.*, **5,** 191.

GRAY, G. W. & WILKINSON, S. G. (1965). The effects of ethylenediamine tetra-acetic acid on the cell walls of some Gram-negative bacteria. *J. gen. Microbiol.*, **39,** 385.

HOENIGER, J. F. M. (1965*a*). Development of flagella by *Proteus mirabilis*. *J. gen. Microbiol.*, **40,** 29.

HOENIGER, J. F. M. (1965*b*). Influence of pH on *Proteus flagella*. *J. Bact.*, **90,** 275.

HOENIGER, J. F. M. (1966). Cellular changes accompanying the swarming of *Proteus mirabilis*. *Can. J. Microbiol.*, **12,** 113.

HUGHES, W. H. (1957). A reconsideration of swarming in *Proteus mirabilis*. *J. gen. Microbiol.*, **17,** 49.

KELLENBERGER, E., RYTER, A. & SECHAUD, J. (1958). Electron microscope study of DNA-containing plasma II. Vegetative and mature phase DNA as compared with normal bacterial nucleoids in different physiological states. *J. biophys. biochem. Cytol.*, **4**, 671.

MARTIN, H. H., LEHMANN, R., HERZOG, J. & KAUL, U. (1974). Altered rigid layer composition in cell envelopes of shape defective forms of *Proteus mirabilis* and *Escherichia coli*. *Ann. N.Y. Acad. Sci.*, **235**, 283.

SMITH, D. G. (1972). The *Proteus* swarming phenomenon. *Sci. Prog., Oxf.*, **60**, 487.

Ultrastructure of Budding and Prosthecate Bacteria

C. S. Dow, D. Westmacott and R. Whittenbury

Biological Sciences, University of Warwick, Coventry, England

Introduction

With the advent of morphogenesis and differentiation as a major growth point in biology, budding and prosthecate bacteria are entering the limelight as models for the study of processes related to development. Although prokaryotic organisms are far removed in many respects from eukaryotic cells they do undergo analogous processes in their growth and morphological cell cycles. The simplicity of their growth media, their rapid generation times and relatively easily manipulated genetic systems and the selective synchronization procedures which can be used, point to several of the advantages which prokaryotes have over eukaryotes as experimental systems. Consequently it is possible that prokaryotes may provide clues about principles, if not the actual processes, which govern differentiation and development in all cells. Even in their own right budding bacteria are proving to be unique organisms which have a lot to offer in the study of microbiology. They vary considerably in their biological properties, some are heterotrophs, some autotrophs and some phototrophs. A point to emphasize is that they are not a tight-knit group of bacteria classifiable into only one or two genera. There is for instance a considerable spread in the guanine plus cytosine ratios of the DNA from different species (Table 1).

Arguments about what the budding process really is and how similar or not reproduction by this method is to conventional binary fission, and the consequences of a budding mode of life, are worth discussion. Several authors have discussed the budding bacteria as a distinct bacterial group (Zavarzin, 1961; Starr and Skerman, 1965; Schmidt, 1971; Hirsch, 1974). We shall consider only those species or genera which highlight the more important exploitable characteristics.

In our opinion "budding" can be simply defined as a process of obligate polar growth resulting in division which is frequently asymmetrical and in which the siblings are not structurally equivalent, i.e. there is clear division of "old" mother and "new" daughter cellular material. Polar growth has

TABLE 1. Guanine plus cytosine ratios of known budding and prosthecate bacteria

Organism	Neutral CsCl buoyant density gcm^{-3}	G + C (mole %)
Caulobacter[1,2]	1.7197–1.7257	61–67
Asticaccaulis[2]	1.7139	55
Prosthecomicrobium[3]	1.7245–1.7280	65.8–69.4
Ancalomicrobium[1]	1.7227	64
Mushroom bacterium[1,4]	1.7257	67.1
Nitrobacter[1]	1.7234	64.7
Hyphomicrobium[1,5]	1.718–1.7255	59.2–66.8
Rhodomicrobium vannielii[1,6]	1.7205–1.7225	61.8–63.8
Rhodopseudomonas palustris[1,6]	1.7235–1.725	64·8–66·3
R. viridis[1,5]	1.7250–1.7300	66.3–71.4
R. acidophila[1,6]	1.7227	64

1. Dow and Whittenbury, pers. comm.; 2. Poindexter (1964); 3. Staley (1968); 4. Whittenbury and Nicoll (1971); 5. Mandel *et al.* (1972); 6. Mandel *et al.* (1971).

been demonstrated in a variety of more conventional rod-shaped organisms, particularly *Escherichia coli* (Donachie and Begg, 1970). It appears most likely that the prime difference between this process and "budding" is simply the obligateness of polar growth of the budding bacteria. In more extreme cases the involvement in reproduction of an integral cellular extension, which is narrower than the diameter of the mature cell, also distinguishes these bacteria from *E. coli* (Fig. 1). The cell wall and cytoplasmic membrane are continuous with the extension.

Such cellular appendages have been referred to as *prosthecae* (Staley, 1968) and although this term is all embracing, a clear distinction must be

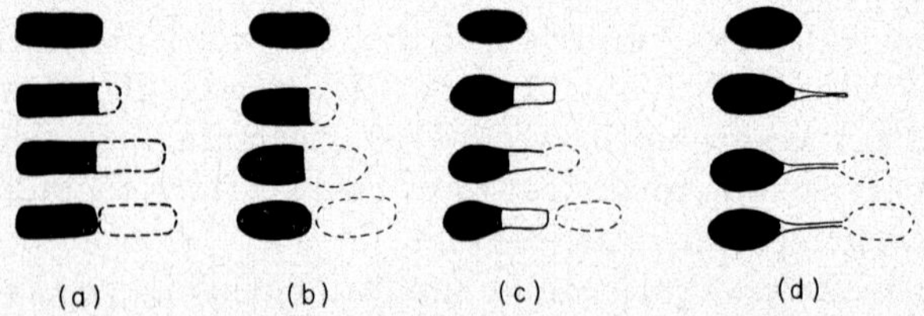

FIG. 1. Comparison of asymmetric polar growth reported for (a) *E. coli* grown on minimal media (Donachie and Begg, 1970) with the obligate polar growth of the budding bacteria, (b) *R. acidophila*, (c) *R. palustris and* (d) *Rm. vannielii.*

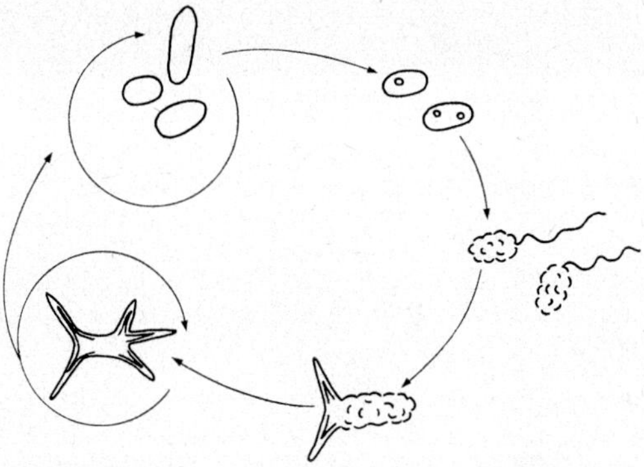

FIG. 2. Phenotypic variation displayed by an *Ancalomicrobium* isolate.

drawn between the appendages of *Ancalomicrobium* and *Prosthecomicrobium* (Staley, 1968) and the stalks of *Caulobacter* species (Poindexter, 1964; Schmidt and Stanier, 1966) which are not directly, if at all, involved in reproduction and the tubular extensions of the budding bacteria which are involved in reproduction processes.

Ancalomicrobium and *Prosthecomicrobium*

These morphologically exotic organisms have been observed by both light and electron microscopy in freshwater, mud and soil samples over several years (Henrici and Johnson, 1935; Nikitin, Vasileva and Lokmacheva, 1966; Hirsch and Rheinheimer, 1968; Orenski, Bystricky and Maramorosch 1966*a* and *b*; Staley, 1968; Volarovich and Terent'ev, 1968). However, only two isolates of *Ancalomicrobium* have been successfully cultured and studied in the laboratory (Staley, 1968; Staley and Mandel, 1973) and although there have been several reported isolations of *Prosthecomicrobium* this organism has not been examined in detail (Hirsch, 1974). Both genera characteristically have several cellular extensions per cell and replication of the isolated species is by budding (polar growth) from the cell body (Staley, 1968).

We have obtained from freshwater an appendaged organism which on initial isolation was very similar to the type species *Ancalomicrobium adeteum* but on sustained laboratory culture displayed considerable phenotypic variation (Dow and Whittenbury, in preparation) (Fig. 2).

On media containing an organic substrate in excess of 200 μg/ml the cell loses its appendages, assumes a Y-shaped morphology (Fig. 3a) and cannot be subcultured successfully, emphasizing the requirement for

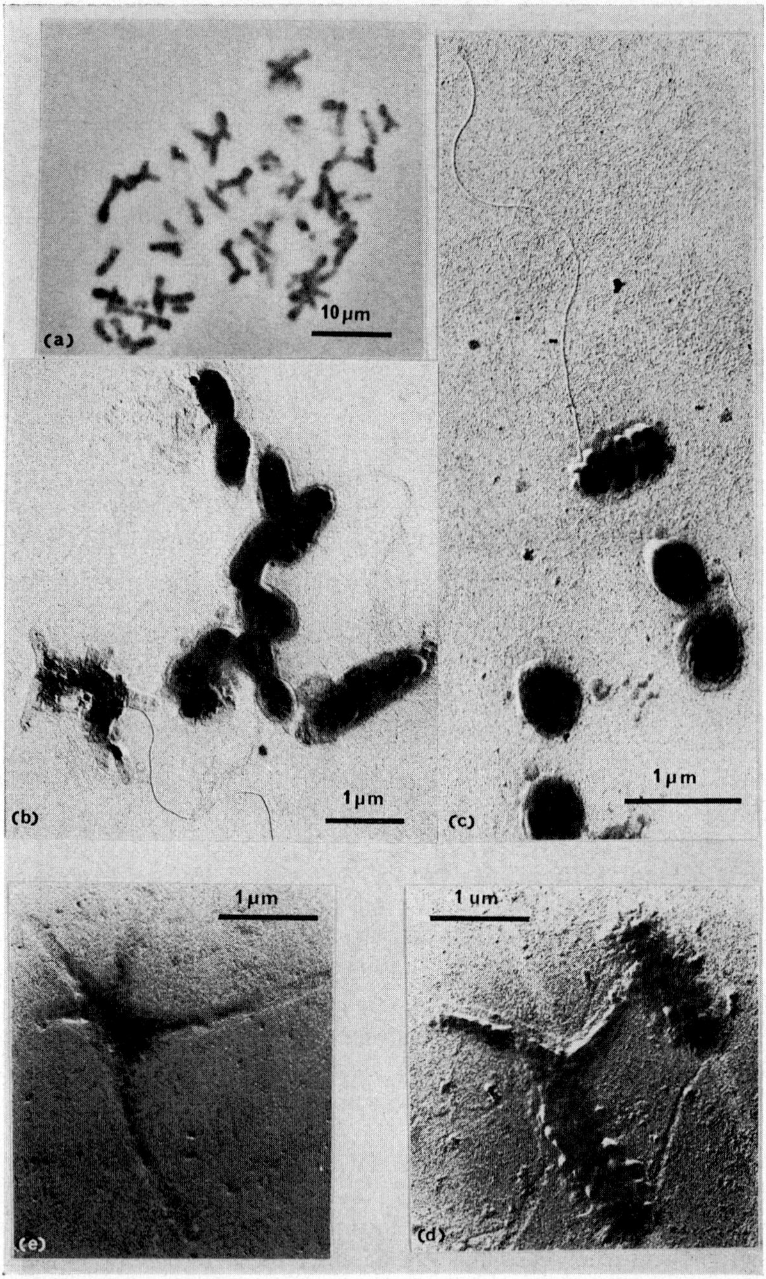

FIG. 3. (a) Phase contrast photomicrograph of a culture of *Ancalomicrobium* grown in the presence of 200 μg/ml peptone. (b), (c), (d), (e) Gold–palladium shadowed electron micrographs of *Ancalomicrobium* at different stages of growth.

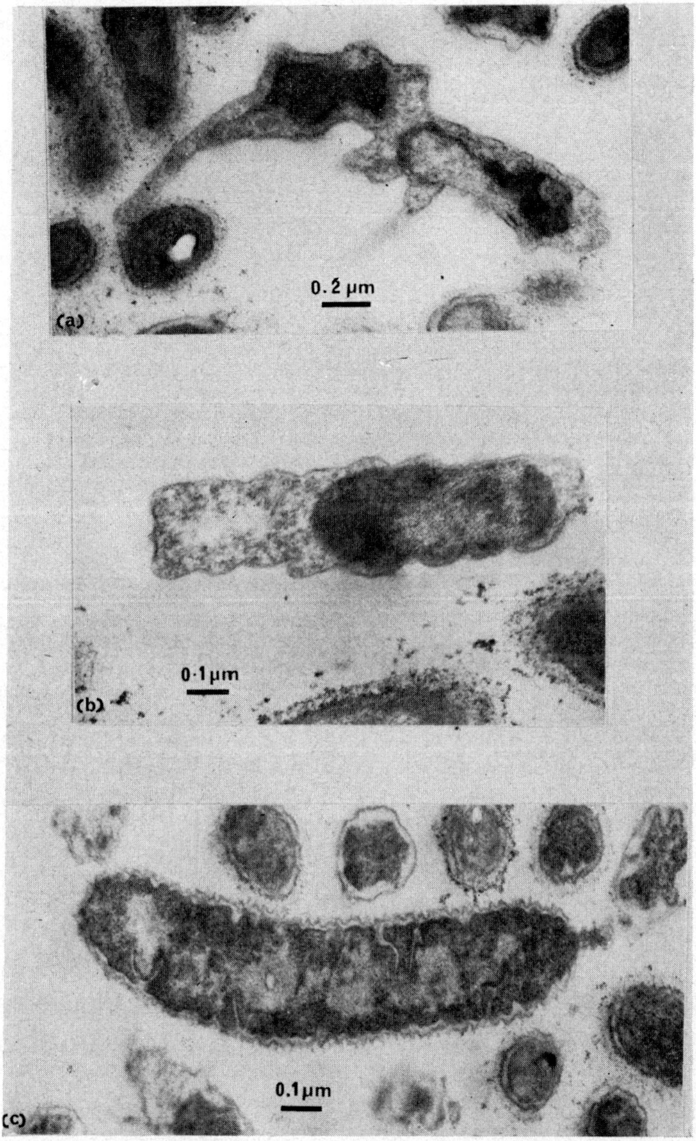

FIG. 4. Ultrathin sections of *Ancalomicrobium* prepared by the method of Ryter and Kellenberger (1958) and embedded in Araldite. (a) Appendaged cell showing continuity of cell wall and cytoplasm with the appendage. (b) Preappendaged cell with characteristic cytoplasmic "condensation". (c) Complex cytoplasmic intrusions observed in a preappendaged cell.

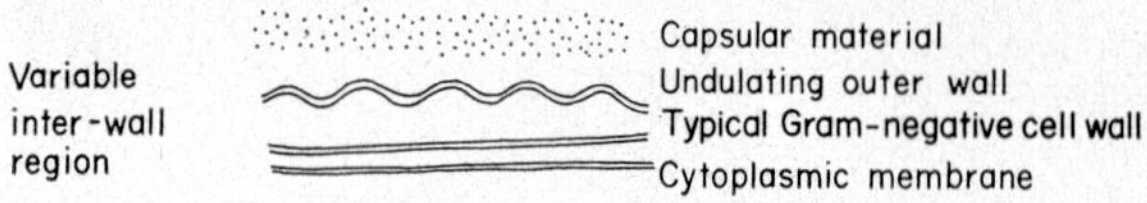

FIG. 5. Ultrastructure of the non-appendaged cell wall.

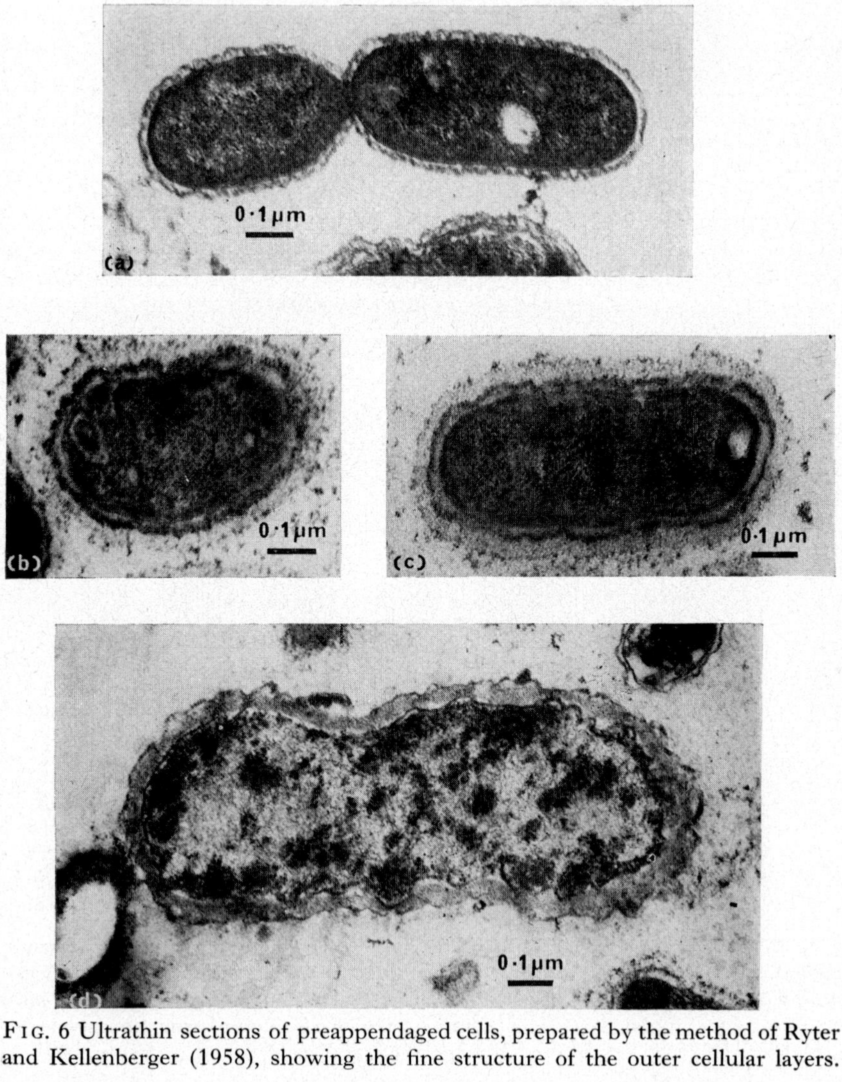

FIG. 6 Ultrathin sections of preappendaged cells, prepared by the method of Ryter and Kellenberger (1958), showing the fine structure of the outer cellular layers.

dilute cultural conditions. On dilute glucose medium the culture is initially characterized by short rod/coccoid cells (Fig. 3b) which upon aging assume a unique cell shape (Fig. 3c) and subsequently develop cellular appendages (Figs. 3d and e). The phenotypic variations are summarized in Fig. 2. It is apparent that the cellular appendages can be induced or repressed in response to cultural conditions. They do not assume a role in replication.

The cell wall and cytoplasm are unquestionably continuous with the appendages (Staley, 1968; Figs. 3d, e; Fig. 4a). In addition the pre-appendaged (Fig. 4b) and the appendaged cell (Fig. 4a) are characterized not only by the cell wall undulations but by the "condensation" of nuclear and adjacent cytoplasmic material. The ultastructural pattern of the cell wall of all non-appendaged cells is shown in Fig. 5. The notable features are the undulating outer wall (Fig. 6) and the variability of the inter-wall region which may be more extensive in some cases (Fig. 6d) than in others (Fig. 6a).

As the culture progresses towards the appendaged stage there is a marked increase in the number of cytoplasmic invaginations. They are initially small and unobtrusive (Fig. 6c) becoming large and complex prior to appendage formation (Fig. 4c).

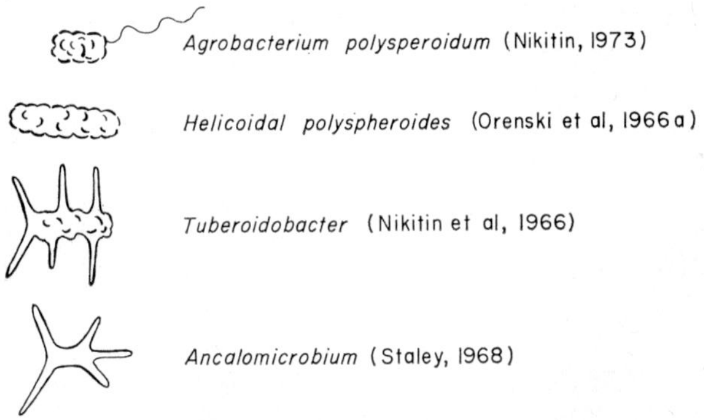

FIG. 7. Classification of *Ancalomicrobium* cultural variants.

The functional significance of these morphological and accompanying ultrastructural modifications is unknown. However, it is obvious that appendage formation can be dissociated from replication. Also such phenotypic variation demands caution in the allocation of generic names to observed but uncultured organisms. The phenotypic variants of this species have been assigned four different generic names depending upon the expressed morphology (Fig. 7).

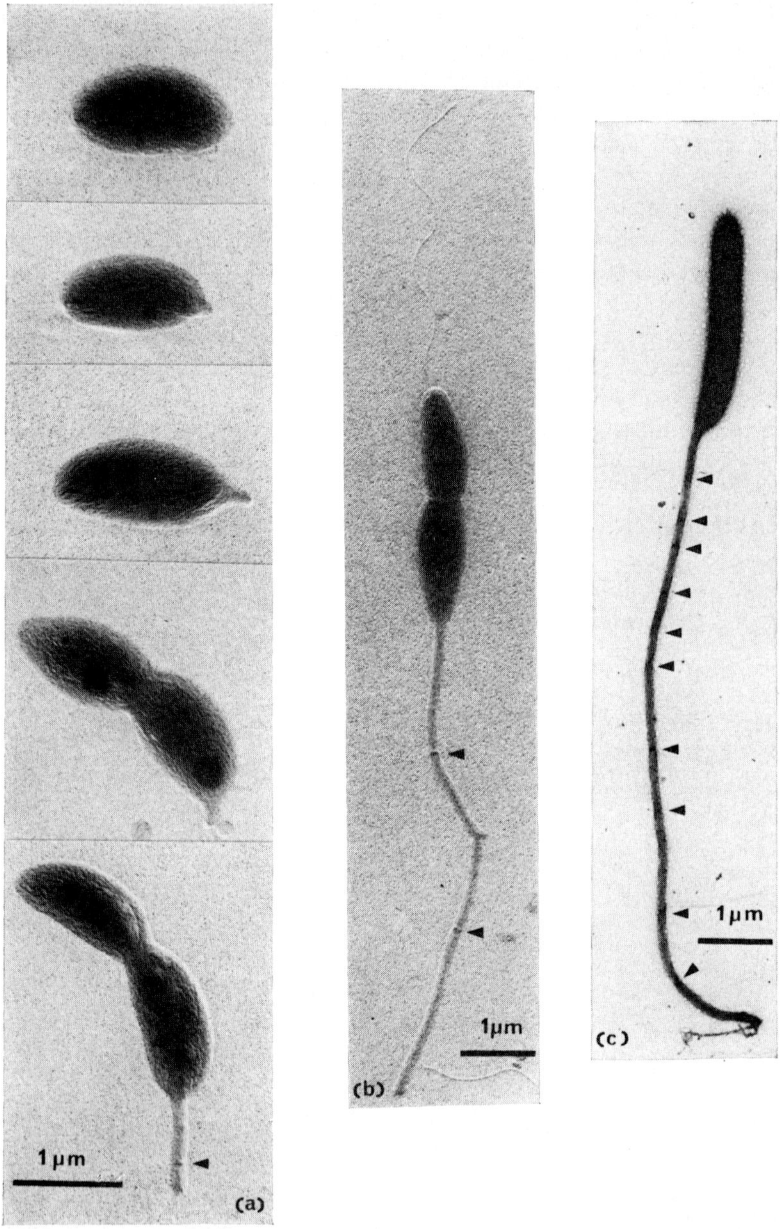

FIG. 8. (a) *Caulobacter* life cycle (Gold–palladium shadows). (b) Mother and flagellated daughter cell on the point of separation (Gold–palladium shadows). (c) Phosphotugstic acid negative stain of a *Caulobacter* mother cell which has undergone twelve generations; as determined from the number of crossbands in the stalk.

Caulobacter

Caulobacter is a Gram-negative, heterotrophic bacterium which undergoes a dimorphic life cycle in which a motile (swarm) cell undergoes obligate sequential differentiation to yield a stalked mother cell (Fig. 8). The stalk length has been shown to be closely related to inorganic phosphate concentration (Schmidt and Stanier, 1966) and it appears that these organisms are specialized for survival in dilute environments. Poindexter (1964) has reviewed the biological properties of the group but it is as a model differentiation system that *Caulobacter* species are now of interest (Shapiro, Agabian–Keshishian and Bendis, 1971; Newton, 1972; Degnen and Newton, 1972). Although environmental factors are influential in determining stalk length it has been recently shown that the mother cell must synthesize new stalk material prior to replication as does the immature swarm cell (Staley and Jordan, 1973). The mother cell produces a crossband within the stalk after each round of replication and so (conveniently for experiments) inadvertently marks each generation (Fig. 9).

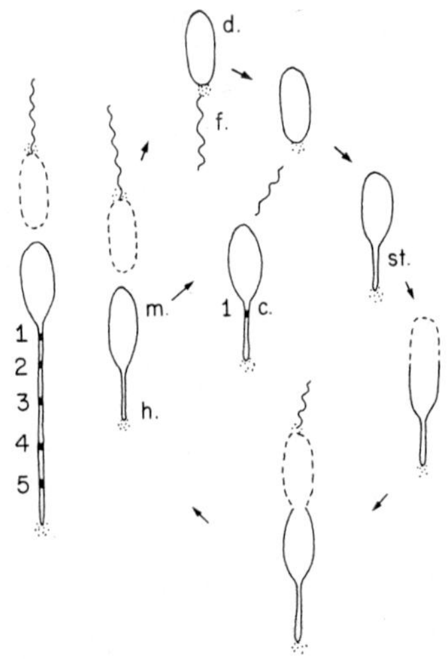

FIG. 9. Life cycle of *Caulobacter* species showing the sequetial synthesis of crossbands with each round of replication. m, mother cell; d, daughter cell; f, flagellum; c, crossband; h, holdfast; st, stalk.

The inter-relationship of stalk synthesis and the replication cycle is of particular interest. Although not physically involved in daughter cell production stalk synthesis is an obligate part of the cycle. The function of the *Caulobacter* stalk therefore would appear to be more complex than a simple environmentally adaptive organelle.

The ultrastructure of *Caulobacter* has been comprehensively studied by Poindexter (1964), Poindexter and Cohen-Bazire (1964), Cohen-Bazire, Kunisawa and Poindexter (1965) and by Schmidt and Stanier (1966). Recent articles have also been published on the ultrastructure of the stalk and crossbands (Jones and Schmidt, 1973; Schmidt, 1973).

As shown, the cellular appendages of *Ancalomicrobium, Prosthecomicrobium* and *Caulobacter* are not directly involved in daughter cell formation. It is those organisms in which the cellular extension is obligatorily involved in daughter cell formation which we now wish to consider. Three species are described here, *R. acidophila, R. palustris* and *R. vannielii* all members of the Rhodospirillaceae (formerly Athiorhodaceae). Their growth cycle "complexity" increases in the order in which they are mentioned and it is apparent that they form a *gradient of differentiation.*

Rhodopseudomonas acidophila

The simplest budding member of the Rhodospirillaceae, *R. acidophila*, was first described by Pfennig (1969). The organism, which grows optimally at pH 5.6 to 6.0, can grow anaerobically in the light and aerobically in the dark on carbon sources such as pyruvate, lactate, malate, fumarate and succinate. Photosynthetic cultures of *R. acidophila* are usually a deep

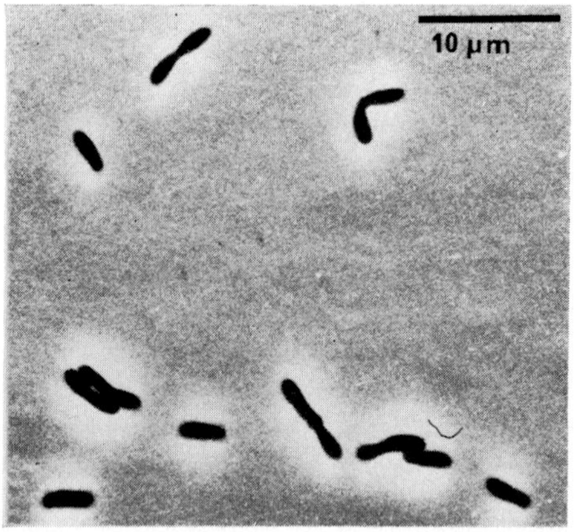

FIG. 10. Phase contrast photomicrograph of *R. acidophila.*

brown-orange colour (carotenoids of the spirilloxanthin series and bacteriochlorophyll *a*). Microscopic examination shows cells of varying length (2 to 6μm in most strains) and 1.0 to 1.5μm wide. Morphology ranges from ovoid cells to dumb-bell shaped cells (Fig. 10), and in many strains "rosettes" or clumps of polarly aggregated cells are formed. Examination of *R. acidophila* on slide culture (Pfennig, 1969) revealed that reproduction was by budding. Figure 11 shows the division cycle revealed by slide cultures and considered alongside data from negatively-stained preparations (Fig. 12) using the electron microscope.

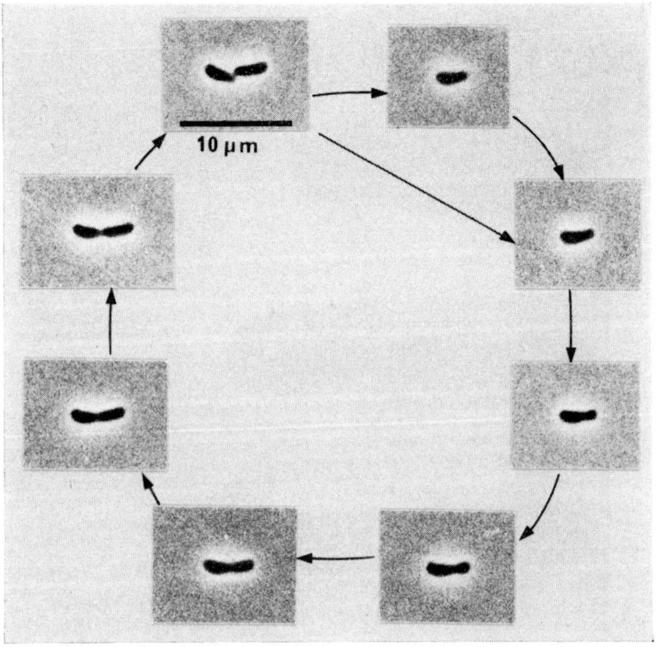

FIG. 11. Division cycle of *R. acidophila*. Phase contrast photomicrographs of a slide culture.

Morphogenesis may best be considered starting at the newly released "daughter" cell. At the pole distal to that at which division occurred the daughter may bear a sub-polar bundle of flagella (Tauschel and Hoeniger, 1974). This is seen most clearly in shadowed preparations (Fig. 13). The daughter, therefore, is often referred to as a swarmer, although growth conditions and high sheer manipulations may give rise to non-motile daughters. During development of the swarmer, flagella are always released and a sticky holdfast material is secreted at the same pole (Fig. 12b) At the same time polar growth at the division pole of the daughter gives rise to a sessile bud. The daughter can now be considered to be a mother cell bearing a daughter bud which gradually enlarges until a symmetrical, dumb-bell shape is produced (Fig. 12b–e). Finally the daughter cell may produce a new bundle of flagella and division by constriction processes

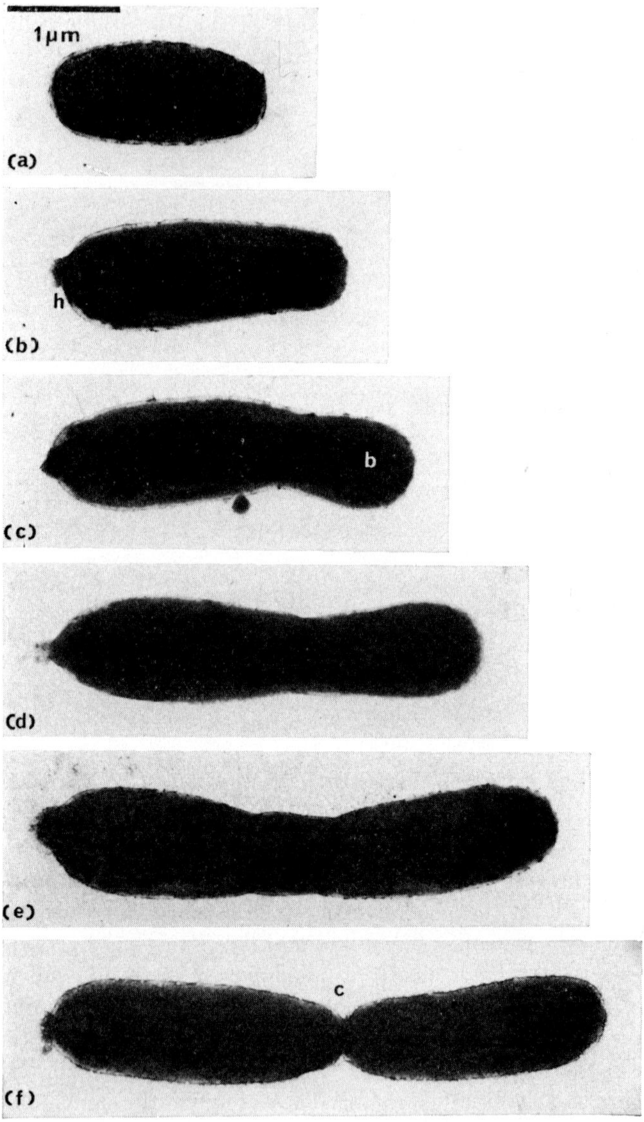

FIG. 12. Electronmicrographs of stages of morphogenesis in *R. acidophila*, negatively stained with uranyl acetate. h, holdfast; b, bud; c, division constriction.

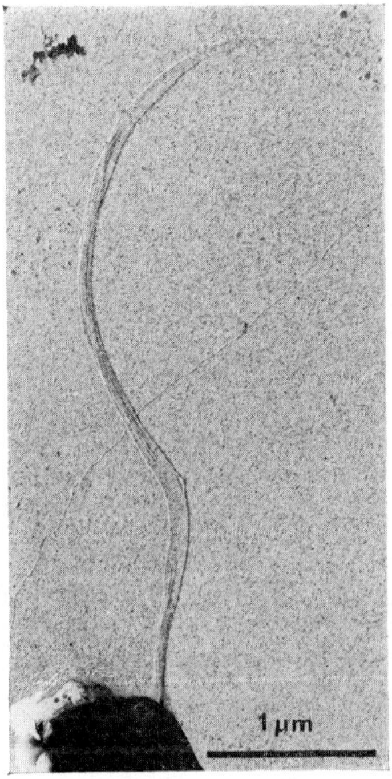

FIG. 13. Sub-polar bundle of flagella on *R. acidophila* shadowed with gold palladium.

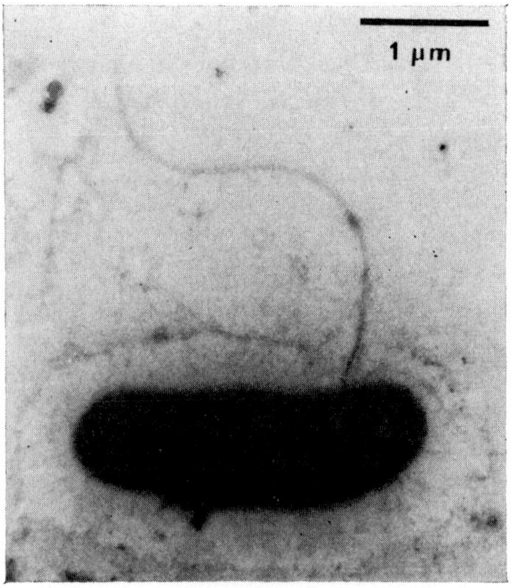

FIG. 14. Unwashed preparation of *R. acidophila*, negatively stained with uranyl acetate, showing capsule and flagella.

separates the two cells (Fig. 12f). This division gives rise to two morphologically dissimilar cells, one bearing a holdfast, the other bearing flagella. However, both continue to develop buds at the division poles, the swarmer cell having first to release its flagella and synthesize a holdfast. At division, therefore, the mother cell may be any number of generations old whilst the swarmer is always only in the first generation.

Tauschel and Hoeniger (1974) showed that *R. acidophila*, under certain conditions, produced an extensive fibrillar capsule. We have been able to show a more clearly defined capsule in unwashed, negatively-stained preparations (Fig. 14). Tauschel and Hoeniger (1974) also noted that formaldehyde-fixed, negatively-stained preparations of *R. acidophila* revealed a delicate surface pattern on the cell wall. The whole cell surface was covered with 9–11 nm diameter rings, of low contrast, arranged in a helical pattern.

The most striking feature of transverse ultrathin sections (Fig. 15a) is

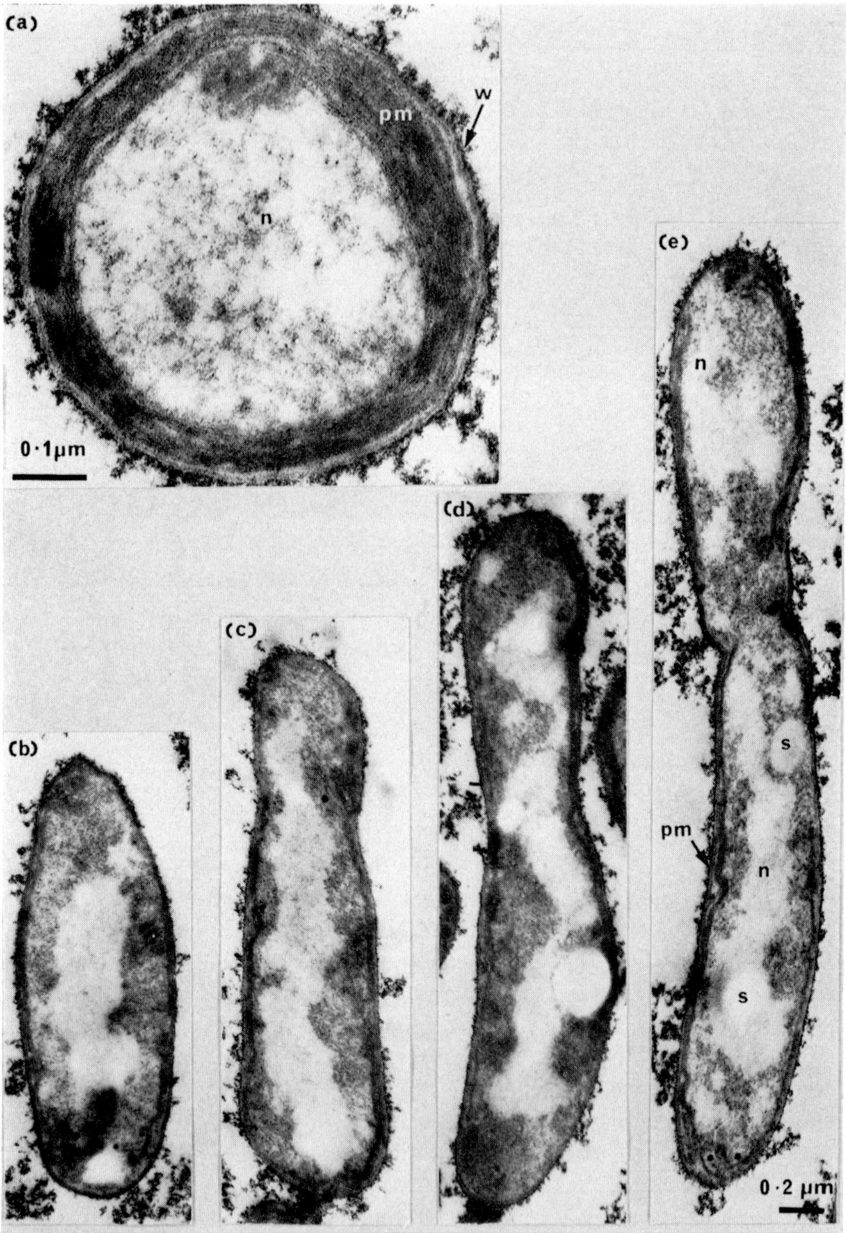

FIG. 15. Ultrathin sections of *R. acidophila*. w, cell wall; pm, photosynthetic membranes; n, nucleus; s, storage granule. (Prepared by the method of Ryter and Kellenberger, 1958).

that the lamellar photosynthetic membranes completely encircle the cytoplasm in that plane. In this way *R. acidophila* resembles *R. vannielii* (Fig. 29d). Longitudinal sections, however, show continuity of photosynthetic membranes only over a major part of the cell periphery (Fig. 15b–e). Although mother cell and daughter photosynthetic membranes seem to be continuous (Fig. 15d) there is a gap near the point where division will occur (Fig. 15c). It is possible that this gap in photosynthetic membranes gives rise to the small translucent area between dividing cells under phase contrast microscopy (Fig. 11). The cytoplasm of *R. acidophila* contains a rather diffuse nucleus. Growth on certain carbon sources gives rise to deposits of poly-β-hydroxybutyrate within the cytoplasm.

Rhodopseudomonas palustris

The diversity of morphology of *R. palustris* when examined by light microscopy and the appearance of rosettes in the cultures (Fig. 16) was first recognised by van Niel (1944'. However, unlike *R. acidophila, R. palustris*

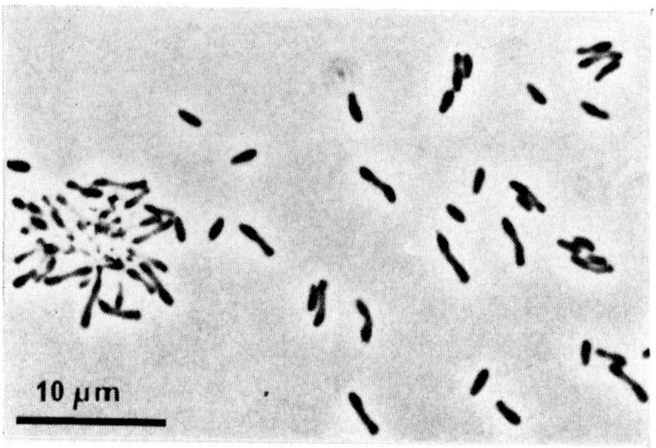

FIG. 16. Phase contrast photomicrograph of *R. palustris*.

grows optimally at pH 6.5 to 7.0 and the cells are considerably smaller; only 0.4 to 0.6μm wide and usually 1.5 to 5μm in length. *R. palustris* is also a facultative photoheterotroph, being able to grow best on carbon sources such as pyruvate, acetate, lactate, malate, fumarate, succinate, butyrate and ethanol. Photosynthetic growth gives rise to a deep orange–red culture, whilst arobic, dark-grown cultures are non-pigmented.

Whittenbury and McLee (1967) showed that *R. palustris* divides by

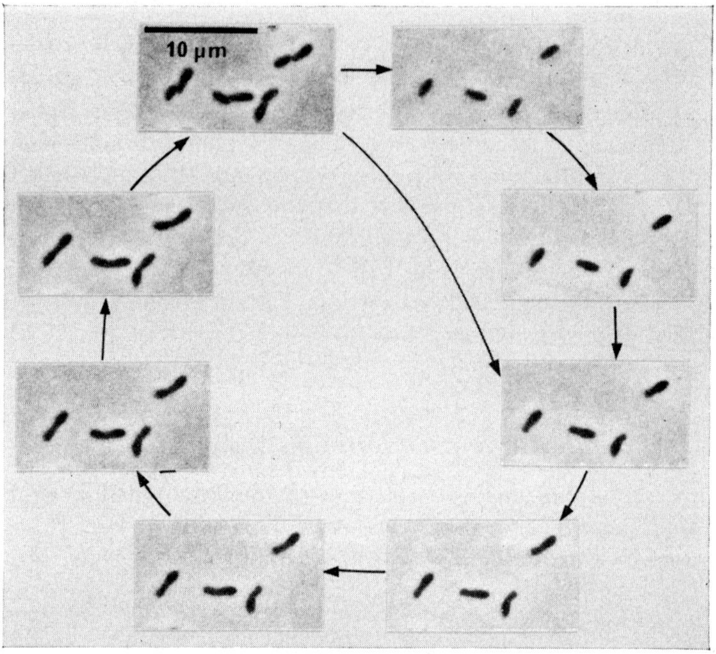

FIG. 17. Division cycle of *R. palustris*. Phase contrast photomicrographs of a synchronous slide culture.

producing a bud on the end of a tube which is slightly narrower than the body of the cell. Figure 17 shows the development of a synchronous slide culture of *R. palustris* as a division cycle, constructed in conjunction with information from shadowed preparations of a synchronous culture (Fig. 18). As with *R. acidophila* the daughters are motile swarmers under suitable conditions, however *R. palustris* only bears one sub-polar flagellum, opposite the division pole of the cell (Fig. 18a). Tauschel (1970) and Tauschel and Drews (1970) showed in their studies of *R. palustris* flagella that the core flagellin is surrounded by a non-proteinous sheath. We have confirmed this observation using a strain isolated in our laboratory (Fig. 20). As the swarmer develops, it synthesizes a holdfast alongside the flagellum and a phase contrast—translucent tube, with a typically blunt end, grows out from the opposite pole and elongates to a finite length (Fig. 20b–d). Meanwhile flagella are shed from the cells. Tube elongation has been found to be inversely proportional to phosphate concentration (C. S. Dow, pers. comm.) in a similar manner to stalk elongation in *Caulobacter* species (Schmidt and Stanier, 1966). Thus strains of *R. palustris* previously described as producing sessile buds (Bosecker, Drews and Tauschel, 1972),

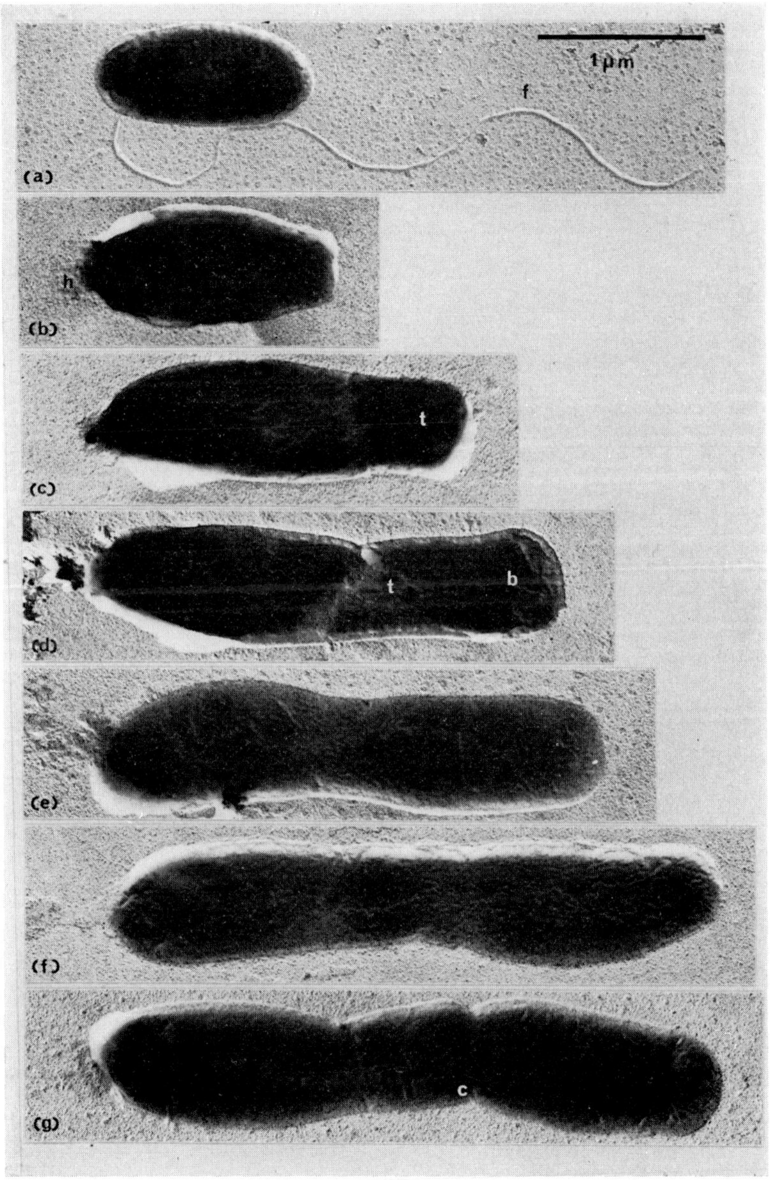

FIG. 18. Electronmicrographs of stages of morphogenesis in *R. palustris* shadowed with gold–palladium. f, flagellum; h, holdfast; t, tube; b, bud; c, division constriction.

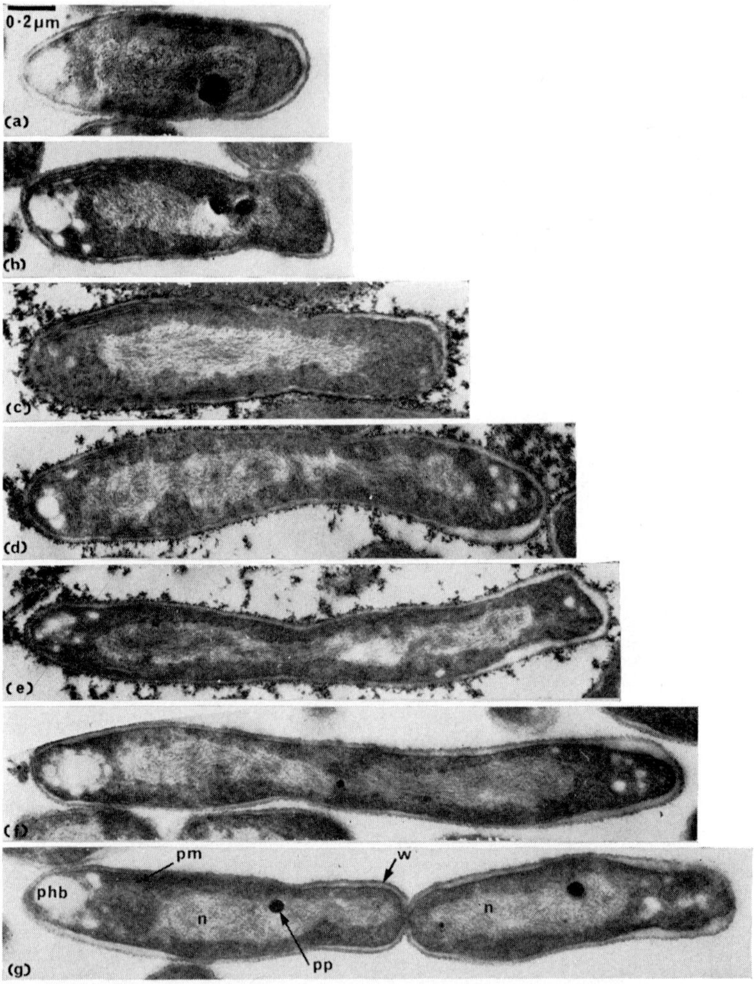

FIG. 19. Ultrathin sections of cells from a synchronous population of *R. palustris*. pm, photosynthetic membrane; w, cell wall; n, nucleus; phb, poly-β-hydroxybutyrate; pp, polyphosphate.

under conditions of phosphate-limited growth produce elongated tubes, confirming another difference between *R. palustris* and *R. acidophila*. Bud development of *R. palustris* takes place at the distal end of the tube until the daughter is of a similar size to the body of the mother cell (Fig. 18d–f). A sub-polar flagellum may then develop on the daughter cell and division occurs by constriction between the distal end of the tube and the newly formed swarmer cell (Fig. 18g). Division gives rise to even more obvious inequality in progeny than in *R. acidophila* since the mother cell not only bears a holdfast and no flagellum but also has a preformed tube on which new buds may develop immediately. Thus the mother cell by-passes a part of the division cycle (Fig. 17) whilst the daughter follows the full cycle, which is unique to first generation progeny.

Ultrathin sections of *R. palustris*, prepared from synchronous cultures at various stages of development (Fig. 19), not only confirm the developmental morphology described above but also show a clear distinction between mother and daughter cell bodies and the "reproductive" tube. In no electron migrographs examined by us have photosynthetic membrane lamellae been found in the tube region, thus explaining the translucent appearance of the tube using phase contrast microscopy. Photosynthetic membranes in longitudinal sections of cells also often appear to be on one side of the cell only (Fig. 19b, e, f and 25) since in these first generation synchronous cells the membranes never totally encircle the cytoplasm in transverse section (Fig. 21). The nucleus of *R. palustris* is very much more compact than in *R. acidophila*. However, just prior to division (Fig. 19f) it extends along a large proportion of the cell. Within the nucleus, granules of polyphosphate are often found, whilst at the poles of the cells are seen inclusions which probably consist of poly-β–hydroxybutyrate storage material.

Examination of thin sections through holdfasts (Fig. 22) indicate that this material, as yet of unknown composition, may originate from the outer layer of the cell wall. The cell wall has a typical Gram-negative structure and where the doubled–tracked, photosynthetic membrane-lamellae lie beneath it there is no discernable cytoplasmic membrane (Fig. 23). Thin sections of aerobically-dark-grown cells (Fig. 24) show no photosynthetic membrane lamellae (confirming the observation of Solov'eva and Fedenko, 1970) whilst close examination shows a typical cytoplasmic membrane inside the cell wall. Figures 25 and 26 show how the lamellar membrane of photosynthetic cells originate from an infolding and modification of the plasma membrane, thus defining the boundary between cell body and tube. A similar infolding may be seen in Fig. 19f on the tube side of the daughter cell. Tauschel and Drews (1967) came to a similar conclusion. However the structure and positioning of membrane lamellae in the strain of *R. palustris*

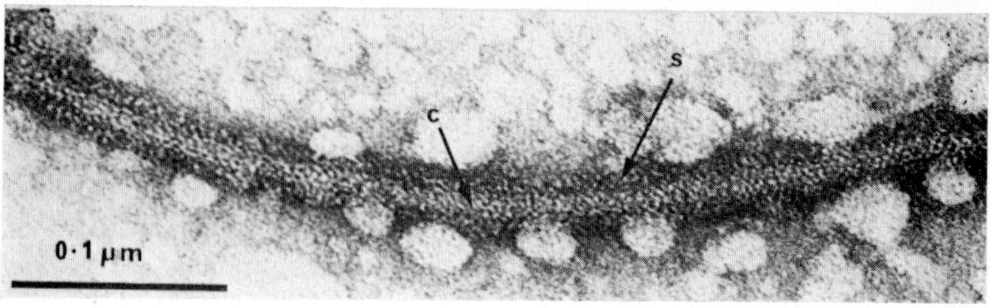

FIG. 20. *R. palustris* flagellum negatively stained with phosphotugstic acid. c, core; s, sheath.

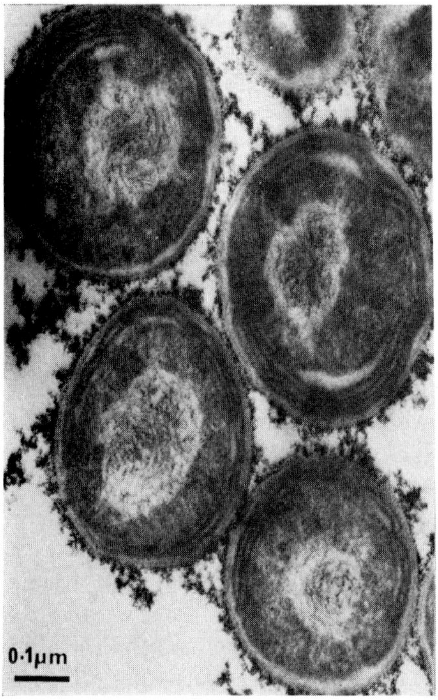

FIG. 21. Tranverse section of *R. palustris* population.

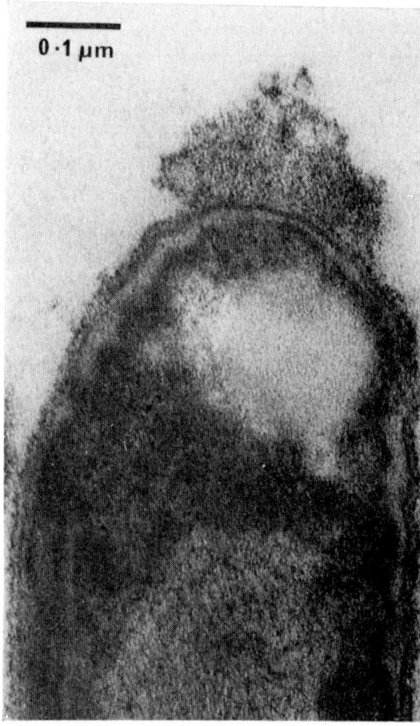

FIG. 22. Longitudinal section of *R. palustris* cell pole and holdfast.

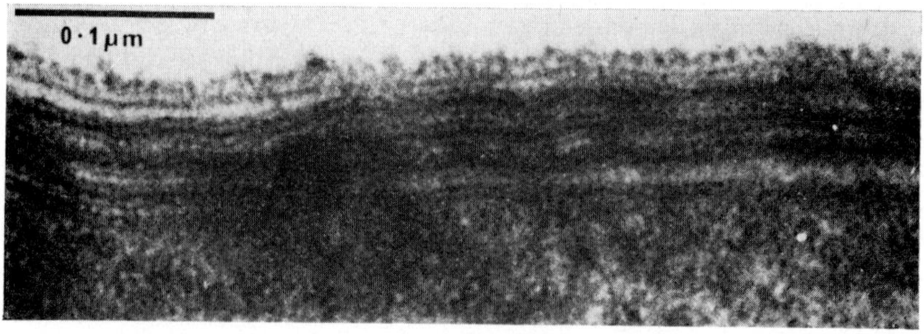

FIG. 23. Cell wall of *R. palustris*, and underlying double-tracked photosynthetic membrane lamellae.

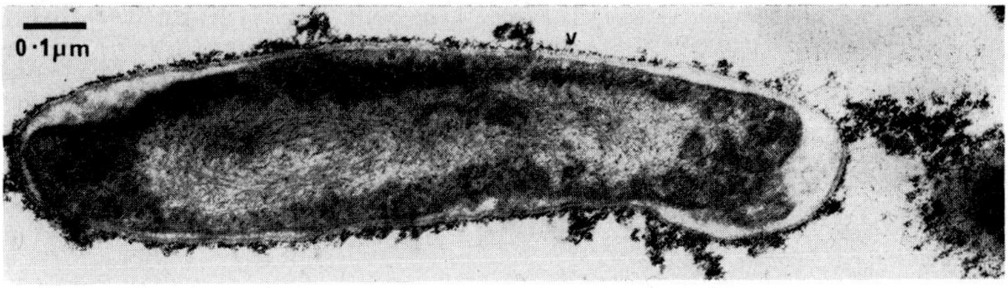

FIG. 24. Ultrathin section of aerobic grown *R. palustris* lacking photosynthetic membranes.

FIG. 25. Ultrathin section of photosynthetically grown *R. palustris* mother cell.

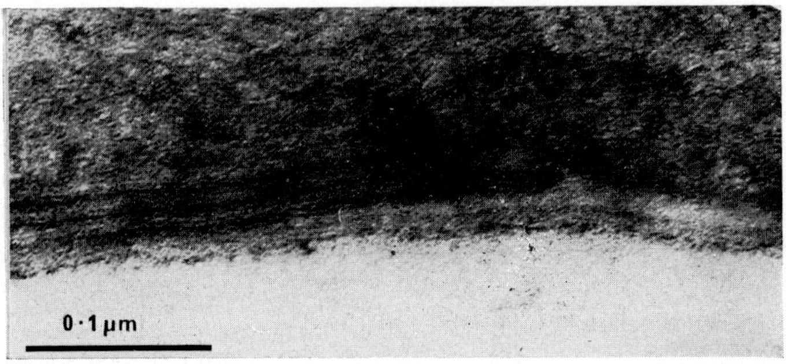

FIG. 26. Detail of photosynthetic membrane origin in Fig. 25.

that they examined was very complex. The cells they studied were, in our experience, probably "old" cells (i.e. having reproduced many times). The high degree of lamellar branching, frequently inter-leaved with broad layers of cytoplasm and periplasm, that they observed is not typical of first generation cells or "young" mother cells.

Rhodomicrobium vannielii

Physiologically *Rm. vannielii* is very similar to *R. acidophila* and *R. palustris* but its developmental cycle is equal in complexity to any found within the prokaryotic kingdom.

Batch culture growth of *Rm. vannielii* was studied by Murray and Douglas (1950) who showed reproduction was by budding. They were, however, unable to offer any explanation of the processes leading to the distribution of inter-cellular links and patterns of the micro-colonies. Actively motile cells were not observed until several years later (Douglas and Wolfe, 1959). Their development and role in the growth cycle was not established at that time.

Vegetative replication, as worked out from slide cultures, is shown as a series of electron micrographs which starts with the peritrichously flagellated swarm cell (Fig. 27a). The first morphologically observable event is the synchronous shedding of the flagella producing a non-motile, non-appendaged (non-tubed) cell (Fig. 27b). This is followed, after a well-defined maturation period, by tube formation from one or both poles of the cell with subsequent bud formation at the tube tip (Fig. 27c–d). When two tubes are formed bud formation only ever occurs on one tube at a time, the other tube is "inactive" until the following generation. On maturation of the daughter cell a plug is synthesized within the tube, physiologically separating mother and daughter cell (Fig. 27f and 29b). This unique division process appears to be under daughter cell control since the plug is always formed at a finite distance from the daughter; in contrast the distance of the plug from the mother cell is variable.

After the first generation subsequent development is dependent on whether or not two tubes have been formed per cell. With the single-tubed cell, branch formation occurs immediately whereas in the bi-tubed cell branching is delayed until the third generation, the second daughter being synthesized on the previously "inactive" tube. This progressive sequence of events is shown diagrammatically in Fig. 28.

Many of the ultrastructural features of *Rm. vannielii* have been established (Boatman and Douglas, 1961; Conti and Hirsch, 1965; Trentini and

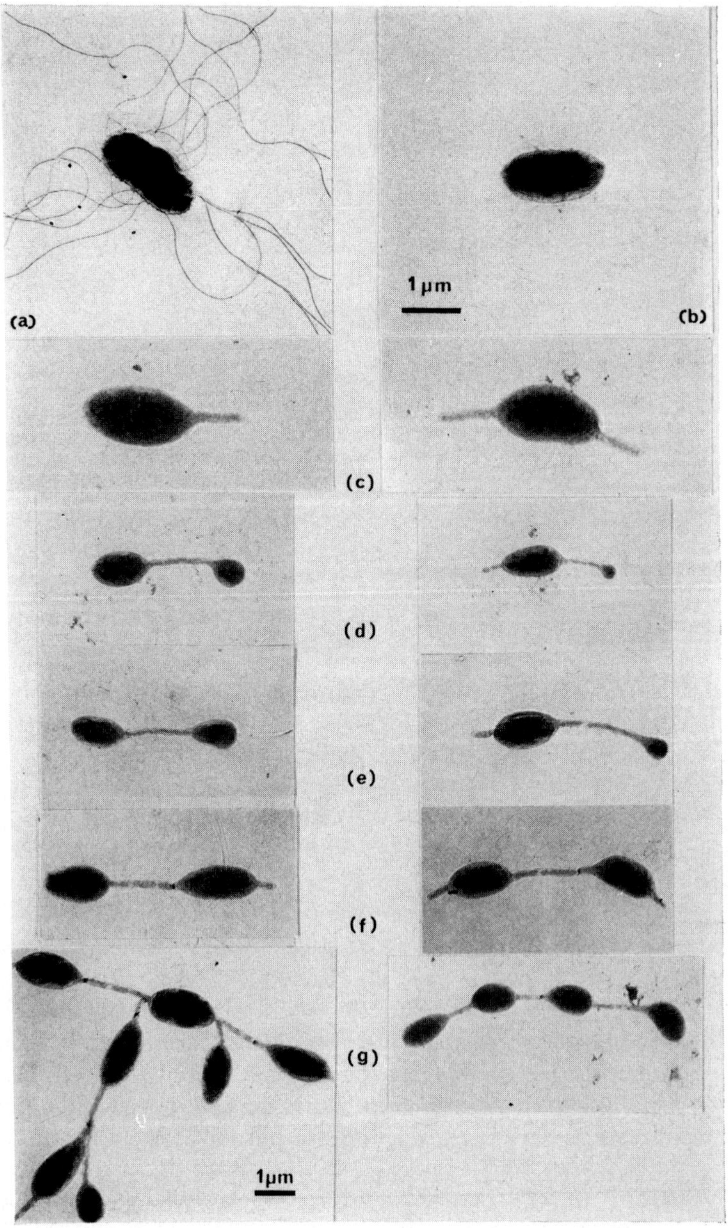

FIG. 27. Developmental cycle of *Rm. vannielii* (Gold–palladium shadows).

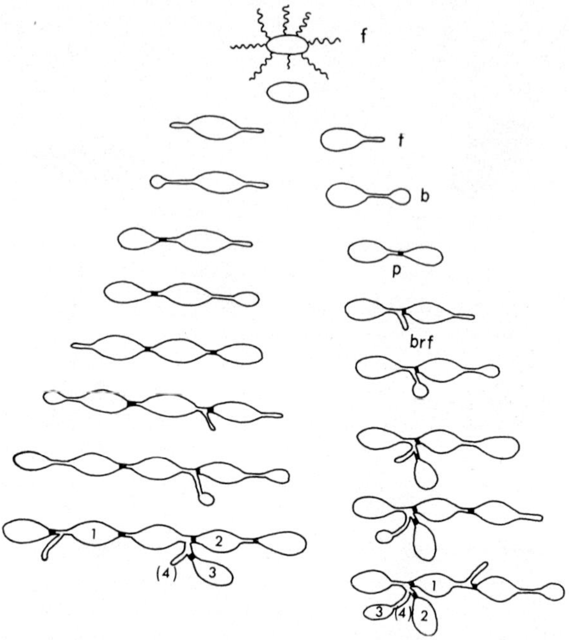

FIG. 28. Diagrammatic representation of progressive sequence of events during the vegetative cell cycle of *Rm. vannielii*. f, flagella; t, tube synthesis; b, bud formation; p, plug formation; brf, branch formation.

Starr, 1967). Thin sections of cells grown at low or moderate light intensities show a peripherally located, symmetrically distributed, lamellate photosynthetic membrane system. (Figs 29a, c and d). When cells are grown under high light intensity, however, the lamellae membrane system is no longer obvious and a considerable degree of "compartmentalization" is apparent (Fig. 30). This has been interpreted as being a reflection of the replication process (Dow and Whittenbury, in preparation).

Slide culture studies apart from revealing the morphology of the cell cycle also indicates that each mature mother cell is limited to forming four daughter cells. When such "old" cultures are examined by electron microscopy, the ultrastructure of a considerable proportion of the cells is found to

FIG. 29. (a) Negatively stained (phosphotugstic acid) preparation of *Rm. vannielii* showing the lamellae membrane arrangement. (b) Ultrathin section of *Rm. vannielii* showing plug formation. (c) Ultrathin section of *Rm. vannielii* showing the lamellae membrane system. (d) Transverse section of *Rm. vannielii*. (Sections prepared by the method of Ryter and Kellanberger, 1958). p, plug.

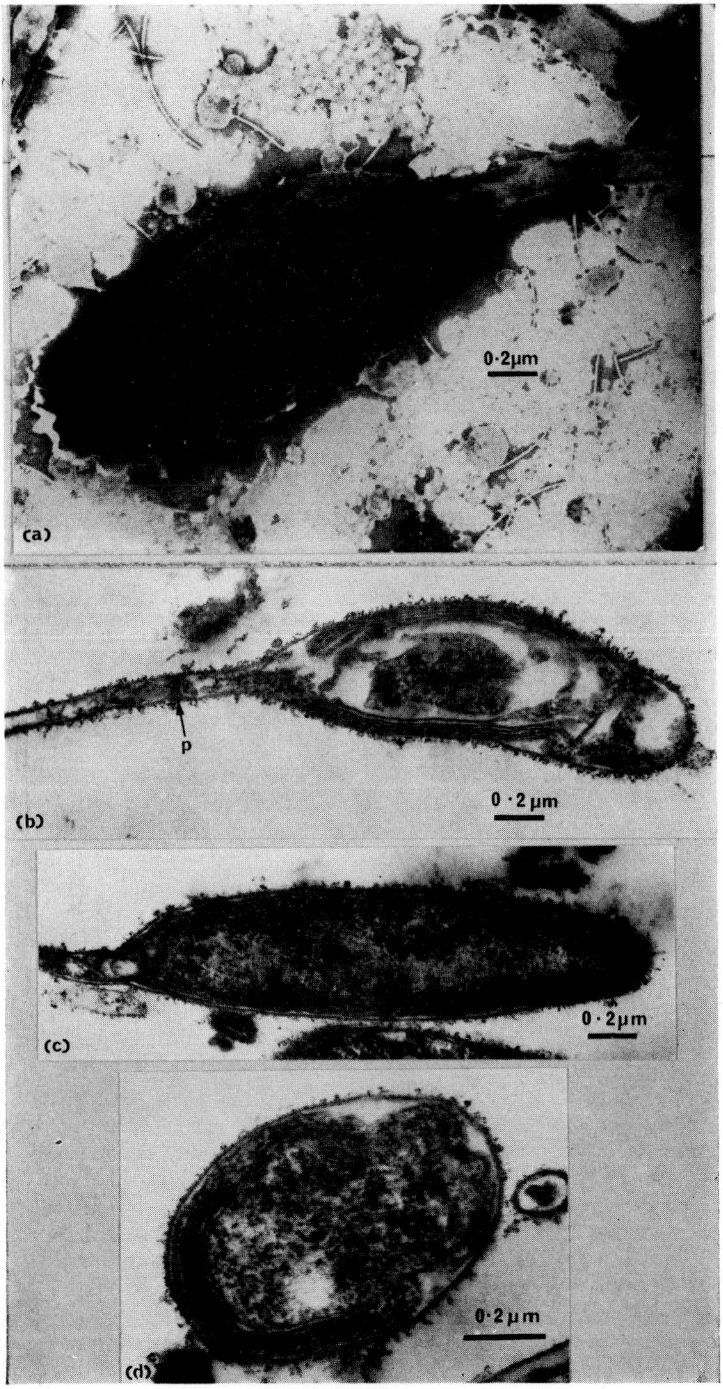

FIG. 29.

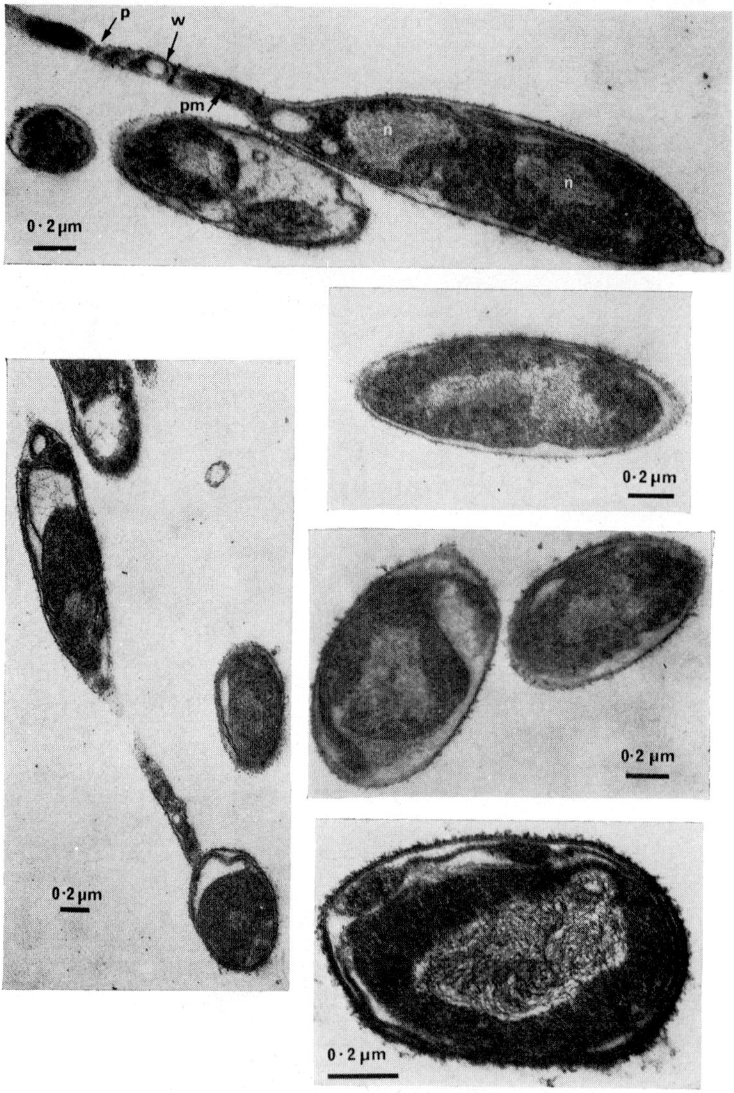

FIG. 30. Ultrathin sections of *Rm. vannielii* grown under high light intensity. There is no lamellar membrane system but a considerable degree of 'compartmentalisation' is apparent (Sections prepared by the method of Ryter and Kellenberger, 1958). p, plug; n, nucleus; w, wall; pm, plasma membrane.

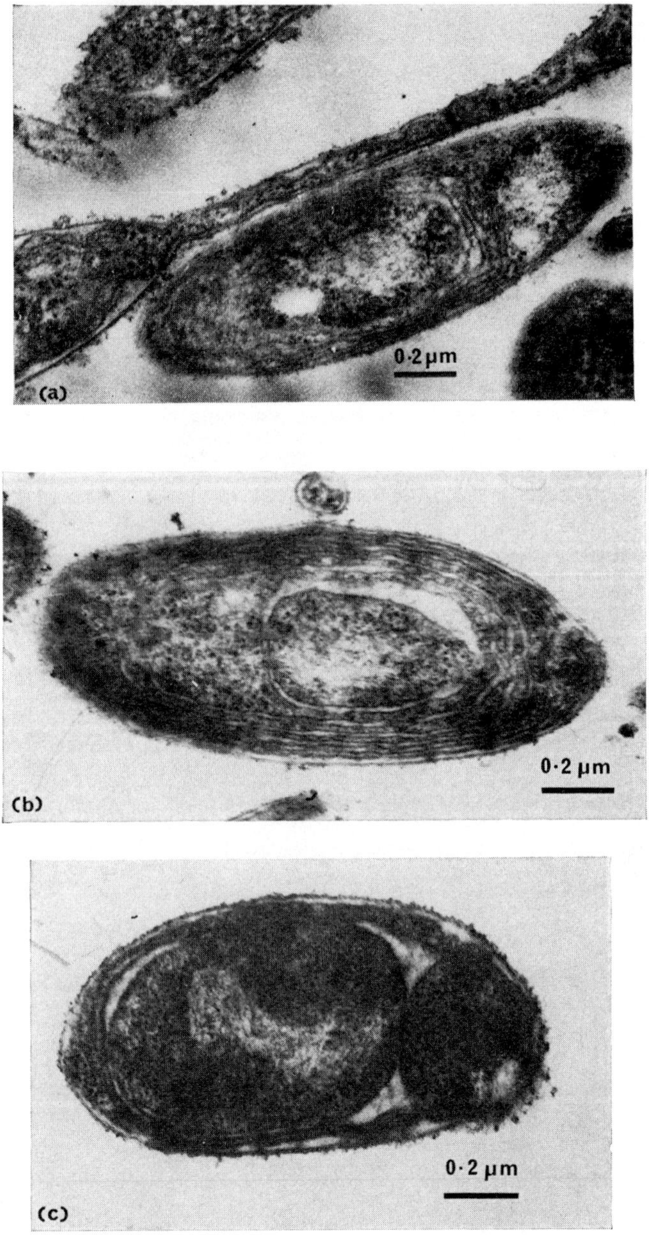

FIG. 31. Ultrathin sections of "aged" *Rm. vannielii* vegetative cells (Prepared by the method of Ryter and Kellenberger, 1958).

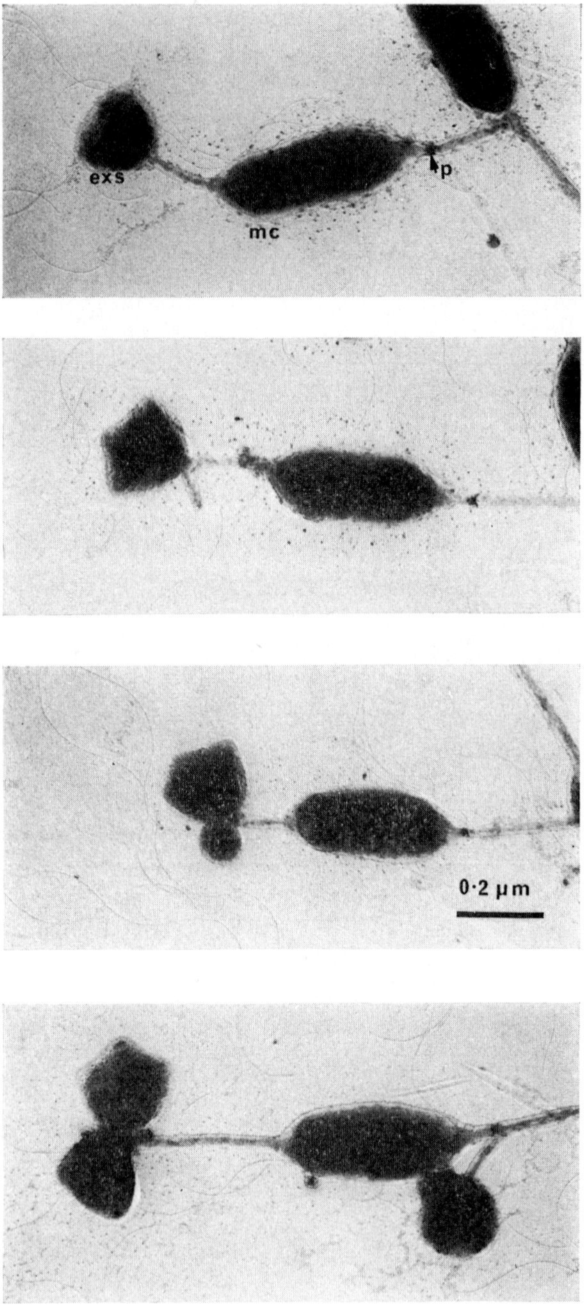

FIG. 32. Gold–palladium shadowed electron micrographs showing *Rm. vannielii* exospore formation. p, plug; exs, exospore; mc, mother cell.

have altered quite drastically (Fig. 31). The functional significance of these modifications is not known but it is tempting to speculate that after the reproductive capacity has been exhausted the mother cell becomes physiologically and hence ultrastructurally adapted to a "new" role—as yet

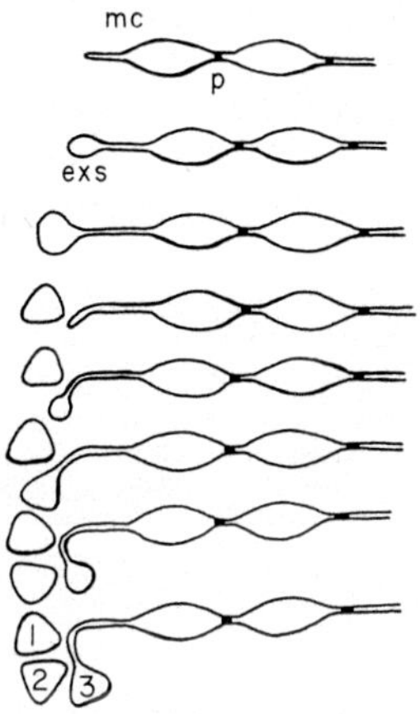

FIG. 33. Exospore formation in *Rm. vannielii.* p, plug; mc, mother cell; exs, exospore.

undefined but possibly of biochemical rather than reproductive significance.

Gorlenko (1969) isolated a strain of *Rm. vannielii* which produced unusual angular cells possessing several resistant properties. The mode of formation or germination was not established. Similarly we have isolated an exospore forming strain of *Rm. vannielii* which is identical to the type strain, *Rm. vannielii* Duchow and Douglas (1949), both morphologically and physiologically. The spores are called "exospores" as they are not formed within the confines of the cell as in the case of the *Bacillus* endospores.

Grown under low light intensity the culture becomes light limited before nutrients are exhausted and under these conditions exospores are formed in abundance. The mode of formation is shown in Fig. 32. Up to four exospores are formed sequentially from one tip (Fig. 33) i.e. each mother cell can produce up to four exospores; there is not a one-to-one relationship as found in the endospore formers.

The angularity of the exospores can be seen from Fig. 32 and Fig. 34. The modified wall structure of the mature exospore is shown in comparison to that of the germinating exospore and to that of the vegetative cell in Fig. 36.

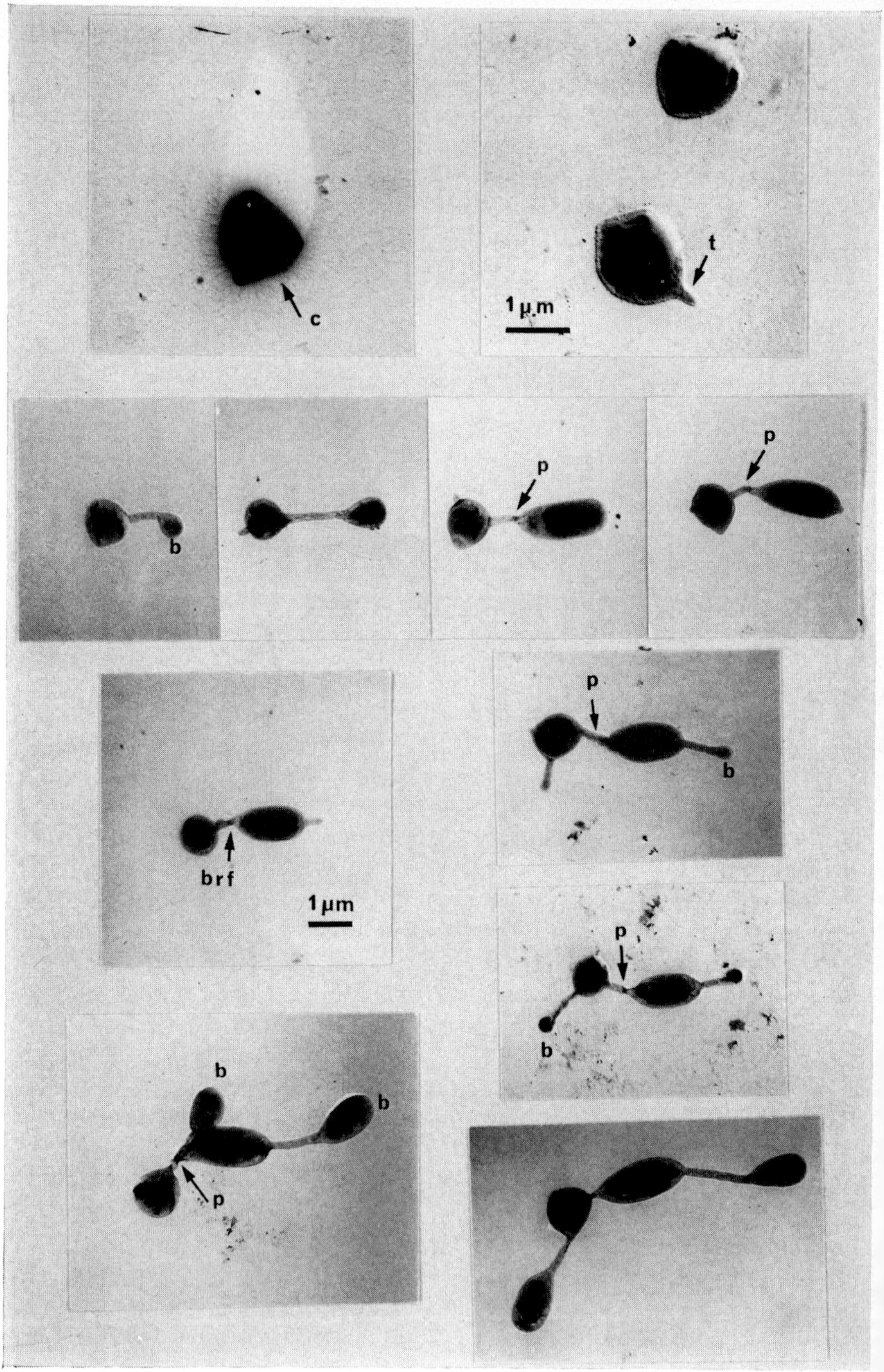

FIG. 34. Gold–palladium shadows showing *Rm. vannielii* exospore germination sequence. c, capsular material; t, tube formation; b, bud formation; p, plug; brf, branch formation.

Germinating exospores of *Rm. vannielii* are unique in that each exospore differentiates to become a "reproductive unit" forming up to four vegetative cells in a distinctive morphological sequence. This is shown as an

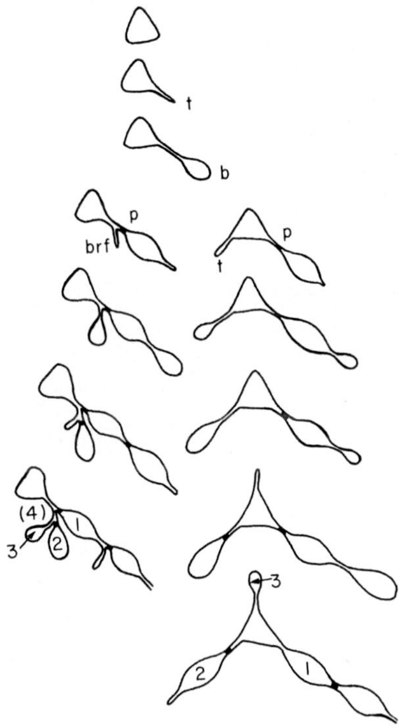

FIG. 35. Germination sequences associated with *Rm. vannielii* exospores. t, tube formation; b, bud formation; p, plug; brf, branch formation.

electron microscopic sequence in Fig. 34 and diagrammatically in Fig. 35. Ultrastructurally the exospore changes considerably on going from a dormant "resting cell" to a viable reproductive unit (Fig. 37). The significance of these events and their temporal expression has not been established.

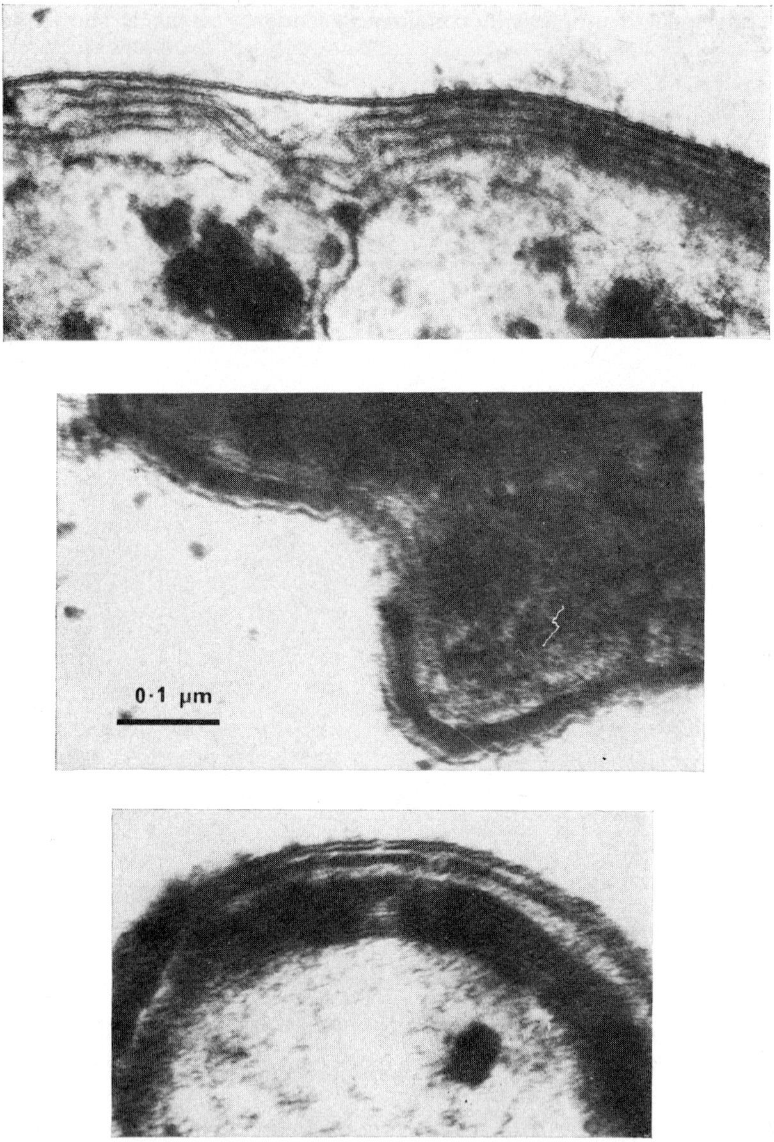

FIG. 36. Envelope structure of the *Rm. vannielii* vegetative cell in comparison to that of the mature and germinating exospore.

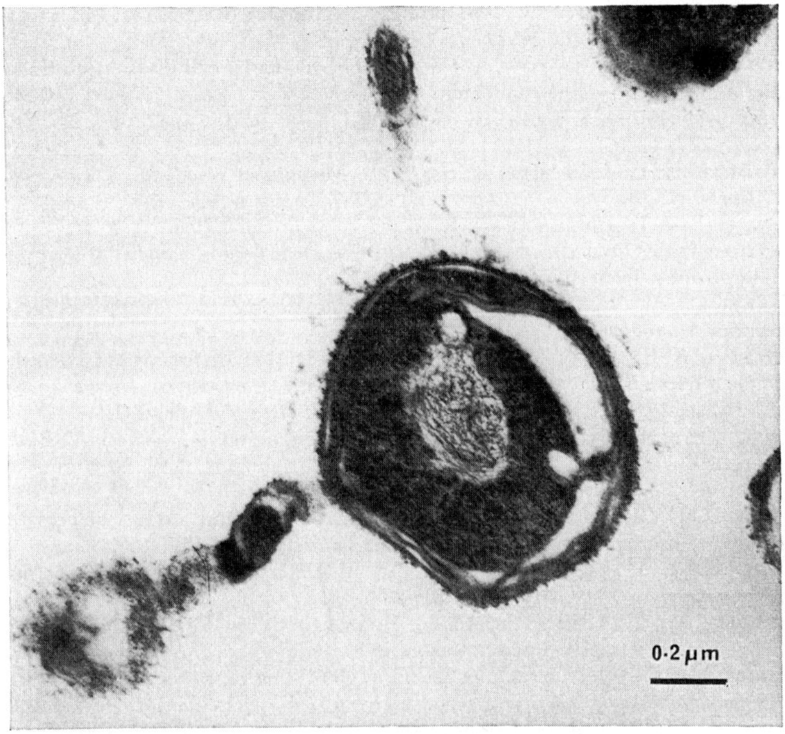

FIG. 37. Ultrastructure of a germinated *Rm. vannielii* exospore.

References

BOATMAN, E. S. & DOUGLAS, H. C. (1961). Fine structure of the photosynthetic bacterium *Rhodomicrobium vannielii*. *J. biophys. biochem. Cytol.* **11,** 469.

BOSECKER, K., DREWS, G. & TAUSCHEL, H. D. (1972). Untersuchugen zur Adsorption des Bacteriophagen Rpl an *Rhodopseudomonas palustris* le5. *Arch. Mikrobiol.* **87,** 139.

COHEN-BAZIRE, G., KUNISAWA, R. & POINDEXTER, J. S. (1965). The internal membranes of *Caulobacter cresentus*. *J. gen. Microbiol.* **42,** 301.

CONTI, S. F. & HIRSCH, P. (1965). Biology of budding bacteria III. Fine structure of *Rhodomicrobium* and *Hyphomicrobium* spp. *J. Bact.* **89,** 503.

DEGNEN, S. T. & NEWTON, A. (1972). Dependence of cell division on the completion of chromosome replication in *Caulobacter crescentus*. *J. Bact.* **110**, 852.

DONACHIE, W. D. & BEGG, K. J. (1970). Growth of the bacterial cell. *Nature, Lond.,* **227**, 1220.

DOUGLAS, H. C. & WOLFE, R. S. (1959). Motility of *Rhodomicrobium vannielii*. *J. Bacteriol.* **78,** 597.

DOW, C. S. & WHITTENBURG, R. in preparation *Bact. Rev.*

DUCHOW, E. & DOUGLAS, H. C. (1949). *Rhodomicrobium vannielii* a new photoheterotrophic bacterium. *J. Bact.* **58,** 409.

GORLENKO, V. M. (1969). Sporulation in a budding photoheterotrophic bacterium. *Microbiology USSR*, **38,** 106.

HENRICI, A. T. & JOHNSON, D. E. (1935). Studies of freshwater bacteria *II*. Stalked bacteria, a new order of Schizomycetes. *J. Bact.* **30,** 61.

HIRSCH, P. (1974). Budding Bacteria. *Ann. Rev. Microbiol.,* **28,** 392.

HIRSCH, P. & RHEINHEIMER, G. (1968). Biology of budding bacteria V. Budding bacteria in aquatic habitats: Occurrence, Enrichment, and Isolation. *Arch. Mikrobiol.* **62,** 289.

JONES, H. C. & SCHMIDT, J. M. (1973). Ultrastructural studies of crossbands occurring in the stalks of *Caulobacter crescentus*. *J. Bact.* **116**, 466.

MANDEL, M., HIRSCH, P. & CONTI, S. F. (1972). DNA base composition of *Hyphomicrobia*. *Arch. Mikrobiol.* **81,** 289.

MANDEL, M., LEADBETTER, E. R., PFENNIG, N & TRUPER, H. G. (1971) DNA base composition of phototrophic bacteria. *Intn. J. Sys, Bact.* **21**, 222.

MURRAY, R. G. E. & DOUGLAS, H. C. (1950). The reproductive mechanism of *Rhodomicrobium vannielii* and accompanying nuclear changes. *J. Bact.* **59**, 157.

NEWTON, A. (1972). Role of transcription in the temporal control of development in *Caulobacter crescentus*. *Proc. Nat. Acad. Sci. USA,* **69,** 447.

NIKITIKIN, D. I., (1973). Direct electron microscopic techniques for the observation of microorganisms in soil. *Bull. Ecol. Res. Comm.,* **17,** 85.

NIKITIN, D. I., VASILEVA, L. V. & LOCKMACHEVA, R. A. (1966). *New and rare forms of soil microorganisms*. p. 70. Moscow, USSR: Science Publishing House.

ORENSKI, S. W., BYSTRICKY, V., & MARAMOROSCH, K. (1966*a*). "Polyspheroids" from American soil. *Nature, Lond.,* **210**, 221.

ORENSKI, S. W., BYSTRICKY, V., & MARAMOROSCH, K., (1966*b*). The occurance of microbial forms of unusual morphology in European and Asian soils. *Can. J. Microbiol.* **12,** 1291.

PFENNIG, N. (1969). *Rhodopseudomonas acidophila,* sp. n., a new species of the budding purple non-sulphur bacteria. *J. Bact.* **99,** 597.

POINDEXTER, J. L. S. (1964). Biological properties and classification of the *Caulobacter* group. *Bact. Rev.* **28**, 423.

POINDEXTER, J. L. S. & COHEN-BAZIRE, G. (1964). The fine structure of stalked bacteria belonging to the family Caulobacteraceae. *J. Cell Biol.* **23**, 587.

RYTER, A. & KELLENBERGER, E. (1958). Étude au microscope électronique de plasmas contenant de l'acide désoxyribonucleicque. Les nucléoides des bactéries en croissance active.

Z. Naturforsch. **13b**, 597.

SCHMIDT, J. M., (1971). Prosthecate bacteria. *Ann. Rev. Microbiol.* **25**, 93.

SCHMIDT, J. M. (1973). Effect of lysozyme on crossbands in stalks of *Caulobacter crescentus*. *Arch. Mikrobiol.* **89**, 33.

SCHMIDT, J. M. & STANIER, R. Y. (1966). The development of cellular stalks in bacteria. *J. Cell Biol.* **28**, 423.

SHAPIRO, L., AGABIAN-KESHISHIAN, N. & BENDIS, I. (1971). Bacterial Differentiation. *Science* **173**, 884.

SOLOV'EVA, SH. V. & FEDENKO, E. P. (1970). Ultrafine cell structure of the parent strain and the pigmented mutant of *Rhodopseudomonas palustris. Mikrobiologiya* (Eng. trans.), **39**, 94.

STALEY, J. T. (1968). *Prosthecomicrobium and Ancalomicrobium:* new prosthecate freshwater bacteria. *J. Bact.* **95**, 1921.

STALEY, J. T. & JORDAN, T. L. (1973). Crossbands of *Caulobacter crescentus* stalks serve as indicators of cell age. *Nature Lond.,* **246**, 155.

STALEY, J. T. & MANDEL, M. (1973). DNA base composition of *Prosthecomicrobium* and *Ancalamicrobium* strains. *Int. J. Syp. Bact.,* **23**, 271.

STARR, M. P. & SKERMAN, V. B. D. (1965). Bacterial Diversity. *Ann. Rev. Microbiol.* **19**, 407.

TAUSCHEL, H. D. (1970). Der Geisselapparat von *Rhodopseudomonas palustris IV.* Isolierung der Geissel und ihrer Komponenter. *Arch. Mikrobiol.* **74**, 193.

TAUSCHEL, H. D. & DREWS, G. (1967). Thylakoidmorphogenese bei *Rhodopseudomonas palustris. Arch. Mikrobiol.* **59**, 381.

TAUSCHEL, H. D. & DREWS, G. (1970). Der Geisselapparat von *Rhodopseudomonas palustris III* Untersuchungen zur Feinstruktur der Geissel. *Cytobiologie,* **2**, 87.

TAUSCHEL, H. D. & HOENIGER, J. F. M. (1974). The fine structure of *Rhodopseudomonas acidophila. Can. J. Microbiol.* **20**, 13.

TRENTINI, W. C. & STARR, M. P. (1967). Growth and ultra-structure of *Rhodomicrobium vannielii* as a function of light intensity, *J. Bact.* **93**, 1699.

VAN NIEL, C. B. (1944). The culture, general physiology, morphology and classification of the non-sulphur purple and brown bacteria. *Bact. Rev.* **8**, 1.

VOLAROVICH, M. P. & TERENT'EV, A. A. (1968). New forms of microorganisms and their distribution in natural peats. *Mikrobiologiya,* **39**, 488.

WHITTENBURY, R. & MCLEE, A. G. (1967). *Rhodopseudomonas palustris* and *Rhodopseudomonas viridis*—photosynthetic budding bacteria. *Arch Mikrobiol.* **59**, 324.

ZAVARZIN, G. A. (1961). Budding bacteria. *Microbiology USSR,* (Eng. trans.) **30**, 774.

Dental Plaque

H. N. NEWMAN

Department of Dental Medicine and Medical Research Council Dental Unit, University of Bristol Dental School, Bristol, England

Dental plaque is the naturally occurring microbial layer found on tooth surfaces. As such it comprises but one part of the complex and abundant flora coating the surfaces of the skin and the greater part of the alimentary canal. There are many such coating floras or plaques, found on a wide variety of natural surfaces, both biotic and abiotic (Newman, 1974*a*). Several of the structural features of dental plaque are common to other microbial films. However, unlike other floras, dental plaque is involved in the causation of the two commonest diseases affecting man—dental caries and chronic inflammatory periodontal disease. Hence the intense interest which has been directed to the study of this particular microbiota over, approximately, the last eighty years.

Plaque Formation

Perhaps the most important factor in the deposition and growth of bacteria on teeth, as on other surfaces, is the manner in which bacteria contact and become attached to the host surface. The primary factor in plaque formation is the degree of stagnation of the relevant portion of the tooth surface (Newman, 1974*b*). The various portions of the tooth surface are not equally exposed to the cleansing action of mastication and saliva. Those most subject to these forces are designated cleansible or self-cleansing, and few bacteria are found on them. In the fissures on the biting surfaces, around the areas where neighbouring teeth are in contact with one another and where the free margins of the gums or gingivae join the teeth, the surfaces are less subject to functional friction and the washing effect of saliva. The problem of stagnation has become more important as the consumption of soft foods has increased (Newman, 1974*b*).

Other factors being equal, some plaque strains adhere more readily to the tooth than others (Gibbons and van Houte, 1973). Nevertheless, no matter what special mechanisms of adherence they possess, such cells will only remain attached to the relatively stagnant portions of the tooth. It may also be observed that there is no evidence to indicate that bacteria without special mechanisms of adherence, which are impacted into a stagnation site, will be removed from that region, other than by artificial means.

The Enamel Cuticle and the Acquired Pellicle

In most instances, the first plaque-forming organisms become attached to an organic layer, either preeruptive cuticle (Fig. 1) or acquired pellicle, rather than directly to the surface enamel crystals (Newman, 1973). The natural enamel surface is not smooth, but contains many microscopic irregularities

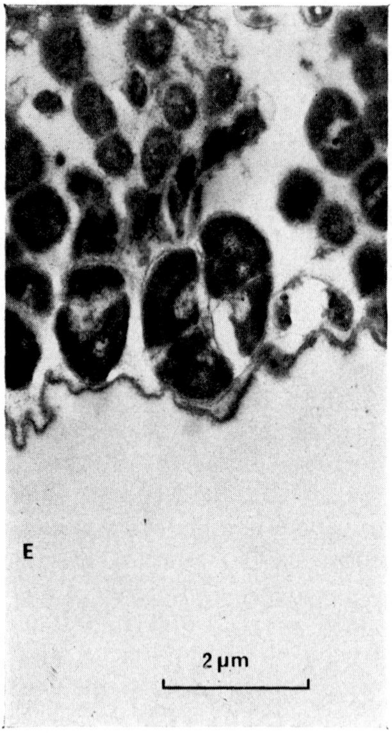

FIG. 1. Transmission electron micrograph (TEM) of relatively thin coccal plaque. Some organisms are linked to one another by fibrils. Note the thin, electron-dense cuticle on which the organisms have been deposited. E=enamel space (× 10 250).

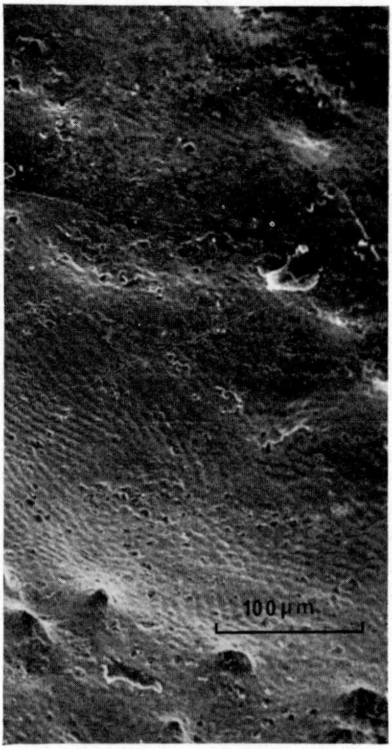

FIG. 2. Scanning electron micrograph (SEM) of portion of natural approximal tooth surface showing prism end depressions and larger pits and elevations (× 200).

in which organisms may lodge after exposure of the tooth to the oral environment (Fig. 2) (Newman and Poole, 1974*a*). The acquired pellicle (Figs 3 and 4) is unique neither in its location nor in its chemical composition. Similar layers may be formed between bacteria and their host surfaces in a wide variety of ecological niches (Newman, 1974*a*). It has been suggested (Winterburn and Phelps, 1972) that sugars are included in protein structures as a means of coding and, consequently, that the great majority of extracellular proteins are glycosylated since such proteins, synthesized by one cell type, are designated for use at another locus. Such a concept may apply to salivary glycoproteins, and it is to be noted that an artificially abraded tooth surface, *in situ* in the mouth, is rapidly covered by salivary

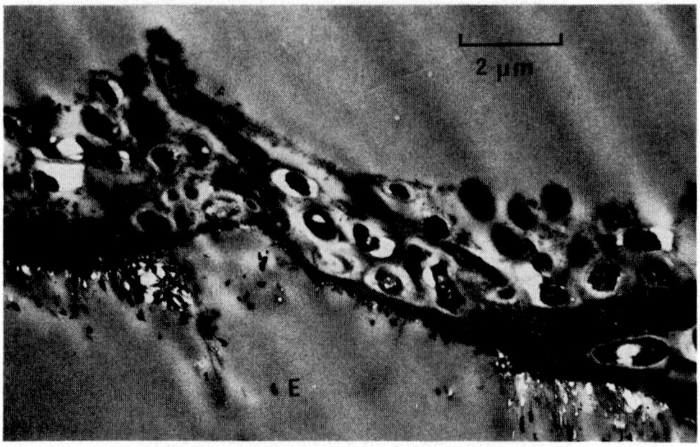

FIG. 3. TEM. Here amorphous pellicular material, considered generally to be of salivary origin, has been deposited on the preeruptive cuticle and encloses some organisms. Note the clear zone around some cells, which may be a zone of matrix lysis, permitting cell growth (× 5400).

glycoprotein. In fact, the instances in which a fresh enamel surface is exposed to saliva are few (McGaughey and Stowell, 1971). It may also be observed that the pellicle may protect against chemical erosion as well as physical wear. It has been shown experimentally that teeth without salivary pellicle develop caries more rapidly under otherwise standard conditions of acid dissolution (Moreno and Zahradnik, 1974).

The structural means of attachment have not been studied thoroughly, but it is clear that many bacteria possess a variety of attachment mechanisms, enabling them to adhere to one another and to a given surface in non-stagnant loci, and these will be reviewed later. In so far as the bacteria/pellicle or cuticle attachment is concerned, it has been suggested that bacterial extracellular or cell wall polysaccharides possess an affinity with the molecules of the acquired pellicle. This would explain attachment on the basis of electrostatic interaction between negatively charged polysaccharide and positively charged protein (Sönju *et al.*, 1974). In addition to salivary glycoprotein and bacterial matter, the acquired pellicle also contains salivary lysozyme and immunoglobulins which may react with bacterial cell-surface determinants and thus influence the adhesion of bacteria to the tooth (Ørstavik and Kraus, 1973).

FIG. 4. Freeze-etch micrograph (FEM) showing palisaded organisms on granular pellicle (P). Note prolongation of this layer into the enamel space, indicating the presence prior to demineralization of an intimate contact between the inner aspect of this layer and the enamel surface crystals. Note also the distorted shape of palisaded cells (× 40 000).

Bacterial Deposition

Organisms are deposited in micro-cavities and then in the troughs of the corrugations or perikymata which form the natural enamel surface (Fig. 5). At the junction with a cleansible area the initial coccal layer may be observed, often in the scalloped acquired salivary pellicle. In a stagnant site, as at the free gingival margin, this layer may be obscured by the presence of large amounts of extracellular matrix (Fig. 6) usually mainly polysaccharide in composition. In health, few bacteria appear to colonize the natural potential crevice between gingiva and tooth, and a plaque-free layer, the preeruptive cuticle described previously, intervenes between the plaque and the cells of the junctional epithelium. Where a contact area between teeth has been established, the most stagnant site in relation to it,

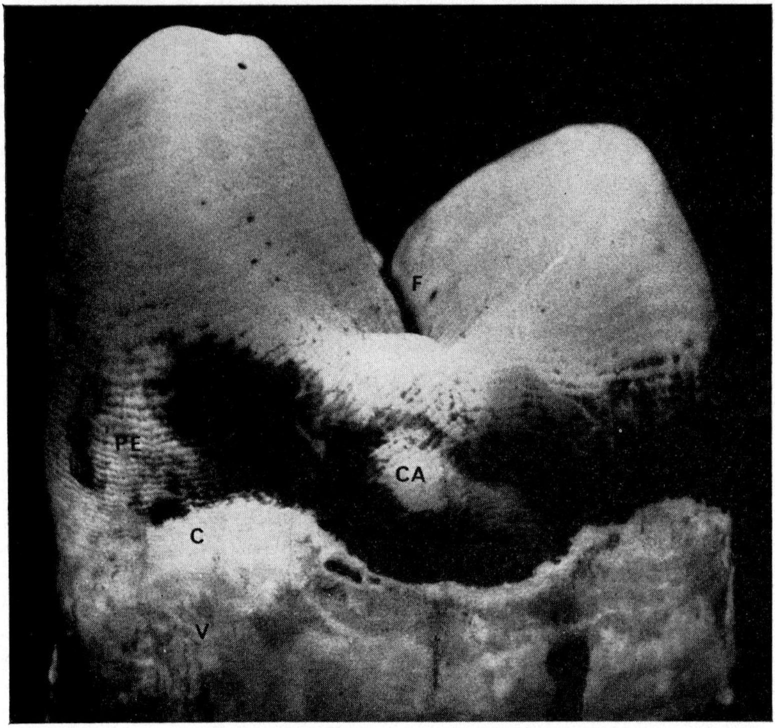

FIG. 5. Approximal surface of upper premolar tooth from child. Stained with osmium tetroxide to disclose organic integument. Note fissure plaque (F), the presence of plaque gingival to the relatively bacteria-free contact area (CA) and the preferential deposition of plaque in the troughs of the perikymata (PE). V = vestigial enamel organ, C = preeruptive cuticle.

that next to the free gingival margin, may contain, as well as coccal forms, segmented or branching bacilli or filaments in its initial layer (Figs 7–12). The deepest sited organisms are usually cell ghosts or cell wall fragments (Fig. 13). Even after the establishment of a crevice plaque, the outer layers may contain many such cell ghosts, indicative of the presence of anti-bacterial factors in the crevicular exudate. These cell ghosts may also be found in developmental or acquired fissures in the enamel (Fig. 14) (Newman and Poole, 1974*a*), and in the deeper portions of thick plaques. In the latter instance it may be that the demise of the ghost cells is due, not

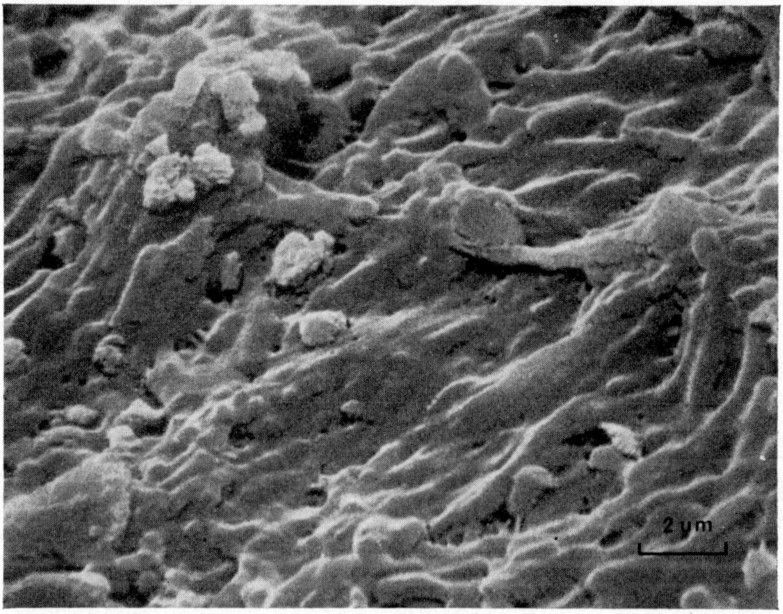

FIG. 6. SEM. The outlines of bacteria at the salivary surface of this portion of approximal plaque are obscured by extracellular matrix, a frequent finding in this stagnant locus (× 6000).

to host antibacterial factors, but to competition for available nutrients or to antagonistic activity on the part of contiguous cells. Apart from chemical antibacterial factors, the high turnover rate of gingival crevicular epithelium and the firmness of apposition of free gingiva to tooth in the crevice region probably also help to prevent bacteria entering the crevice. It has been observed (Holmberg and Hallander, 1973) that *Streptococcus sanguis* is the dominant organism in early plaque and it is of interest that, while positive identification has yet to be made, the predominant structurally intact cells in the crevice region would seem to be Gram-positive or negative cocci and to contain granules of carbohydrate polymer, and to be present (Fig. 15) in a largely carbohydrate matrix. In addition to the crevice region, bacterial colonization has been observed on the oral surface of the gingiva (Takeuchi *et al.*, 1974) as well as on other oral epithelial surfaces (MacCallum, 1974).

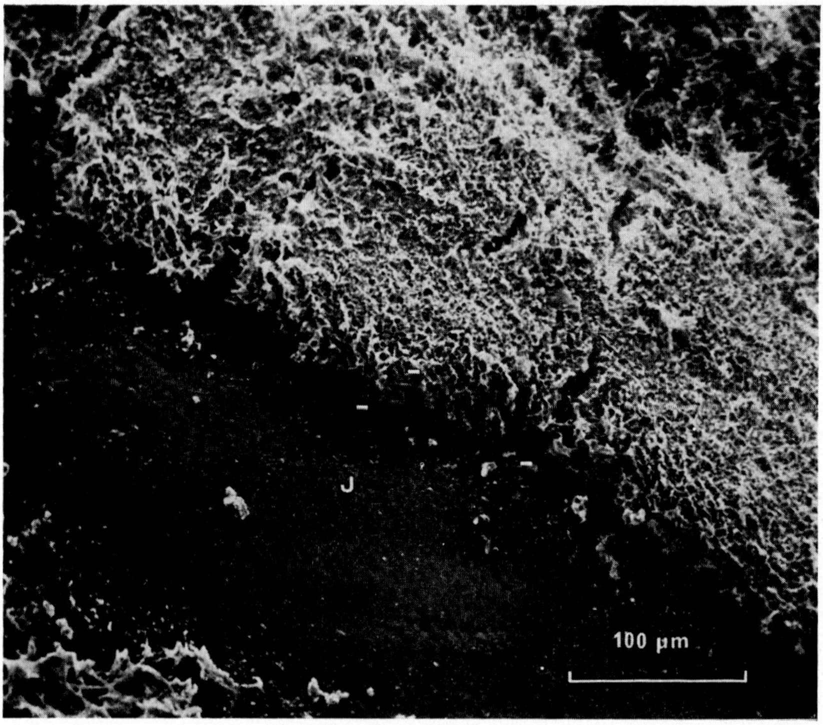

FIG 7. Plaque has collected at what was *in vivo* the free gingival margin where it adjoined the enamel surface. The crevice region is relatively free of bacteria (× 250).

FIGS. 7–12. SEM. Series illustrating variations in plaque topography in its most gingival portion, in the site of onset of chronic inflammatory periodontal disease, and close to the initiating site of approximal dental caries.

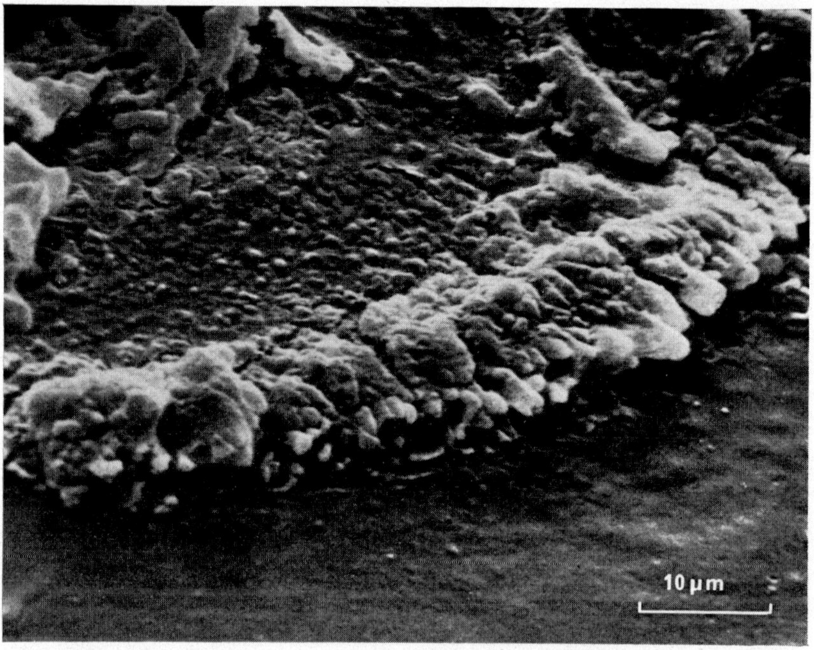

FIG. 8. This micrograph depicts an agglomeration of apparently coccal forms in an abundant matrix gingival to the contact area ($\times$ 1800).

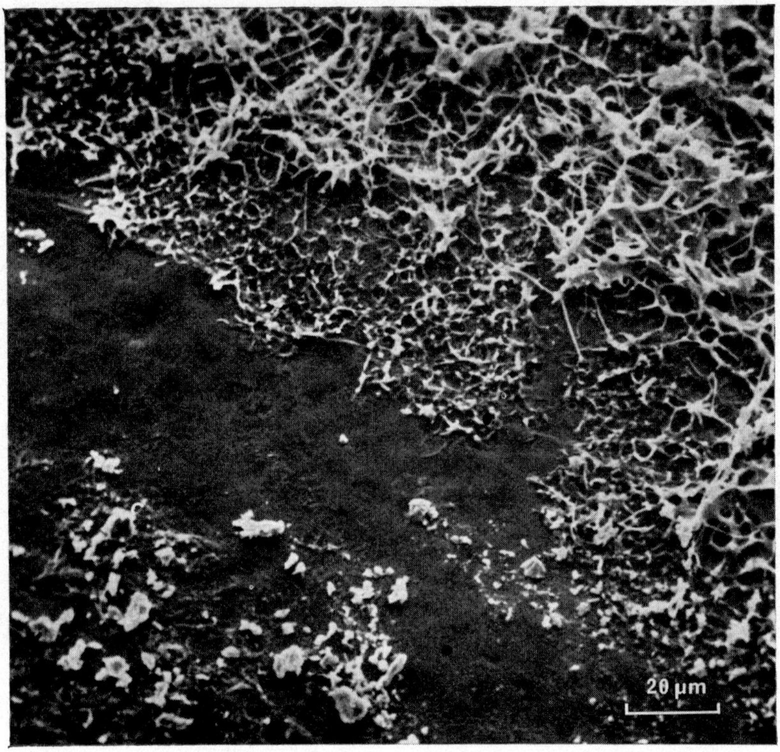

FIG. 9. In this micrograph organisms have begun to extend into the gingival crevice region ($\times$ 650).

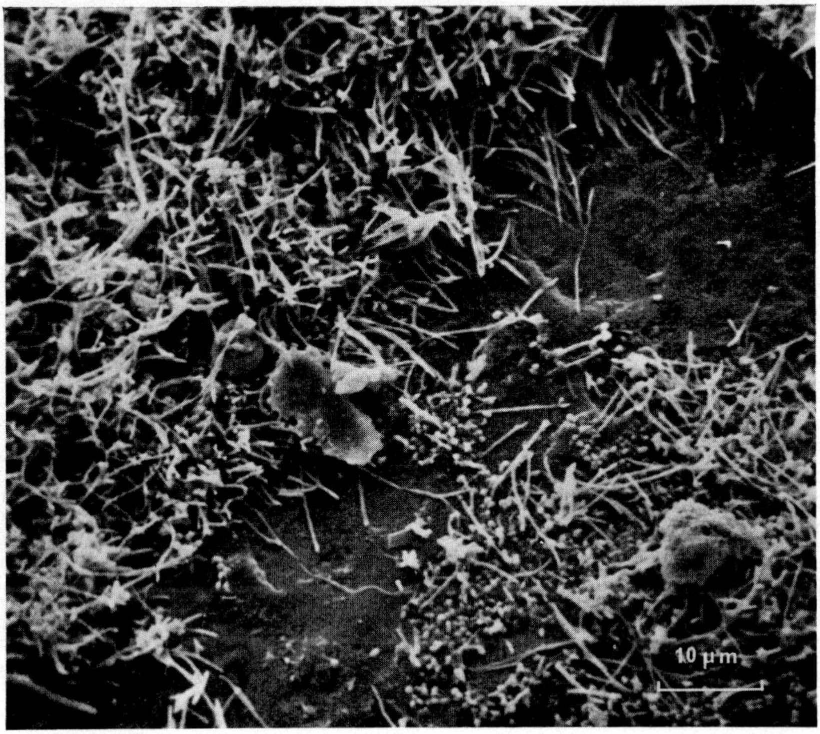

FIG. 10. In this portion of crevice the predominant organisms are filaments, some of which appear to be branching, which have grown over a primary layer of coccal forms (× 1400).

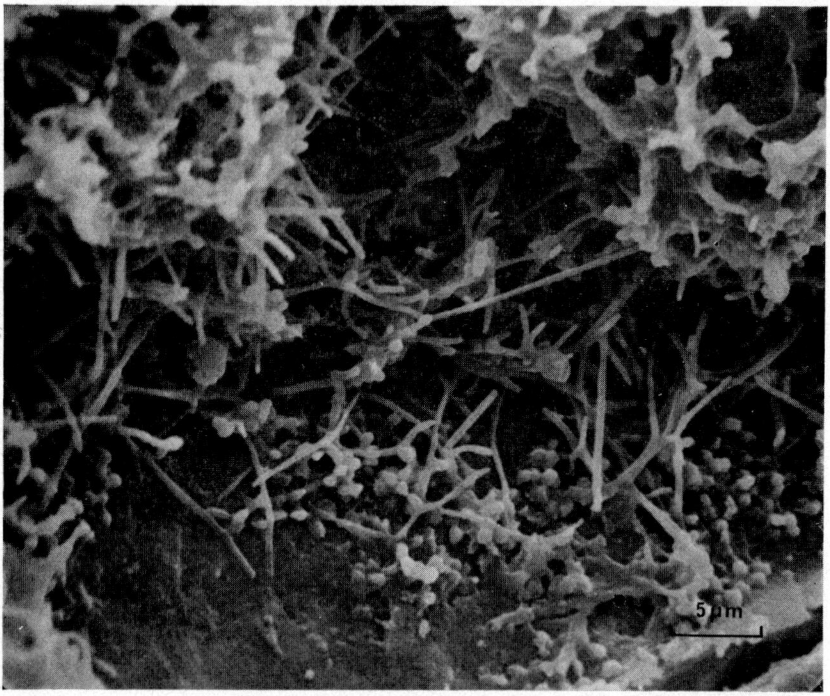

FIG. 11. This higher power micrograph depicts the association between filaments and coccal forms in the initial plaque layer in the crevice region (× 2700).

FIG. 12 (*opposite*). Although the predominant cells in this plaque portion are coccal organisms embedded in a dense matrix, a branching and septate (S) filament has extended ahead of the coccal cells into the crevice (× 2100).

FIG. 13 (*opposite*). TEM. The most gingivally located plaque often consists, as in this micrograph, of palisaded, polysaccharide-containing cells with an outer layer in which many cell ghosts are present. E = enamel (× 4080).

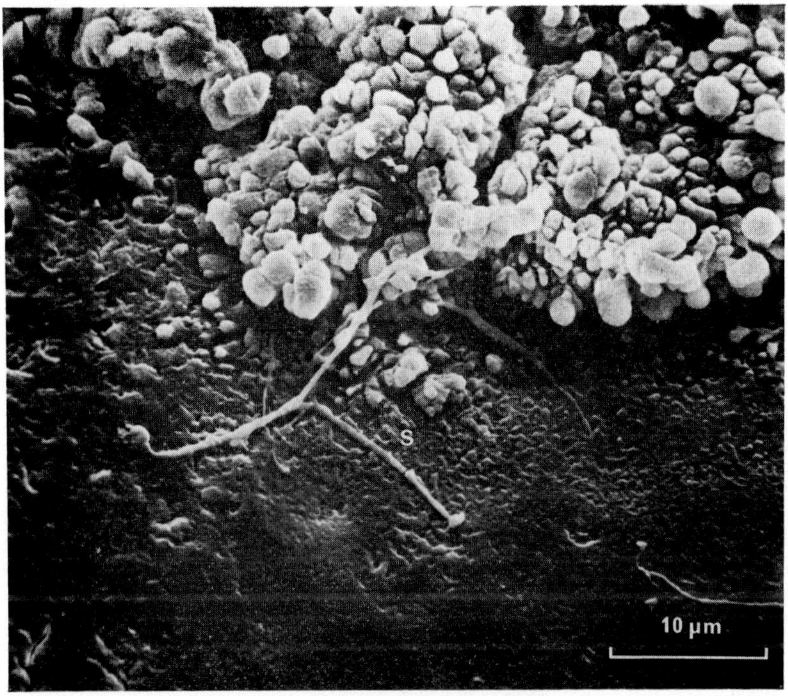

FIG. 12.

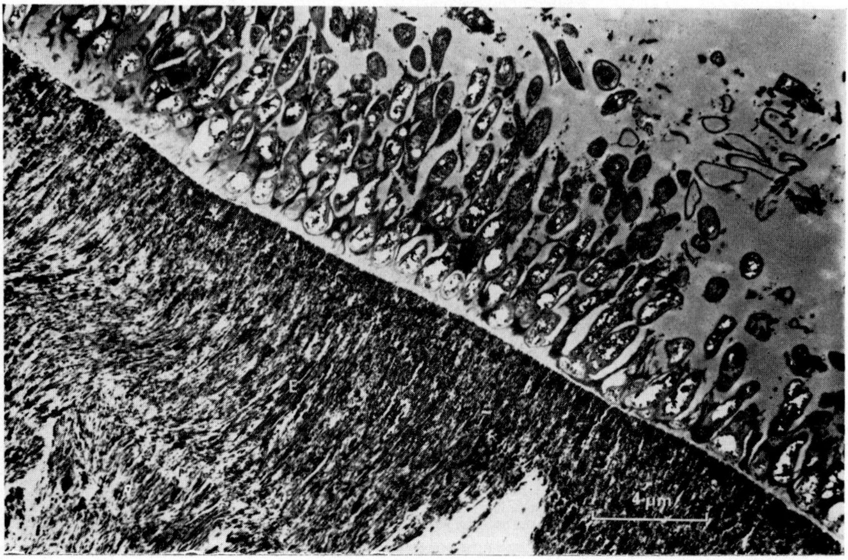

FIG. 13.

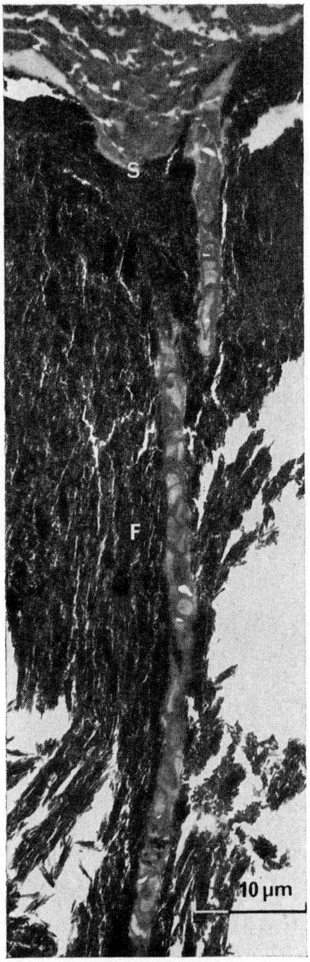

Fig. 14. TEM. This fissure (F) in the enamel surface (S) is packed with bacterial cell ghosts ($\times$ 1550).

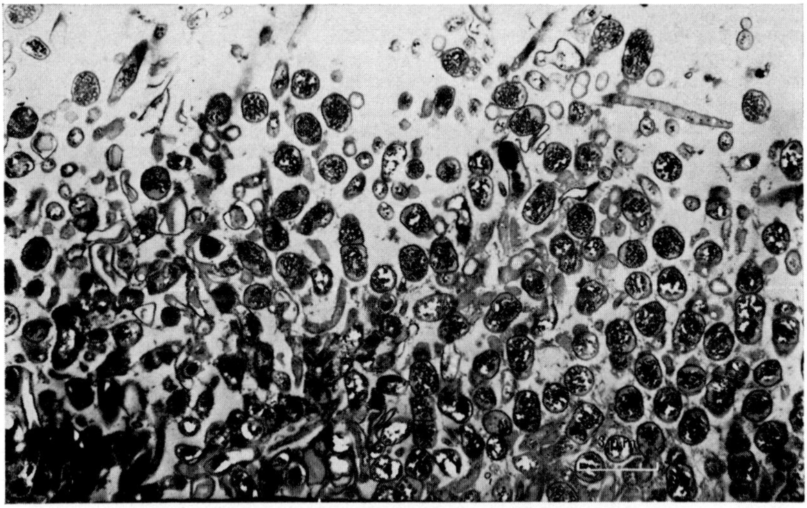

FIG. 15. TEM. This outer portion of the most gingivally located plaque indicates that the predominant organisms in this region are polysaccharide-containing, apparently coccal cells, and many cell ghosts. Some filaments are also present. The distribution of cell ghosts suggests that antibacterial factors in the crevicular exudate may have some effect on the survival of organisms in this region. The presence of many cells containing polysaccharide suggests that the latter may have adapted to unfavourable growth conditions in this region (× 3525).

Plaque Growth

As plaque increases in thickness, organisms at the deep surface of the film next to the enamel become aligned in parallel (Fig. 16). Palisading has also been observed in a number of pure cultures (Henneberg, 1964) and in other microbiotas (Brock, 1966; Takeuchi and Zeller, 1972). It is not at present clear how cells become aligned in this manner, nor is it known if the

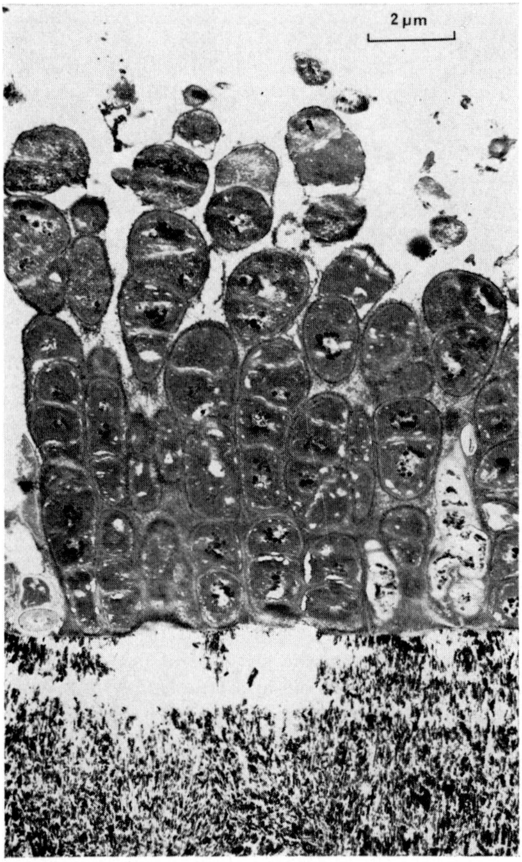

FIG. 16. TEM. Palisaded coccoid cells containing polysaccharide and located in the most gingival portion of plaque. Note cell ghosts in the deep plaque and fibrillar matrix between organisms. The "ghosts" in this location may have been produced by antagonistic activity on the part of neighbouring organisms rather than by crevicular antibacterial factors. (Reproduced from *Brit. dent. J.*, **135**, 106, 1973, by kind permission of the Editor.) (× 6225).

arrangement has some underlying biological significance. It is evident that close packing of cells is related to palisading, since the palisaded cells usually show a marked distortion of their rigid cell walls (Fig. 17), even on relatively hydrated freeze-etched specimens (Newman and Britton, 1974). Where organisms are so densely concentrated, palisading may facilitate cell growth in the line of least resistance to movement, and towards nutrients at the outer surface. Palisading may also aid diffusion of nutrients, perhaps by capillary action (Henneberg, 1964). It is also possible that nutrients taken up by one cell may be transferred to others in the same or contiguous palisades. It may be noted that the bulk of natural plaque is composed of palisaded cells, either filaments, cocci, bacilli, or even spirochaetes, though the latter are found mainly in the deep gingival crevices produced by inflammatory periodontal disease (Soames and Davies, 1974).

Colonial Groupings

Organisms in plaque are usually grouped in a number of contiguous micro-colonies which each seem to be composed of morphologically similar cells. Many coccal and filamentous colonies occur, but anaerobic bacteria such as spirochaetes are found in the gingival crevice only after the onset of inflammation. The environment within each micro-colony may differ from that prevailing in adjoining micro-colonies, so that the biochemical activity of a given group of cells in plaque does not depend solely on its position but on the specific metabolism of each micro-colony and on the relationships which may develop between adjoining colonies. A number of antagonistic relationships have been observed between plaque isolates *in vitro* (Donoghue and Tyler, 1974) but it is not yet certain that these exist *in situ* in natural plaque.

Interbacterial Attachment

In the late nineteenth century a curious microbial configuration was observed in dental plaque. This was found subsequently to consist of a central filamentous organism coated by smaller, usually coccal, though occasionally bacillary, cells. The coccal version bears a marked resemblance to maize and has, therefore, been described as a "corn cob" (Schroeder, 1969), whereas the configuration in which bacillary elements coat the central cell has been likened to a test-tube brush. In spite of careful work by Goadby (1900), which showed clearly that the phenomenon was due to the attachment of cells of one species to those of another, and that it was not a feature of specific strains, investigators have come to regard each of the configurations as being composed of the cells of a single strain or species,

FIG. 17. FEM. Portion of palisaded plaque. Note distortion of cell outlines, even in this non-dehydrated preparation (× 21 000).

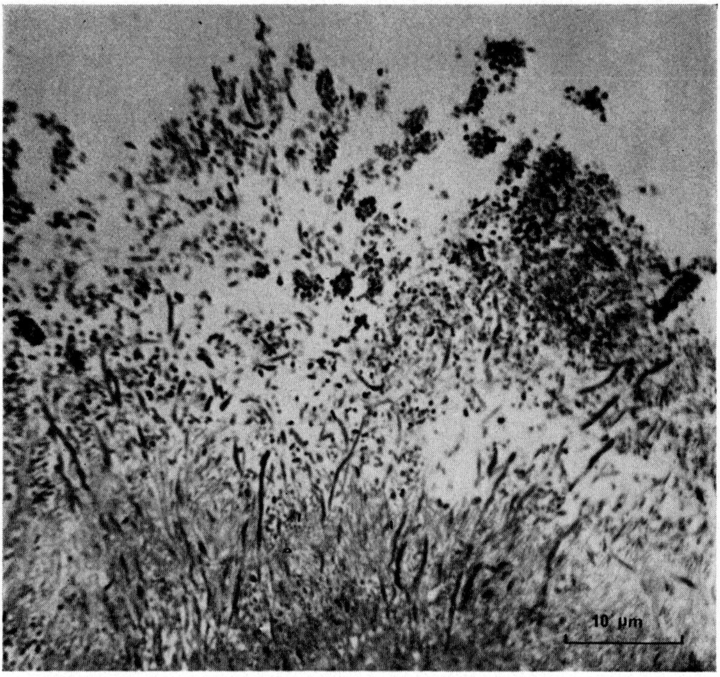

FIG. 18. Araldite section, stained with crystal violet. Rosettes are present in the outer portion of this section. Note the relative lack of organization of cells in the outer plaque (× 1600).

the corn cob being named *Leptothrix racemosa* and the "test-tube brush" organism *L. falciformis*. The coating cells were considered as having budded off the central filament. Recently, using immunofluorescence techniques in conjunction with routine transmission electron microscopy, it has been possible to provide further evidence that coating cells and central organism may consist of different species (Newman and McKay, 1973). In section the configurations resemble rosettes (Fig. 18) and, in this form, analogous configurations have been observed in several other microbiotas. Although they occur commonly, they constitute but a small proportion of the plaque on a given tooth surface. No direct commensal relationship has yet been observed between the constituent cells, although it seems that the central filament may, in its young form, be Gram-positive and intact, while at a later stage it becomes Gram-variable and usually has the appearance of a ghost cell as seen in transmission electron micrographs (Newman and McKay, 1973). Coating cells are found usually with their divisional septa at right angles to the central filament, to which they may be attached by their cell walls, by extracellular amorphous material or by fine fibrils. Where the filament tip is not obscured by coating cells, it may be seen to be expanded by comparison with its proximal portion. While positive identification has yet to be made, available evidence indicates that the central organism may be a yeast or a species of *Leptotrichia* (Goadby, 1900; Newman and McKay, 1973; Vreven and Frank, 1973). If it is a yeast, the central filament must be degenerate, since yeasts are strongly Gram-positive and the filament is usually Gram-variable or negative.

Morphology, Division and Growth of Organisms in Plaque

Early studies implied that coccal cells such as streptococci, veillonellae and neisseriae made up the bulk of the plaque viable count. More recent investigations suggest that filamentous organisms account for the major proportion of both the viable count (Hardie and Bowden, 1974) and the volume (Newman and Poole, 1974*b*). This finding may result from the difficulties of isolating every viable cell. It must also be remembered that a single filament occupies a volume comparable to that of an equilinear chain of cocci of equivalent cross-sectional area. A further problem arises since it is by no means certain that the morphology of an organism *in situ* in plaque corresponds to that of the same strain in artificial culture. Close-packing of cells seems to be a further modifying factor, since many such organisms deep in plaque, especially in palisaded regions, have a box-like morphology. In non-palisaded regions coccal cells whose division is physically impeded may also septate randomly to produce cake diagram-like arrangements (Fig.

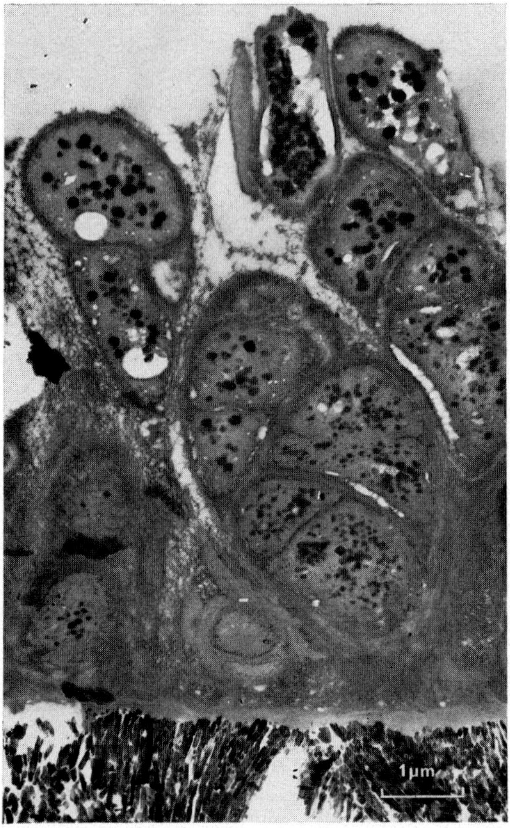

FIG. 19. TEM. Uneven cell division can produce odd configurations such as the central cake diagram-like appearance. Note fibrillar matrix between cells. Again these organisms in the crevice region contain polysaccharide granules. The relatively large size of coccal cells in this region may, in addition to other features described, be an indication of slow growth rate and interference with cell division (× 11 840).

19). It has also been shown that some strains which have *in situ* a coccal morphology become filamentous in pure culture (Hurst, 1950) and, conversely, that few plaque filaments grow as such on pure culture (Snyder, Bullock and Parker, 1967). Such changes in morphology may be related to nutritional imbalances in the plaque interior. Satellitism may also induce changes in morphology. A recent study of symbiotic streptococci has shown that cells from the immediate vicinity of the host colony consisted of Gram-positive cocci in chains, whereas organisms from the outer edge of the zone of satellitism were Gram-negative and pleomorphic, with filamentous and globular forms. The presence of pleomorphic forms indicates the absence of essential growth factors (George, 1974). Involution forms develop at an early stage on media poised initially at a low or high pH, but at a later stage on initially neutral media which contain a fermentable sugar and so become acid in the course of growth (Duguid and Wilkinson, 1961), a factor which may be important in relation to the pleomorphism of plaque strains in a cariogenic locus. DNA precursors such as thymidine (Jeener and Jeener, 1952) or thymine (Cohen and Barner, 1954) may be growth factors, the lack of which leads to microbial pleomorphism. This may explain palisading and filamentous growth of organisms from the deep surface of plaque as it increases in thickness, and emphasizes the significance of the slow rate of penetration of plaque by tritiated DNA precursors (Newman and Wilson, 1975).

As regards the formation of palisaded chains of smaller forms, it would seem that cell separation after septation in such instances is inhibited, either by the physical restriction of cell separation in such loci, or by inhibition of the relevant cell separation enzyme (Duguid and Wilkinson, 1961). On a medium deficient in thymidine or thymine some bacilli may grow into very long filaments. When supplied with DNA, these filaments show rapid nuclear multiplication and then segmentation into bacilli (Jeener and Jeener, 1952). Filament formation is adopted by many micro-organisms under unfavourable nutritional circumstances (Brock, 1966). However, these forms do occasionally develop in conditions which appear favourable for normal growth (Duguid and Wilkinson, 1961). In regard to plaque strains, Erikson (1954) showed that the production of a filamentous mycelium by a species of *Nocardia* was favoured by the presence of extra-cellular polysaccharide which presumably restricted diffusion of growth substrates to the organism. Such a phenomenon may be of some importance in plaque, which often contains large amounts of extracellular polymeric carbohydrate matrix. The filamentous mode of growth also maintains a constant ratio of cell surface area to volume, thus allowing for adequate rate of nutrient uptake. The thickness of the cell wall, which increases in sites of growth limitation (van Houte and Saxton, 1971) is a further restrictive factor in cell growth and division (Chung, 1971).

It is not yet clear how filamentous organisms grow, although it would appear that growth takes place throughout the length of the filament. Septa form in some strains and, in such instances, growth may start from the septum, and possibly continue in previously formed segments. In non-segmented and, possibly, septate, forms, nutrients absorbed at one end of the cell may, conceivably, be transported throughout the length of the organism. Whether or not palisades are present, cell ghosts are found frequently in the deeper portions of thick plaques, and intact cells may show several features indicative of slow growth rate, including thick cell walls and an abundance of intra- and extracellular carbohydrate polymers. The presence of cross-septa in a given preparation is not necessarily indicative of an active division process and it may be noted that, even under starvation conditions, several plaque strains will only partly utilize their nucleic acids and proteins (Williams, 1972).

Structure and Nutrition of Organisms in Plaque

It is evident from what has been discussed that substrates may penetrate given regions of plaque at different rates. Thus, portions of plaque of similar thickness, even composed of morphologically similar organisms, either randomly arranged or in palisades, may show varying rates of uptake of different and the same substrates. Sucrose appears to be taken up more quickly than other substrates and neither plaque structure nor thickness appear to limit its absorption (Fig. 20). In contrast, even in thin regions of plaque, tritiated DNA precursors are absorbed at a slow rate (Fig. 21), and, while outer organisms may become labelled within one or two hours, the

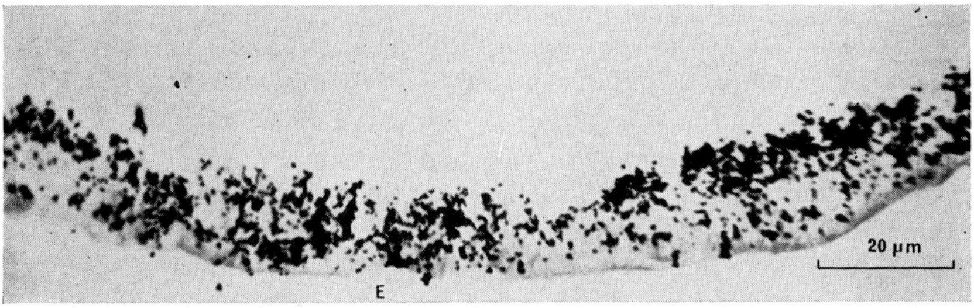

FIG. 20. Autoradiograph of thin plaque incubated for 15 min in tritiated sucrose. Note the relatively even and rapid distribution of label (× 750).

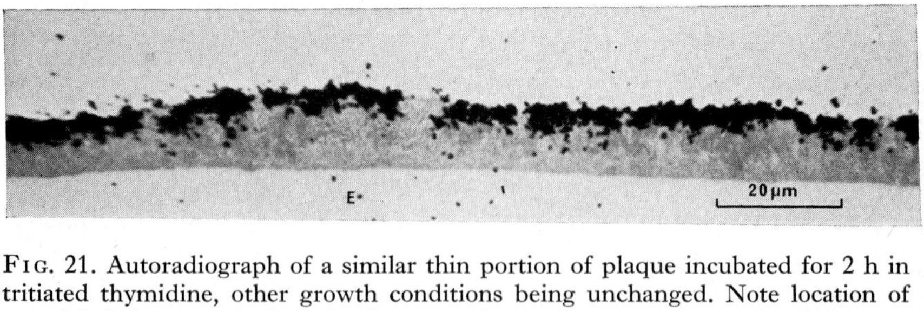

FIG. 21. Autoradiograph of a similar thin portion of plaque incubated for 2 h in tritiated thymidine, other growth conditions being unchanged. Note location of label almost exclusively in the outer plaque. E=enamel space (× 750).

substrate may take a day or longer to reach the deep surface of thick regions of plaque. In view of evidence indicating a slow rate of growth of cells in deep plaque (van Houte and Saxton, 1971), it would seem that growth rate in such circumstances is dependent mainly on rate of diffusion of labelled substrate through plaque, which may, possibly, be favoured by palisading and a reduction in the amount of diffusion-limiting extracellular matrix. It has been suggested that the breakdown of dextran around *S. mutans* may allow unrestricted cell division, and that the production of dextranases by organisms in a dense carbohydrate matrix may be related to the growth of the producer cells in such an unfavourable environment (Guggenheim and Burckhardt, 1974). This phenomenon may also explain the presence of non-staining or clear zones around some organisms in electron micrographs of plaque.

The Matrix of Dental Plaque

As plaque increases in thickness, the organisms, although densely packed and occasionally having contiguous cell walls (Newman and Britton, 1974), are often separated by a complex interbacterial matrix of mainly bacterial (Frank and Houver, 1970) and salivary (Belcourt, 1973) origin. The carbohydrate polymer component of the matrix has received the most attention. The initial attachment of bacteria to cuticle or pellicle seems to depend on an affinity between extracellular bacterial matter and the pellicle protein. The presence of protein on the enamel mineral, hydroxyapatite, seems to reduce the ability of unencapsulated strains of *S. mutans* to adhere, but increases the binding of encapsulated strains of the latter (Kornman, Kreitzman and Clark, 1972), indicating that a possible mode of adherence, perhaps fundamental to bacteria in many habitats (Newman, 1974*a*), may depend on an affinity between coating glycoprotein and bacterial extracellular polysaccharide (see also Wagner and Barrnett, 1974). The polysaccharide may be in the form of amorphous matter or granules, a capsule, a holdfast, usually polar or arising from one site on the cell wall (Halhoul and Colvin, 1974) or fine, fimbria-like fibrils arising from the cell wall. It has been proposed that extracellular slimes and capsular polysaccharides secreted by bacteria consistently possess a fibrillar ultrastructure, suggesting that they are secreted initially as individual polymeric macromolecular filaments at specific loci on the cell surface (Ridgway and Lewin, 1973). Fine axial filaments originate from goblet-like bases on some marine organisms, the goblets being subcellular membrane-bound organelles responsible, it is suggested, for filament synthesis. Where goblet-like sub-units have not been detected, analogous wall-bound structures may be present to perform such a secretory function (Ridgway and Lewin, 1973).

Johnson, Bozzola and Schechmeister (1974) have studied the extracellular polysaccharides produced by several strains of *S. mutans*, and they have found both globular and fibrillar forms, the fibrils being composed each of two protofibrils to which amorphous electron-dense material is often adherent. The fibrillar component appeared to predominate towards the centre of a given colony.

It is well known that animal and bacterial cell walls possess a coat which in typical electron micrographs appears blurred. Recent work has made it clear that such a "fuzzy coat" may be found in relation to many organisms in the plaque interior (Schroeder, 1970) as well as to organisms adhering to oral epithelium (Barnett, 1973). On closer examination the fuzzy coat has a distinct fibrillar appearance (Schroeder and de Boever, 1970). Carbohydrate polymers in plaque frequently have a granular or amorphous appearance, especially in the outer layers, but fibrillar polysaccharides are more prominent in the deeper plaque (Schroeder and de Boever, 1970). The fibrils possess frequently a beaded appearance (Newman and Britton, 1974) and would seem to correspond to the fibrils observed in pure cultures of plaque strains by Newbrun, Lacy and Christie (1971) and by Johnson, Bozzola and Schechmeister (1974). In plaque the fibrillar matrix appears to originate as branching networks from extracellular globular units on the cell wall or in the matrix (Figs 22–24). Some cells in plaque have a distinctly fimbriate form (Newman and Britton, 1974) and it may be that matrix fibrils arise from or are continuous with these fibrils of cell wall origin, or form as a result of enzyme activity in the interbacterial matrix. Analogous fibrils in pure cultures of some plaque streptococcal strains have been described as being composed of distinct protofibrils, two to each fibril (Johnson, Bozzola and Schechmeister, 1974). While a similar finding has not been made as yet in natural plaque, organisms in the latter may be so densely packed, or the matrix stain so heavily for carbohydrate that details of its ultrastructure may be obscured. It may be observed that cells in such an abundant carbohydrate matrix often contain correspondingly large amounts of intracellular polysaccharide.

Fibrillar and other matrix components may subserve other than attachment functions, as reviewed recently by Newman and Poole (1974*b*). An illustration of current trends in work on such structures is provided by the study of Wagner and Barrnett (1974) on the fine structural characteristics of the junctional complexes formed between rat colon bacteria and mucosal epithelial cells. They suggested that, since bacterial capsule was antigenic, and on reaction with homologous antibody produced a microprecipitation in the capsule with an accompanying alteration in its refractive index, the presence of filaments or fine fibrils only in the region of contact between the bacterium and the mucosal cell might result from such a microprecipitation

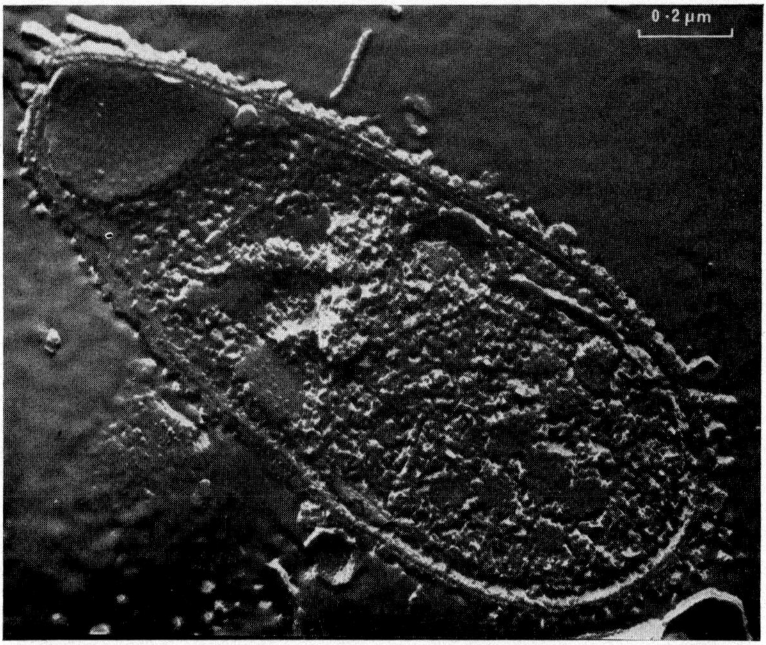

FIG. 22. FEM. In this micrograph globular material may be observed arising from the cell wall of this Gram-negative organism present in the superficial portion of a sample of natural plaque (× 64 000).

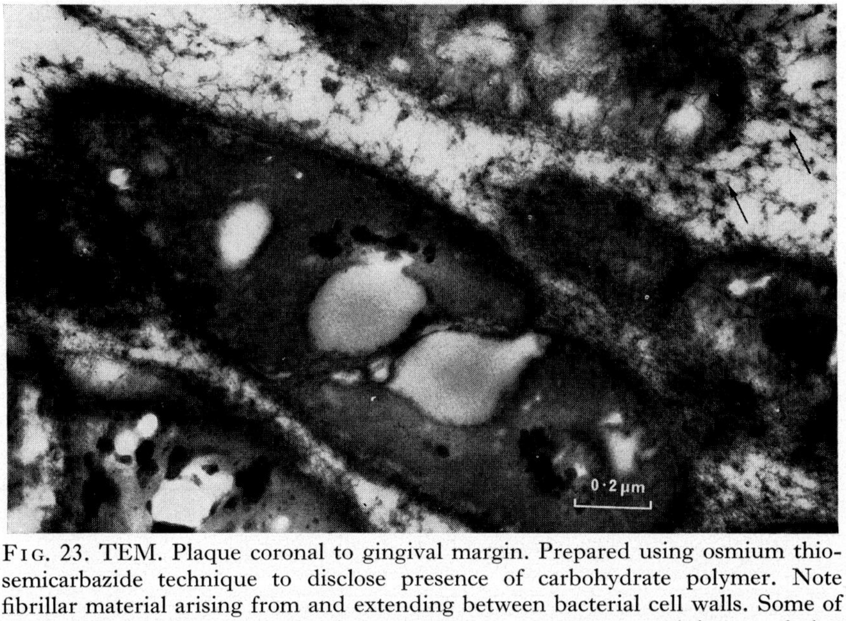

FIG. 23. TEM. Plaque coronal to gingival margin. Prepared using osmium thio-semicarbazide technique to disclose presence of carbohydrate polymer. Note fibrillar material arising from and extending between bacterial cell walls. Some of the matrix fibrils have a double protofibrillar appearance reminiscent of that described by Johnson *et al.* (1974) in relation to pure isolates of *S. mutans*, and appear to arise from central globular units (arrows) (× 48 750.)

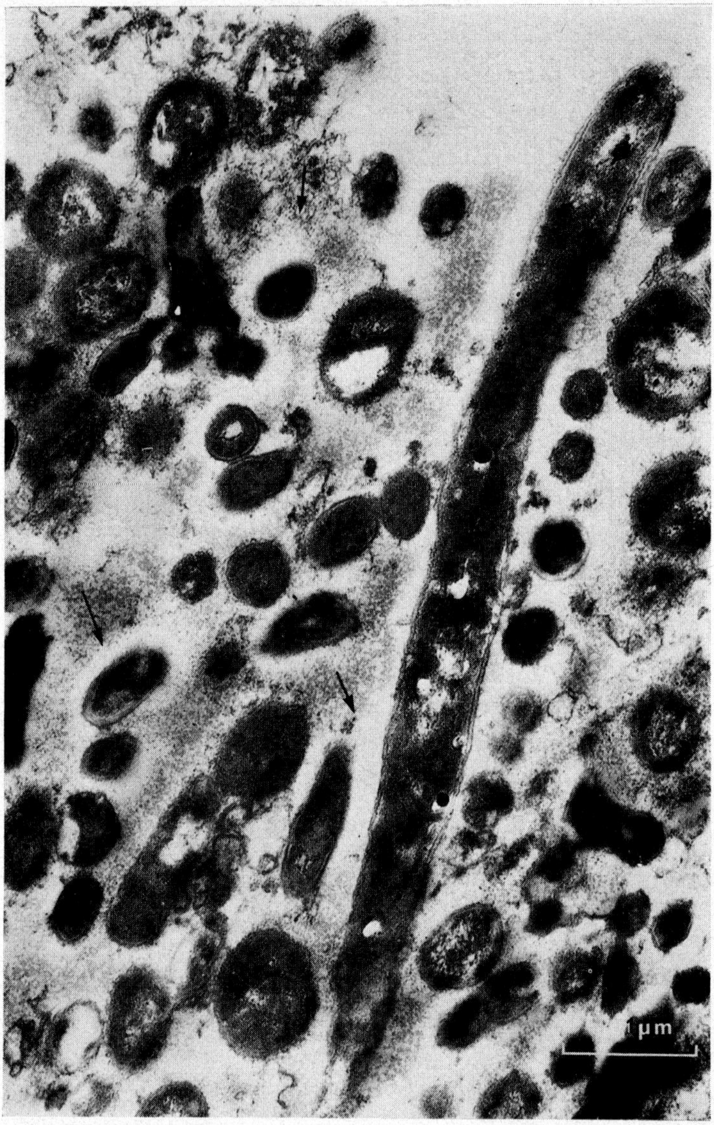

FIG. 24. TEM. Treated with uranyl acetate and phosphotungstic acid. Organisms in sections of natural plaque often have a clear zone between cell walls and stainable matrix (arrows). This may correspond to the zone of dextran lysis permitting cell growth, as described by Guggenheim and Burckhardt (1974). (× 18 600).

reaction elicited by antibody. This observation was supported in their study by the absence of fibrils on the surface of the bacterial cell distal to its point of contact with the mucosal cell. Plaque organisms occasionally show this preferential location of fibrils and, as mentioned above, many plaque organisms possess fine fibrils arising, apparently, from the outer surface of the cell wall, and often extending between contiguous cells. Any consideration of the biological significance of these fibrils must therefore include the possibility of the appearance being due to a microprecipitation reaction of the kind described above.

It is evident that the presence of a dense carbohydrate matrix may provide a partial explanation of intercellular aggregation and adherence in plaque. Yet, since plaque will only accumulate in stagnation sites, it may be that the matrix subserves functions additional to those suggested above. Recent studies make it clear that only cariogenic strains of *S. mutans* produce a dense, sticky extracellular carbohydrate polymer (de Stoppelaar *et al.*, 1971). It may be that, in plaque as in other microbiotas, the production of such polymers helps to prevent the cell being brought into contact with toxic levels of acid or sucrose, a function which would be aided by a dense, relatively inert and highly branched polymer structure. As a corollary it may be observed that the restricted diffusion resulting from a cell being surrounded by such a matrix would impede its access to essential nutrients and that producer cells are, therefore, adapted to the high concentrations of acid and sucrose and the low levels of growth substrates in such sites (van Houte and Saxton, 1971).

There are usually marked differences between the superficial and deep portions of the plaque matrix. In the outer plaque the matrix is generally amorphous or granular and partially removed by treatment with ethylene-diamine-tetra-acetate. It has also been demonstrated *in vitro* that aggregation of all strains of *S. mutans* (except one mutant which did not produce a dense extracellular polysaccharide) depended on the presence of calcium or magnesium ions which bound to protein receptors on the bacterial cell wall (Kelstrup and Funder-Nielsen, 1974). The attachment of bacteria to cuticle or acquired pellicle may be an analogous phenomenon.

In the deeper layers the matrix contains more carbohydrate (Critchley, Saxton and Kolendo, 1968) (Fig. 25) and may be aligned in parallel strands between palisaded cells (Fig. 26). As cells are generally closer to one another in this region, it follows that the accompanying matrix is less abundant. In plate colonies, or in pure cultures of artificial plaques of carbohydrate polymer producers, where growth is not physically restricted, the organisms may become more widely spaced due to polymer production. Indeed, so much matrix may be formed that microbial outlines may be distinguished clearly only at the edge of the colony. Contrarily, the adherence of

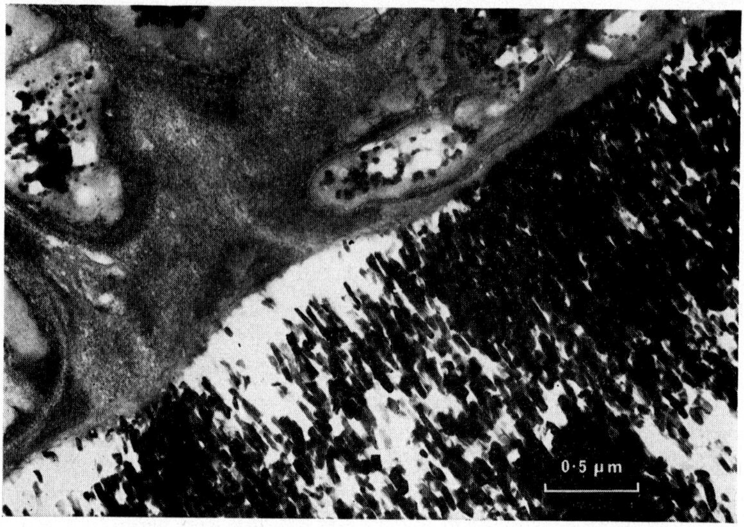

FIG. 25. TEM. Plaque on enamel surface near gingival margin. Prepared by osmium thiosemicarbazide technique. Fibrillar matrix predominates in the deeper plaque (× 25 280).

FIG. 26. FEM. Palisading of organisms is apparently related to the concomitant alignment of matrix fibrils in parallel between cell columns (arrows) (× 30 000).

plaque is not always associated with detectable polysaccharide synthesis (Kelstrup *et al.*, 1974).

Features of Bacterial Cells in Plaques

Intracellular polysaccharide

Several morphologically dissimilar cells in natural plaque contain vesicles which, with an appropriate electron-histochemical technique, stain for polysaccharide, usually in the form of round, electron-dense granules which may occupy a large proportion of the cell cytoplasm. Producer cells are often associated with large amounts of extracellular polysaccharide, although the latter tends to be in fibrillar rather than granular form. Intracellular polysaccharide (IPS)-containing cells are a common feature of intact organisms in the most gingivally located plaque on teeth with little or no associated clinical gingivitis. Cell ghosts are another common feature of this region of plaque. It would seem on this basis that IPS-producing cells are among the first organisms to colonize successfully the crevice region.

Freeze-etched preparations show that many cells in this region and elsewhere contain collections of round vesicular bodies similar in dimension to the granules seen by conventional electron microscopy (Figs 27 and 28).

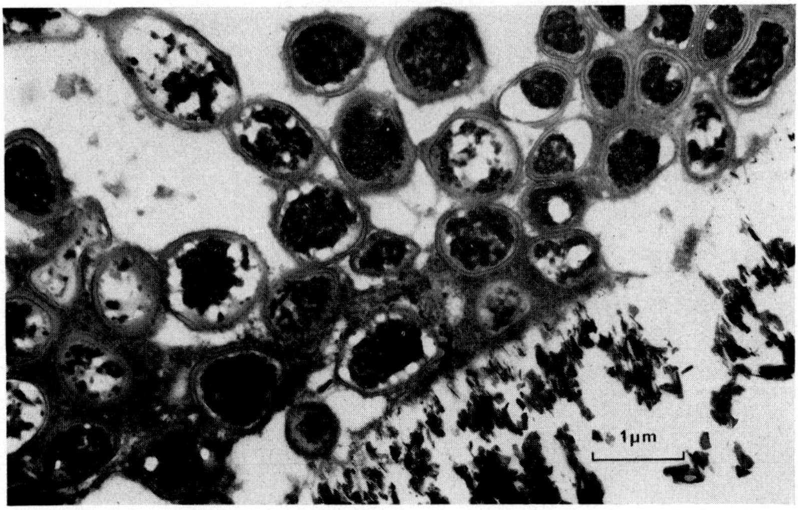

FIG. 27. TEM. Plaque in relation to gingival margin, and stained for carbohydrate polymer. Note predominance of Gram-positive cells packed with intracellular polysaccharide granules (× 11 200).

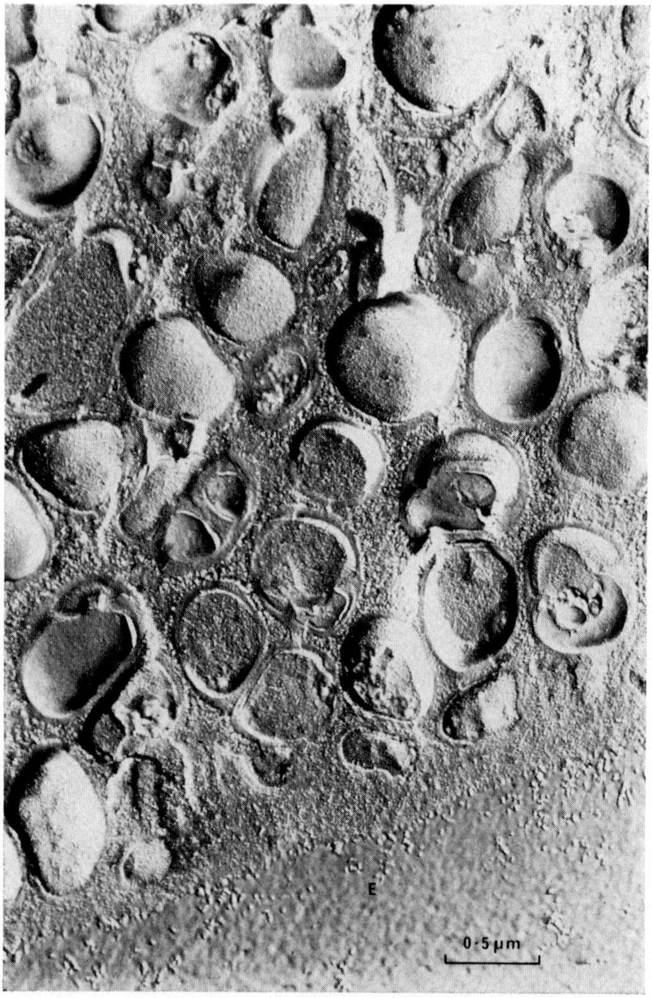

FIG. 28. FEM. Several cells contain agglomerations of vesicular bodies in thick-walled, Gram-positive cells which may be analogous to the granules in similar cells seen in Fig. 27. E=enamel space (× 24 000).

These may represent vesicular mesosomes or polysaccharide granules. A recent study has shown that actively growing cells of the rumen isolate *Megasphaera elsdenii* contain few polysaccharide inclusions, while older cells accumulate so much of this material that the normal cytoplasmic components are displaced to one locus at the edge of the cell (Cheng *et al.*, 1973), a phenomenon which may be observed in relation to the vesicles found in freeze-etched plaque specimens (Newman and Britton, 1974). In the rumen locus, such cells may lyse due to their lacking control of their polysaccharide production (Cheng *et al.*, 1973). This would suggest an association between unbalanced growth and IPS formation in producer cells.

As yet, there has been little attempt to explain the biological significance of this IPS in plaque bacteria. The conventional view is that it acts as a carbon and energy reserve. Yet high concentrations of sucrose and other sugar and acid concentrations (Ingram, 1957). In this they resemble closely the cariogenic streptocci. The production of polymeric intracellular constituents—polysaccharide, polymetaphosphate or poly-β-hydroxybutyrate—is, however, associated with the absence of a growth nutrient (Duguid and Wilkinson, 1961). It is thought that the resultant unbalanced growth is related to the cessation of cell division and even to alterations in cell morphology, as discussed previously. It is difficult to prove a storage function, and the only satisfactory way would be by experiments with mutants unable to synthesize these compounds in excess (Duguid and Wilkinson, 1961). Since IPS in a given cell is limited in amount, it is unlikely to be adequate for cell growth in the absence of an exogenous carbon and energy source. The association of IPS production with unbalanced growth rather than with a storage function in plaque is further supported by the finding that IPS-containing plaque organisms, including *S. mutans*, are limited in their growth (Ellwood, Hunter and Longyear, 1974), and by the association of cell ghosts with IPS-containing cells.

The question of the importance of the part played by the binding of small molecular weight solutes by large molecular weight constituents of the cytoplasm, and by the cohesion of the large molecular weight constituents to one another, in resisting osmotic stress, can be answered only by more quantitative measurements of permeability and osmotic pressure (Mitchell and Moyle, 1956). Observations on plasmolysis and swelling of bacteria provide evidence for the existence of an osmotic barrier, relatively impermeable to sucrose (and some electrolytes) in both Gram-positive and Gram-negative organisms. The presence of large amounts of sucrose in the medium, as in a cariogenic situation, may cause plasmolysis of Gram-negative cells, with retraction of the cytoplasmic membrane from the cell wall. In the case of Gram-positive organisms the cells contract when placed in a hypertonic sucrose medium without separation of cytoplasmic membrane and cell wall. In general it would seem that bacteria are not freely but only selectively permeable to many of their nutrients and waste products (Mitchell and Moyle, 1956). IPS granules may, therefore, possess the capacity to bind solutes and may also prevent cell damage due to changes in extracellular osmotic pressure brought about by excess sucrose.

Cell wall

Organisms in the deeper plaque may have thicker cell walls than their more superficially located counterparts. Freeze etching discloses the existence of fibrillar connections between cell wall and cytoplasmic membrane in cells at various levels (Fig. 29), although these may be a freeze-etching artefact rather than true fibrils (Sleytr, personal communication). Conventional electron microscopy discloses sites of polysaccharide formation in the same locus, as well as on the outer aspect of the cell wall, where fimbriae and fine matrix fibrils may be seen. These appearances are noticeable especially in relation to Gram-positive cells (Fig. 30). The fine intermembrane, possibly artefactual, connections appear to arise from particles which give the convex aspect of the cytoplasmic membrane disclosed by freeze etching its rough appearance. The outer cell wall of plaque organisms may be smooth or covered by fibrils (Newman, 1972*a*; Newman and Britton, 1974), corresponding in appearance to the cell walls of several plaque isolates examined by freeze etching (Newman and Britton, 1973; Girard and Jacius, 1974).

Branching

Branching of organisms is occasionally seen in the deeper layers, but is more commonly observed, because of close packing in the latter, in superficial plaque (Newman, 1972*b*) and in filamentous forms colonizing the gingival crevice. It cannot, however, be said that plaque organisms form mycelia in the manner of higher forms, although natural plaque consists frequently of a dense, palisaded meshwork of filaments and other forms, with smaller cells enclosed between them.

Topographical Variations in Plaque Structure

Some reference has already been made to changes occurring in plaque structure from its initial stages to the formation of a thick layer. Variations may be associated with its location on the tooth and with the presence of

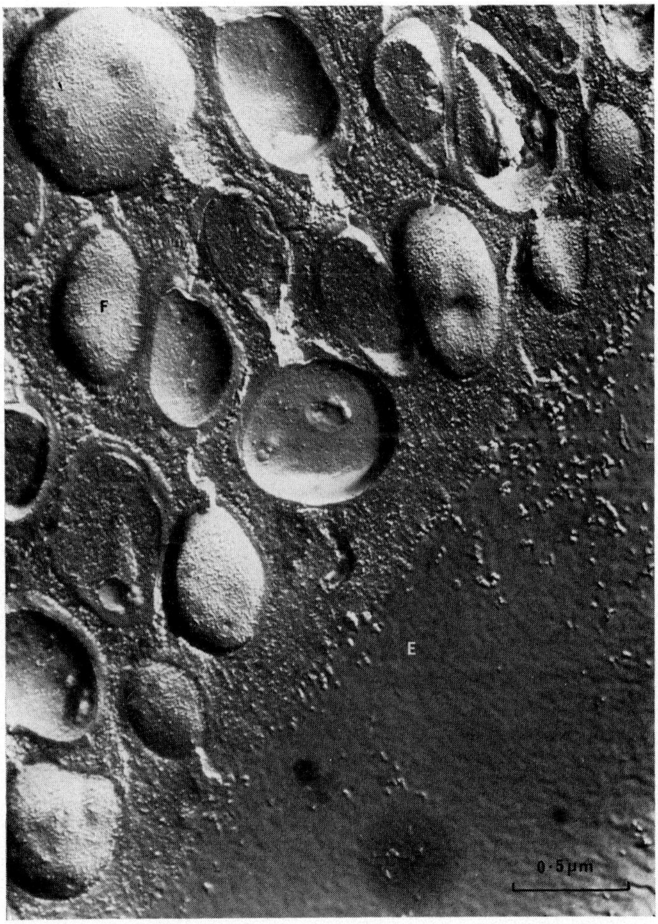

FIG. 29. FEM. Note fibrillar extensions (F) between cell wall and cytoplasmic membrane. It has yet to be demonstrated whether these are true membrane interconnections or an artefact of the technique (× 32 000).

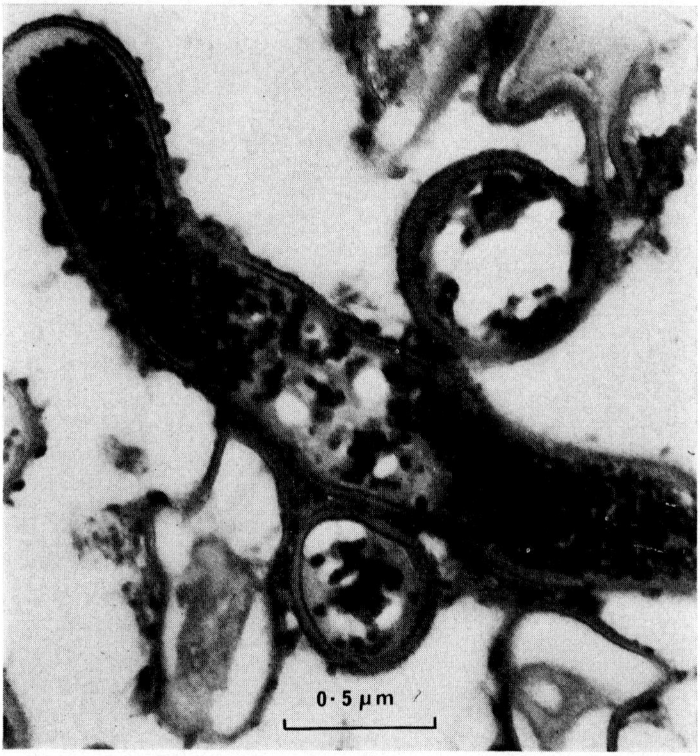

FIG. 30. TEM. Outer gingival margin portion of plaque stained for carbohydrate polymer. Note staining of both inner and outer surfaces of cell walls (× 40 000).

disease processes in which plaque is involved, which may be exacerbated by systemic complications. The nearer the location to a stagnation site, the thicker the plaque layer. The occlusal limit on a smooth surface is generally less obscured by matrix than the gingival border of a given plaque (Fig. 31).

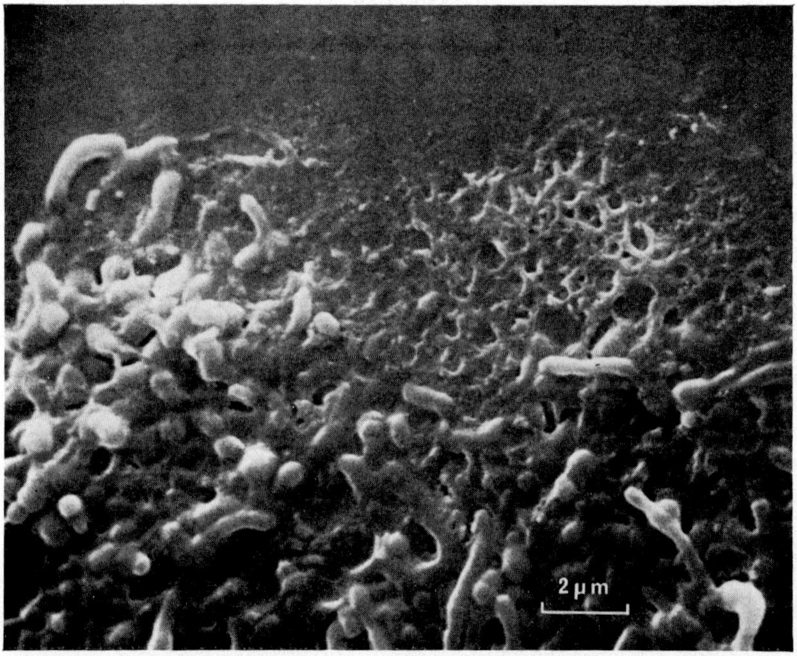

FIG. 31. SEM. Occlusal border of approximal surface plaque. Coccal, bacillary and filamentous forms are evident. Depressions in acquired pellicle may represent sites occupied by coccal cells, viz. the scalloped appearance of this layer in transmission electron micrographs (× 5500).

Fissure Plaque

The structure of natural fissure plaque has not received as much attention as that on the smooth surface, but, from available evidence, the general features of both portions of plaque are similar (Schroeder and de Boever, 1970). Coccal, rod and filamentous forms are found, the cocci reaching the deepest part of the natural occlusal fissures in from one to two days (Folke, Sveen and Thott, 1973). Intermicrobial matrix is abundant (Loe, Karring and Theilade, 1973). Palisading is seen, and cell wall thickness increases in the deeper layers (Wilson, 1973), but unstructured plaque is also common in fissures. The abundance of cell ghosts relates to the slight labelling of fissure plaque that occurs in specimens incubated with tritiated DNA precursors (Newman and Wilson, 1975).

Changes in Plaque in Chronic Inflammatory Periodontal Disease

It is doubtful whether bacteria actually invade the crevice or gingival tissues. Few plaque strains are motile, and such organisms, mainly vibrios and spirochaetes are found in this region only after the onset of inflammation, and only enter the tissues rapidly in acute ulcerative gingivostomatitis. Penetration of epithelium is believed to be consequent on widening of intercellular spaces which is an early feature of chronic inflammatory periodontal disease. This occurs in the junctional and oral sulcular epithelium. Some organisms may become attached to cells, for instance, desquamated oral epithelial cells (Fig. 32) and the mainly neutrophil component of the oral leucocyte population. These neutrophils may also form a thick layer on the outer plaque in relation to inflamed gingival tissue (Fig. 33). With advancing periodontal disease the dental plaque comes to

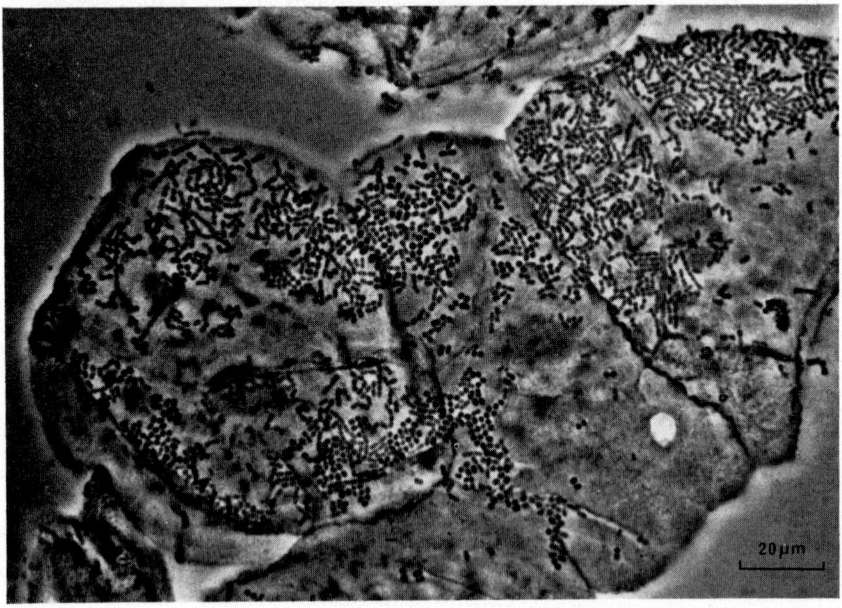

FIG. 32. Phase-contrast micrograph. Human oral epithelial cells coated with coccal forms. (Courtesy of Mr. M. S. Gillett, M. R. C. Dental Unit, Bristol.) (× 576).

FIG. 33. FEM. Round cells containing many cytoplasmic inclusions, present in a superficial portion of plaque. The cell in the top right of the micrograph appears to be in a state of lysis. N=nucleus ($\times$ 6000).

occupy an ever deepening crevice between tooth and surrounding soft tissues. In this region mineralization may begin around organisms in the plaque matrix and continue intracellularly after death of the enclosed organism. This mineralization does not appear to be cell-mediated, since it occurs whether or not the associated cells are viable. Calcified plaque may be firmly attached to the tooth surface, especially when it extends on to the root cementum. For an excellent review of dental calculus, the reader is referred to the monograph by Schroeder (1969).

Note on Specimen Preparation

The material illustrated in the photomicrographs was prepared from plaque specimens obtained from the approximal surfaces of clinically healthy premolar and molar teeth extracted for orthodontic reasons from 8 to 15 year old children. Specimens for examination by scanning electron microscopy were freeze-dried, coated with gold–palladium and examined and photographed in a Stereoscan II scanning electron microscope, at an accelerating voltage of 10 kV. Some specimens for transmission electron microscopy were prepared using buffered glutaraldehyde and osmium tetroxide fixatives as described by Newman (1973*a*): others were stained for carbohydrate polymer using the osmium thiosemicarbazide technique (Critchley *et al.*, 1967). Material for freeze-etch microscopy was prepared in a Balzer's 360M freeze-etch device using methods described previously (Newman and Britton, 1974). Specimens for autoradiography were treated as described by Newman and Wilson (1975).

Acknowledgements

Acknowledgements are due to Mr. M. S. Gillett, Mr. A. B. Britton, Mr. K. Robbins and to the staff of the Long Ashton Stereoscan Unit for their kind cooperation in several aspects of the work encompassed in this review, and to Dr H. D. Donoghue for helpful comments.

References

Barnett, M. L. (1973). Adherence of bacteria to oral epithelium *in vivo*: electron microscopic observations. *J. dent. Res.*, **52,** 1160.

Belcourt, A. (1973). Etude des protéines de la matrice acellulaire de la plaque dentaire et de la salive filtrée humaines. *J. Biol. Buccale,* **1,** 261.

Brock, T. D. (1966). *Principles of Microbial Ecology*. Englewood Cliffs, New Jersey: Prentice-Hall.

Cheng, K.-J., Hironaka, R., Roberts, D. W. A., & Costerton, J. W. (1973). Cytoplasmic glycogen inclusions in cells of anaerobic Gram-negative rumen bacteria. *Canad. J. Microbiol.,* **19,** 1501.

Chung, K. L. (1971). Thickened cell walls of *Bacillus cereus* grown in the presence of chloramphenicol: their fate during cell growth. *Canad. J. Microbiol.,* **17,** 1561.

Cohen, S. S., & Barner, H. D. (1954). Studies on unbalanced growth in *Escherichia coli. Proc. Nat. Acad. Sci., Wash.,* **40,** 885.

Critchley, P., Saxton, C. A., & Kolendo, A. B. (1968). The histology and histo-chemistry of dental plaque. *Caries Res.,* **2,** 115.

Critchley, P., Wood, J. M., Saxton, C. A., and Leach, S. A. (1967). The polymerisation of dietary sugars by dental plaque. *Caries Res.,* **1,** 112.

Donoghue, H. D., & Tyler, J. E. (1975). Antagonisms amongst streptococci isolated from the human oral cavity. *Archs. oral Biol.,* **20**, 381.

Duguid, J. P., & Wilkinson, J. F. (1961). Environmentally induced changes in bacterial morphology. In *Microbial Reaction to Environment.* 11th. Symp. Soc. Gen. Microbiol. 69. Cambridge: The University Press.

Ellwood, D. C., Hunter, J. R., & Longyear, V. M. C. (1974). Growth of *Streptococcus mutans* in a chemostat. *Archs oral Biol.,* **19,** 659.

Erikson, D. (1954). Factors promoting cell division in a 'soft' mycelial type of Nocardia: *Nocardia turbata* n.sp. *J. gen. Microbiol.,* **11,** 198.

Folke, L. E. A., Sveen, O. B., & Thott, E. K. (1973). A new methodology for study of plaque formation in natural human fissures. *Scand. J. dent. Res.,* **81,** 411.

Frank, R. M., & Houver, G. (1970). An ultrastructural study of human supragingival plaque formation. In *Dental Plaque* (W. D. McHugh, ed.) 85. Edinburgh and London: E. and S. Livingstone.

George, R. H. (1974). The isolation of symbiotic streptococci. *J. med. Microbiol.,* **7,** 77.

Gibbons, R. J., & van Houte, J. (1973). On the formation of dental plaques. *J. Periodont.,* **44,** 347.

Girard, A. E., & Jacius, B. H. (1974). Ultrastructure of *Actinomyces viscosus* and *Actinomyces naeslundii. Archs oral Biol.,* **19,** 71.

Goadby, K. W. (1900). Micro-organisms in dental caries. *Dent. Cosmos,* **42,** 210.

Guggenheim, B., & Burckhardt, J. J. (1974). Isolation and properties of a dextranase from *Streptococcus mutans* OMZ 176. *Helv. odont. Acta,* **18,** 101.

Halhoul, N., & Colvin, J. R. (1974). The novel structure of a microorganism of human gingival plaque. *Canad. J. Microbiol.,* **20,** 1307.

Hardie, J. M., & Bowden, G. H. (1974). The normal microbial flora of the mouth. In *The Normal Microbial Flora of Man.* (Skinner, F. A. and Carr, J. G., eds) 47. London and New York: Academic Press.

Hennenberg, G. (1964). Sagittal sections of bacterial colonies. In *Pictorial Atlas of Pathogenic Microorganisms.* (Henneberg, G. ed.) 26. Band 1. 2. Auflage. Stuttgart: Gustav Fischer Verlag.

Holmberg, K., & Hallander, H. O. (1973). Production of bactericidal concentrations of hydrogen peroxide by *Streptococcus sanguis. Archs oral Biol.,* **18,** 423.

Hurst, V. (1950). Morphologic instability of actinomycetes associated with enamel. *J. dent. Res.,* **29,** 571.

Ingram, M. (1957). Micro-organisms resisting high concentrations of sugars or salts. In: *Microbial Ecology.* 7th Symp. Soc. Gen. Microbiol. 90. Cambridge; The University Press.

Jeener, H., & Jeener, R. (1952). Cytological study of *Thermobacterium acidophilus* R. 36 cultured in absence of deoxyribonucleosides or uracil. *Exp. Cell Res.,* **3,** 675.

Johnson, M. C., Bozzola, J. J., & Schechmeister, I. L. (1974). Morphological study of *Streptococcus mutans* and two extracellular polysaccharide mutants. *J. Bact.,* **118,** 304.

Kelstrup, J., & Funder-Nielsen, T. D. (1974). Adhesion of dextran to *Streptococcus mutans. J. gen. Microbiol.,* **81,** 485.

Kelstrup, J., Theilade, J., Poulsen, S., & Møller, I. J. (1974). Bacteriological, electron microscopical, and biochemical studies on dento-gingival plaque of Moroccan children from an area with low caries prevalence. *Caries Res,* **8,** 61.

Kornman, K. S., Kreitzman, S. N., & Clark, W. B. (1972), Elution profiles of streptococcus from hydroxyapatite: influence of capsule and protein. *J. dent. Res.*, **51,** (Suppl.) Abs. 75, p67 (Amer. Div., I.A.D.R.).

LÖE, H., KARRING, T., & THEILADE, E. (1973). An *in vitro* method for the study of the microbiology of occlusal fissures. *Caries Res.*, **7,** 120.

MACCULLUM, D. K. (1974). *In situ* localization of proline in oral bacteria and on lingual epithelium. *J. dent. Res.*, **53,** 138.

MCGAUGHEY, C., & STOWELL, E. C. (1971). Adsorption of salivary proteins by hydroxyapatite: relations between the effects of calcium ions, hydrogen ions, temperature, and exposure time. *J. dent. Res.*, **50,** 542.

MITCHELL, P., & MOYLE, J. (1956). Osmotic function and structure in bacteria. In *Bacterial Anatomy*. 6th. Symp. Soc. Gen. Microbiol. 150, Cambridge: The University Press.

MORENO, E. C., & ZAHRADNIK, R. T. (1974). Chemistry of enamel subsurface demineralization *in vitro*. *J. dent. Res.*, **53,** 226.

NEWBRUN, E., LACY, R., & CHRISTIE, T. M. (1971). The morphology and size of the extracellular polysaccharides from oral streptococci. *Archs oral Biol.*, **16,** 863.

NEWMAN, H. N. (1972*a*). Freeze-etching and dental research. *J. periodont. Res.*, **7,** 91.

NEWMAN, H. N. (1972*b*). Structure of approximal human dental plaque as observed by scanning electron microscopy. *Archs oral Biol.*, **17,** 1445.

NEWMAN, H. N. (1973). The organic films on enamel surfaces. 2. The dental plaque. *Brit. dent. J.*, **135,** 106.

NEWMAN, H. N. (1974a). Microbial films in nature. *Microbios*, **9,** 247.

NEWMAN, H. N. (1974*b*). Diet, attrition, plaque and dental disease. *Brit. dent. J.*, **136,** 491.

NEWMAN, H. N., & BRITTON, A. B. (1973). Ultrastructure of selected bacteria isolated from dental plaque as revealed by freeze-etching. *J. dent. Res.*, **52,** 1194.

NEWMAN, H. N., & BRITTON, A. B. (1974). Dental plaque ultrastructure as revealed by freeze-etching. *J. Periodont.*, **45,** 478.

NEWMAN, H. N., & MCKAY, G. S. (1973). An unusual microbial configuration in human dental plaque. *Microbios*, **8,** 117.

NEWMAN, H. N., & POOLE, D. F. G. (1974*a*). Observations with scanning and transmission electron microscopy on the structure of human surface enamel. *Archs oral Biol.*, **19,** 1135.

NEWMAN, H. N., & POOLE, D. F. G. (1974*b*). Structural and ecological aspects of dental plaque. In *The Normal Microbial Flora of Man*. (Skinner, F. A. and Carr, J. G., eds.) 111. London and New York, Academic Press.

NEWMAN, H. N., & WILSON, C. M. (1975). Thymidine and thymine uptake in human dental plaque: an autoradiographic study. *Caries Res.*, **9,** 405.

ØRSTAVIK, D., & KRAUS, F. W. (1973). The acquired pellicle: immunofluorescent demonstration of specific proteins. *J. oral Path.*, **2,** 68.

RIDGWAY, H. F., & LEWIN, R. A. (1973). Goblet-shaped sub-units from the wall of a marine gliding microbe. *J. gen. Microbiol.*, **79,** 119.

SCHROEDER, H. E. (1969). *Formation and Inhibition of Dental Calculus*. Berne, Stuttgart and Vienna: Hans Huber.

SCHROEDER, H. E. (1970). The structure and relationship of plaque to the hard and soft tissues: electron microscopic interpretation. *Int. dent. J.*, **20,** 353.

SCHROEDER, H. E., & DE BOEVER, J. (1970). The structure of microbial dental plaque. In *Dental Plaque*. (McHugh, W. D, ed.) 49. Edinburgh. E. and S. Livingstone, and London:

SNYDER, M. L., BULLOCK, W. W., & PARKER, R. B. (1967). Morphology of Gram-positive filamentous bacteria identified in dental plaque by fluorescent antibody technique. *Archs oral Biol.,* **12,** 1269.

SOAMES, J. V., & DAVIES, R. M. (1974). The morphology and distribution of spirochaetes in the gingival crevice of the Beagle dog. *J. dent. Res.,* **53,** (Suppl.), 1075 (Br. Div., I.A.D.R.).

SONJU, T., CHRISTENSEN, T. B. KORNSTAD, L., & ROLLA, G. (1974). Electron microscopy, carbohydrate analyses and biological activities of the proteins adsorbed in two hours to tooth surfaces *in vivo. Caries Res.,* **8,** 113.

DE STOPPELAAR, J. D., KÖNIG, K. G., PLASSCHAERT, A. J. M., & VAN DER HOEVEN, J. S. (1971). Decreased cariogenicity of a mutant of *Streptococcus mutans. Archs oral Biol.,* **16,** 971.

TAKEUCHI, H., SUMITANI, M., TSUBAKIMOTO, K., & TSUTSUI, M. (1974). Oral micro-organisms in the gingiva of individuals with periodontal disease. *J. dent. Res.,* **53,** 132.

TAKEUCHI, A., & ZELLER, J. A. (1972). Scanning electron microscopic observations on the surface of the normal and spirochete-infested colonic mucosa of the Rhesus monkey. *J. Ultrastruct. Res.,* **40,** 313.

VAN HOUTE, J., & SAXTON, C. A. (1971). Cell wall thickening and intra-cellular polysaccharide in microorganisms of the dental plaque. *Caries Res.,* **5,** 30.

VREVEN, J., & FRANK, R. M. (1973). Cythochimie ultrastructurale de la phosphatase acide et de la pyrophosphatase dans la plaque dentaire humaine. *J. Biol. Buccale,* **1,** 63.

WAGNER, R. C., & BARRNETT, R. J. (1974). The fine structure of procaryotic-eukaryotic cell junctions. *J. Ultrastruct. Res.,* **48,** 404.

WILLIAMS, R. A. D. (1972). Use of bacterial polymers by *Streptococcus sanguis. J. dent. Res.,* **51,** 1236 (Abs., Br. Div., I.A.D.R.).

WILSON, D. F. (1973). Ultrastructure of human tooth fissure plaque. *J. dent. Res.,* **52,** (Suppl.) 581 (Amer. Div., I.A.D.R.).

WINTERBURN, P. J., & PHELPS, C. F. (1972). The significance of glycosylated proteins. *Nature, Lond.,* **236,** 147.

Microfungi, Yeasts and Yeast-like Organisms

R. R. Davenport, Barbara Bole, Beverley McLeod and Elizabeth Parsons

University of Bristol, Department of Agriculture and Horticulture, Research Station, Long Ashton, Bristol, England

During the course of academic and industrial microbiological investigations it was found that certain features of micro-organisms could be more clearly observed using scanning electron microscopy techniques rather than by light microscopy. The examples given here are of known micro-organisms, which are part of a reference collection, used for identification and taxonomic studies. A gold–palladium alloy was applied to all the specimens some of which had first been dehydrated by the critical point method.

During the investigations it was shown that micro-organisms on plant organs could be readily distinguished from debris; furthermore morphological characteristics of the micro-organisms, as observed with scanning electron microscopy could be used as a means for their probable identification.

Methods

Growth of micro-organisms

The filamentous fungi were grown on the surface of apple-juice yeast-extract agar plates and the yeasts cultivated either in the same way or in liquid apple-juice extract (Beech and Davenport, 1969). This is a routine isolation medium; although originally prepared for apple product micro-organisms it is a good general purpose medium for most fungi, including yeasts and yeast-like organisms, but one can equally use other rich fungal media such as wort agar (Oxoid, 1971). In all cases the incubation temperature was 25° and colonies and cells were prepared and examined after one week using the following methods.

Preparation of fungi (*after Greenhalgh and Evans, 1971*)

Specimen stubs were covered with double sided adhesive tape (Sellotape) and the fungal specimens mounted by gently brushing the sticky stub across the surface of the colony. This procedure resulted in minimum damage to the cells and could avoid overcrowding on the stub. The specimens were then coated with gold–palladium alloy (Johnson Mathley Metals Ltd, London) to a thickness of about 20 nm in an Edwards 12E6 Coating unit modified to allow rotation and tilting of the specimen to ensure even distribution of the metal. The coated specimens were examined in a Cambridge Scientific Instruments "Stereoscan" Mk 11A at an accelerating voltage of 10 KV (Figs 1–6).

Preparation of yeast colonies—critical point dry procedure (Parsons et al., *1974)*

Yeast colonies were cut from the agar plate, fixed for 30 min. in 3% glutaraldehyde, washed twice in distilled water and then dehydrated through a graded series of ethanol/water (25% to 100% ethanol), leaving the specimens in each solution for 20 min. The specimens were then passed through a similar graded amyl acetate/ethanol series and then critical-point dried using liquid carbon-dioxide in a Polaron E300 apparatus. (Polaron Equipment Ltd, Watford). The dried colonies were attached to aluminium specimen stubs with Durofix and coated with gold (about 25 nm thick) in a Polaron sputter coater (Model E5000) and then examined in the scanning electron microscope at 10 KV (Figs 7–12).

Preparation of cells from liquid cultures (after Lung, 1974)

Yeast cultures were centrifuged and washed in sterile glass distilled water and adjusted to a standard concentration (+2 on scale devised by Wickerham, 1951), of *c* 10^6 cells/ml. This was pipetted directly on to a clean aluminium stub and the cells allowed to settle by leaving the suspension undisturbed for five minutes. The specimens were then dehydrated through an ethanol series with ascending concentrations from 25–100% at intervals of five minutes, transferred through a similar graded concentration series of amyl acetate at the same time intervals and finally dried in liquid CO_2 in the

---

FIG. 1. *Cunninghamella echinulata*—spiny conidium (×3150).

FIG. 2. *Zygorrhychus moelleri*—mature zygospores, with deep sculptural walls (×400).

FIG. 3. *Aspergillus carbonairus*—conidiophore: a=conidiophore head, b=phiolides, c=conidia (×440).

FIG. 4. *Penicillium notatum*—conidiophore: a=conidiophore head, b=phiolides, c=conidia (×500).

FIG. 5. *Paecilomyces variotii*—conidia (×5500).

FIG. 6. *Geotrichum candidum*—arthrospores (×6000).

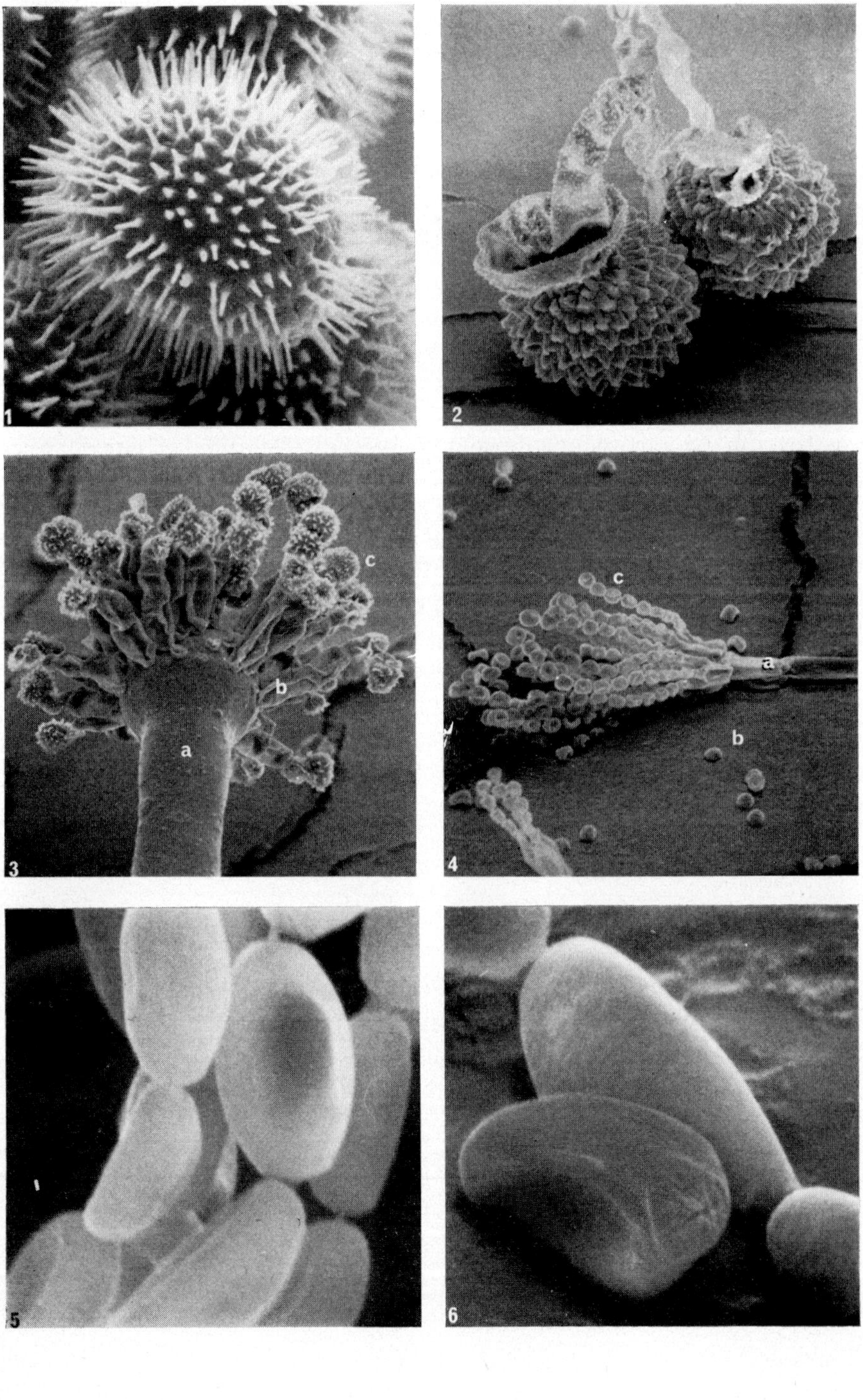

1
2
c
b
a
3
c
a
b
4
5
6

critical point drying apparatus. Coating procedures and scanning electron microscope examinations were carried out as described above for the yeast colonies.

Results and Discussion

Figs 1–6 illustrate some fungi which were coated without being subjected to any elaborate preparations. It can be seen that, in general, the softer parts of the organisms were somewhat distorted but the firmer areas remained intact; this is similar to the findings of Greenhalgh and Evans (1971). In the case of yeasts some degree of cellular distortion is due to the condition of the cells in a particular environment (Davenport, 1976).

However, Figs 7–12 illustrate that more delicate organisms such as yeasts can be successfully treated to prevent distortion of either yeast colonies (Fig. 7) or cells (Figs 8–12). Elongated and oval cells (Fig. 8) and oval-circular cells (Figs 9 and 10) of yeasts which reproduce by multipolar budding (note the bud scars) can be seen. Figure 11 shows non-budding cells (i.e. cylindric fission cells) and Fig. 12 the cells of a bipolar budding yeast. Thus one can obtain scanning electron micrographs of yeasts which are not only a useful complement to light photomicrographs but also act as reference pictures for ecological studies (Davenport, 1976).

FIG. 7. *Hansenula anomals* var. *anomala*—colony grown on apple juice yeast extract agar, one week at 25°c ($\times$34).

FIG. 8. *H. anomals* var. *anomala*—cells from edge of colony (Fig. 7) ($\times$600).

FIG. 9. *Saccharomyces cerevisiae*—cells: a = bud scars from parent cell, i.e. multipolar budding ($\times$1200).

FIG. 10. *Debaryomyces hansenii*—cells: a = multipolar budding ($\times$4900).

FIG. 11. *Schizosaccharomyces pombe*—cells: a = cylindric fission cells, b = annulation formed by previous fission process ($\times$3500).

FIG. 12. *Kloeckera apiculata*—cells: a = bipolar budding cell (mother), b = beginning of bud of one daughter cell, c = second daughter bud, just before separation by fission ($\times$3250).

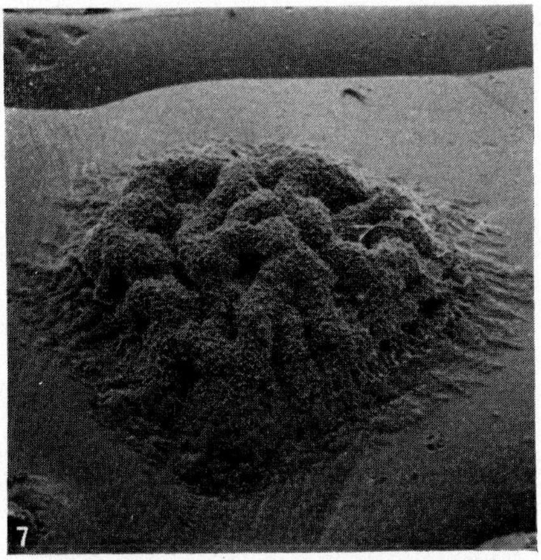
7

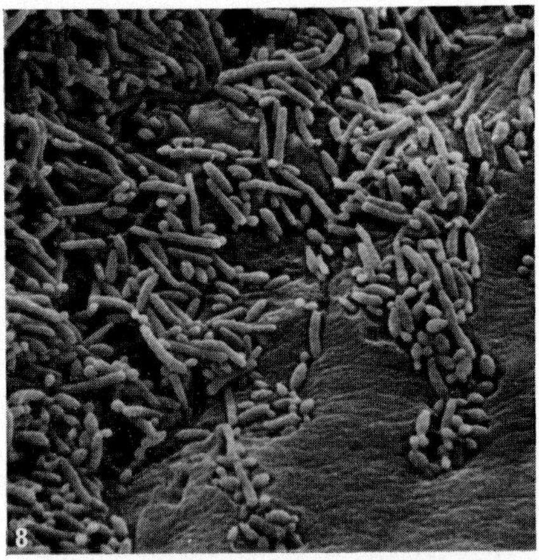
8

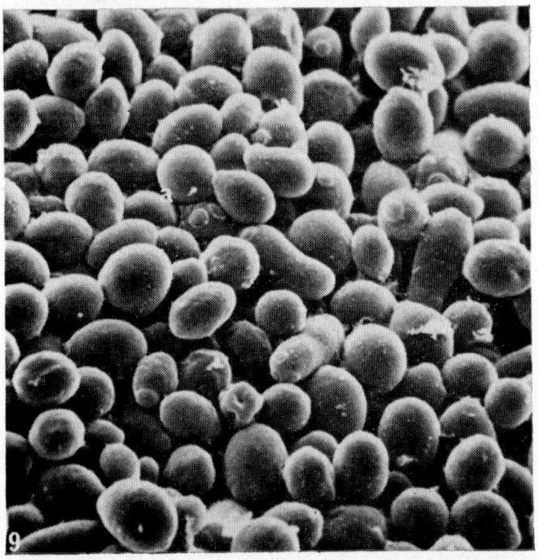
a
9

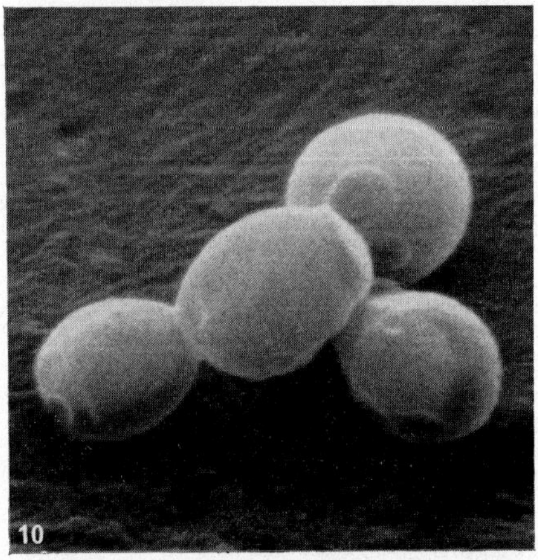
10

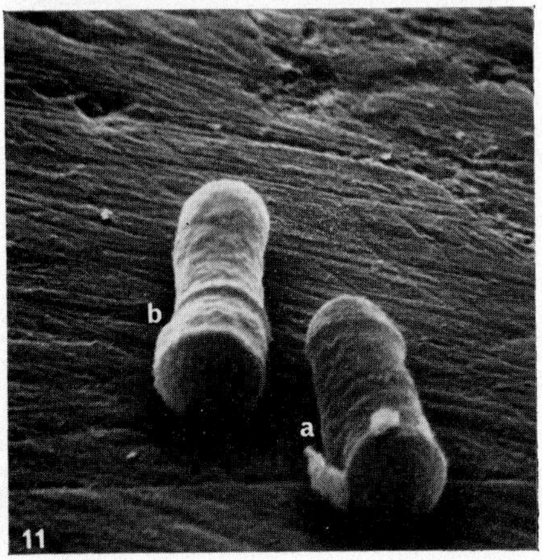
b
a
11

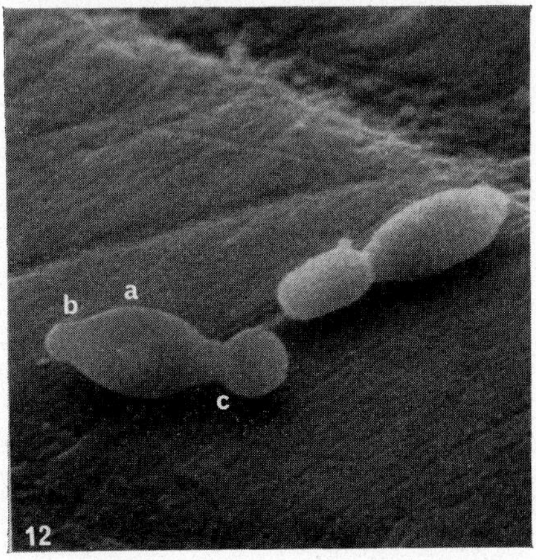
b
a
c
12

References

BEECH, F. W. & DAVENPORT, R. R. (1969). The isolation of non-pathogenic yeasts. In *Isolation Methods for Microbiologists.* Society of Applied Bacteriology, Technical Series 3. (Simpson, D. A. and Gould, G. W. eds). London and New York: Academic Press.

DAVENPORT, R. R. (1976). Experimental ecology and identification of micro-organisms. In *Microbial Ultrastructure—The Use of the Electron Microscope.* Society of Applied Bacteriology, Technical Series 10. (Fuller, R. and Lovelock, D. W., eds) London: Academic Press.

GREENHALGH, G. N. & EVANS, L. V. (1971). Electron Microscopy. In *Methods in Microbiology,* Vol. 4 (Norris, J. R., Ribbons, D. W., and Booth, C., eds) London and New York: Academic Press.

LUNG, B. (1974). The preparation of small particulate specimens by critical point drying: application for scanning electron microscopy. *J. Microsc.* **101,** 88.

OXOID (1971). Oxoid manual (3rd Ed.) London.

PARSONS, E. BOLE, B. HALL, D. J. & THOMAS, W. D. E. (1974). A comparative survey of techniques for preparing plant surfaces for the scanning electron microscope. *J. Microsc.,* **101,** 59.

WICKERHAM, L. J. (1951). Taxonomy of Yeasts. *Tech. Bull. No.* **1029** *Washington: U.S. Dept. Agric.*

Experimental Ecology and Identification of Micro-organisms

R. R. Davenport

University of Bristol, Department of Agriculture and Horticulture Research Station, Long Ashton, Bristol, England

Detailed studies of cider apple orchards and a vineyard at the Research Station (Davenport, 1968; 1970; 1973) have shown the presence of many microbial forms on plant and animal surfaces as well as in beverages made from the processed fruits. The use of scanning electron microscopy (SEM) has been used to demonstrate (a) the occurrence and distribution and (b) the significance and identification of selected propagules. Emphasis has first been placed on those organisms of possible importance in beverage industries, i.e. species *Saccahromyces, Schizosaccharomyces* and *Saccharomycodes*. In the second instance examples have been given of propagules occurring on natural substrates.

Methods

Morphological characteristics are widely used for the delimitation of genera of fungi. Although many yeasts and yeast-like organisms show few morphological features for generic separation (Lodder, 1970), it is possible to recognize some characteristics, i.e. types of vegetative reproduction. These characteristics can be used to form groups or genera of organisms; thence, with details of their source, one can select tests for further identification (Davenport, 1974). The use of SEM is a valuable supplement to existing identification methods. Figures 1–6 show a selection of organisms from cider (an alcoholic apple beverage) obtained by membrane filtration (Beech and Davenport, 1969; 1971*a*). Part of the membrane was aseptically cut off and glued to a specimen stub with "Durofix" and coated with gold–palladium alloy (Greenhalgh and Evans, 1971; Davenport *et al.*, 1976). The remainder of the membrane was placed on a nutrient pad soaked in liquid apple-juice yeast-extract medium to allow the organisms to grow. Subsequently these were isolated and identified (Beech and Davenport, 1969 and Davenport, 1973; and 1974).

Micro-organisms and other propagules on surfaces, such as leaves and fruits, can be viewed directly by conventional microscopy preceded by the Sellotape impression technique (Davenport, 1967; Beech and Davenport, 1971*b*). The SEM technique has been used to view micro-organisms *in situ*, on a wide variety of substrates. This procedure not only demonstrates the position of micro-organisms in relation to other propagules and surface topography, but also shows microbial forms which are either not readily grown in culture or are as yet unculturable, as well as non-biological propagules. Figures 7–12 are some examples illustrating propagules on surfaces.

Comments on Organisms in Cider (Figs 1–6)

The organisms in Figs 1–6 were taken from part of an experiment to determine whether or not known species could be recognized after remaining in a cider sample at 25° for three to six months (i.e. bad storage conditions over a long shelf-life period). It can be seen that the micro-organisms were easily distinguished from debris; moreover, there was little morphological change. Hence it is possible to make use of some of these results as an example for the rapid indentification of yeast and yeast-like organisms. Thus tables (in appendix) show how the combined use of SEM and selected features can give help in rapidly identifying yeasts. For species differentiation one would have to follow more detailed schemes, i.e. classical taxonomy (Lodder, 1970) or a simplified system (Davenport, 1974).

Comments on Figs 7–12

Figures 7–11 show some features of various surfaces on which micro-organisms were present, and these were subsequently identified by comparing with known fungi (Davenport *et al.* 1976). Thus Figs 7, 8 and 10 are of *Cladosporium herbarum*, *Kloeckera apiculata* and *Penicillium chrysogenum*. Figure 9 shows how sometimes bacteria reside along the veins of leaves, while Fig. 11 illustrates the occurrence of micro-organisms as well as

---

FIG. 1. *Saccharomyces cerevisiae*. a, budding cells (×1400).

FIG. 2. *Schizosaccharomyces pombe*. a, fission cells; b, fission walls; c, chain of cylindric fission cells (×1750).

FIG. 3. *Saccharomycodes ludwigii*. a, bud-fission scar; b, annulations formed by successive buds (×900).

FIG. 4. *Lactobacillus malii*. a, rod shaped bacterial cells. (×1650).

FIG. 5. *Sacch. uvarum*. a, budding cells (×1500).

FIG. 6. *Sacch. bailii* var, *bailii*. a, budding cells; b, ascospores released from ascus (×2800).

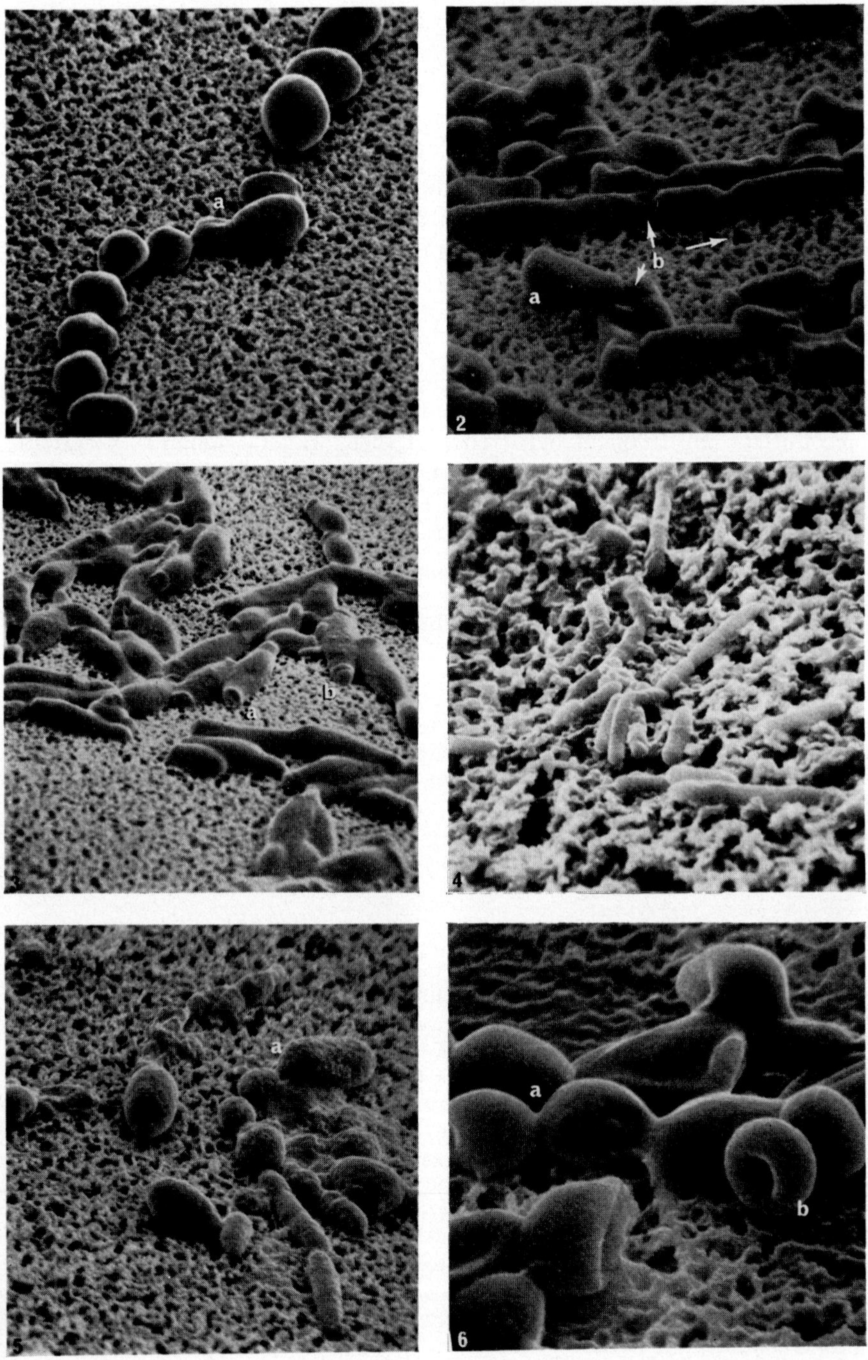

1
a
2
a
b
3
a
b
4
5
a
6
a
b

mineral propagules. Some further examples of this type of propagula are shown on Fig. 12. Studies using XES (X-Ray Energy Dispersive Analyser) showed that non-biological particles could be of various minerals including iron, silicon and a complex of iron-aluminium-silicate; furthermore, it is suggested that these particles are probably from industrial sources (Allen *et al.*, 1975). Two points arise from the presence of these: firstly, their size and shape could possibly allow them to be mistaken for circular yeasts or similar smooth-walled micro-organisms; secondly, the occurrence of these minerals poses questions of micro-ecological interest. For example, do these mineral particles have any interacting effect with micro-organisms (see Fig. 11), are there any ill effects produced by them on plant surfaces, and do some plants have an electric charge on their hairs whilst the mineral spheres have a dissimilar polarity (see Fig. 12)? If this is true then it would account for the frequent occurrence of these mineral spheres on certain leaf hairs where many micro-organisms (including plant pathogen conidia) can be found, thus illustrating a possible defence mechanism for some leaf surfaces.

Conclusion

The application of SEM, using simple methods, can be a useful supplement to existing microbiological methods, thus aiding early identification and assessing the ecological significance of certain propagules.

FIG. 7. *Cladosporium herbarum*. Conidium on a grape leaf. (×5500).

FIG. 8. *Kloeckera apiculata*—bipolar budding yeast, on a grape pedicle, a, mother cell; b, bud fission scar of one daughter cell; c, second daughter cell (×3350).

FIG. 9. Bacteria on a grape leaf vein. a, leaf vein; b, leaf stomata; c, bacteria cells. (×450).

FIG. 10. *Penicillium chrysogenum* isolated from cider; growing on a membrane filter. (×250)

FIG. 11. Mineral spheres and micro-organisms on a grape leaf. a, iron sphere; b, fungal hypha; c, micro-organisms (×200).

FIG. 12. Particles on a grape leaf hair. a, leaf hair; b, silicon spheres; c, mineral debris (×2750).

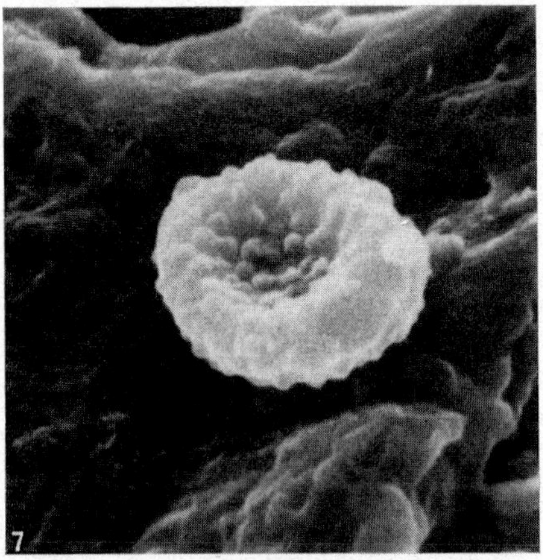
7
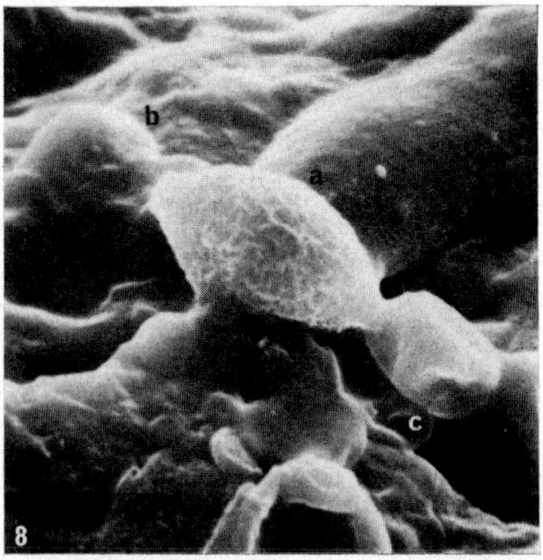
b
a
c
8
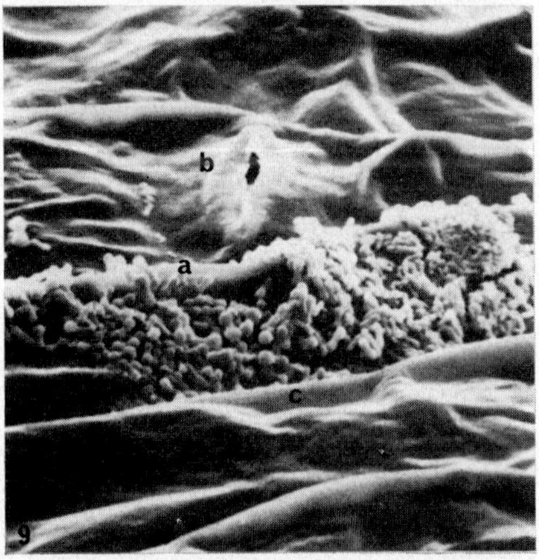
b
a
c
9
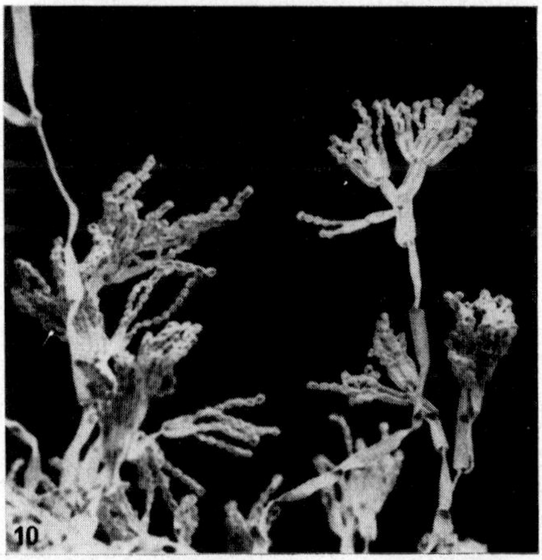
10
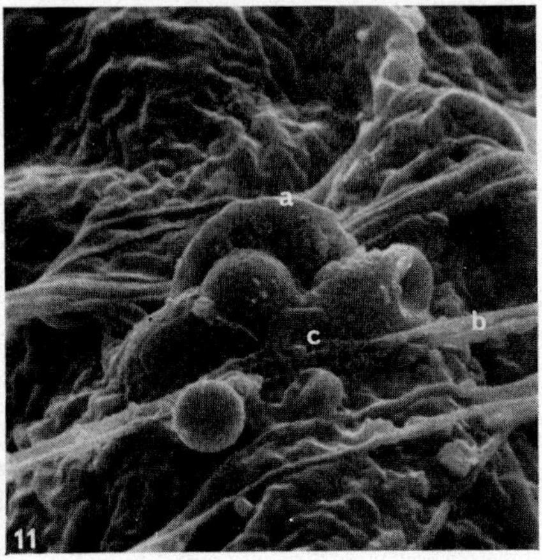
a
c
b
11
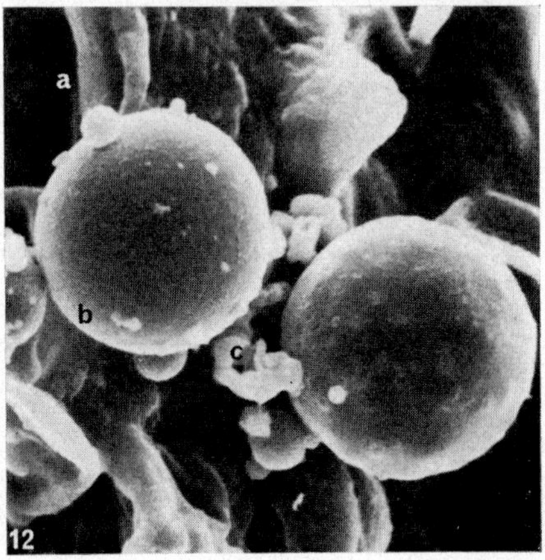
a
b
c
12

Appendix

TABLE 1. An example of separating the organisms on Fig. 1–6 into groups and possible genera as an aid to rapid yeast identification.

Groups	Possible genera
A. Mode of vegetative reproduction	
Budding	
Fission C	
B. Multipolar budding, Figs 1, 5 and 6	*Saccharomyces, Hansenula, Torulopsis* or *Candida*
Bipolar budding (or bud fission) Fig. 3	*Saccharomycodes, Hanseniaspora* or *Kloeckera*
C. Fission, small cells, i.e. > 1 μ dia., Fig. 4	Lactic acid bacteria
Fission, large cells, Fig. 2	*Schizosaccharomyces*

TABLE 2. Further separation of the bipolar budding yeasts

Genera	Cells (up to 23μ and 8μ wide)	Resistant to > 50 p/m SO_2	Ascospores formed
Saccharomycodes	+	+	+ circular only
**Hanseniaspora*	—	—	+ various shapes
Kloeckera	—	—	—

* N.B. Imperfect forms of *Hanseniaspora* are placed in the genus *Kloeckera*.

Acknowledgments

The author gratefully acknowledges the assistance of Mrs. Barbara Bole and Miss Beverley Mcleod with SEM preparations.

References

ALLEN, G., NICKLESS, G., WIBBERLEY, B. & PICKARD, J. (1975) Heavy metal particle characterisation. *Nature, Lond.*, **252,** 571.

BEECH, F. W. & DAVENPORT, R. R. (1969). The isolation of non-pathogenic yeasts. In *Isolation Methods for Microbiologists*. Society of Applied Bacteriology, Technical Series 3, (Simpson, D. A., and Gould, G. W., eds), London and New York. Academic Press.

BEECH, F. W. & DAVENPORT, R. R. (1971*a*). Isolation, purification and maintenance of yeasts. In *Methods in Microbiology*, Vol. 4 (Norris, J. R., Ribbons, D. W. and Booth, C., eds), London and New York: Academic Press.

BEECH, F. W. & DAVENPORT, R. R. (1971*b*). A survey of methods for the quantitative examination of yeast flora of apple and grape leaves. In *Ecology of leaf surface micro-organisms* (Preece, T. F. and Dickenson, C. H., eds), London and New York: Academic Press.

DAVENPORT, R. R. (1967). The microflora of cider apple fruit buds. *Rep. Long Ashton Res. Stn for 1966*, 246.

DAVENPORT, R. R. (1968). The origin of cider yeasts. Thesis, London, Institute of Biology.

DAVENPORT, R. R. (1970). Epiphytic yeasts associated with the developing grape vine. M.Sc. Thesis, University of Bristol.

DAVENPORT, R. R. (1973). Vineyard yeasts—an environmental study. In *Sampling —Microbiological Monitoring of Environments* (Board, R. G. and Lovelock, D. W. eds), London and New York: Academic Press.

DAVENPORT, R. R. (1974). A simple method, using stripdex equipment for the assessment of yeast taxonomic data and identification keys. *J. appl. Bact.*, **37**, 269.

DAVENPORT, R. R., BOLE, B., MCLEOD, B. & PARSONS, E. (1976). Experimental ecology and identification of micro-organisms. In *Microbial Ultrastructure—The Use of the Electron Microscope*. Society of Applied Bacteriology, Technical Series 10, (Fuller, R. and Lovelock, D. W., eds). Academic Press: London and New York.

GREENHALGH, G. N. & EVANS, L. F. (1971). Electron microscopy. *Methods in Microbiology*, Vol. 4 (Norris, J. R., Ribbons, D. W., and Booth, C. eds). London and New York: Academic Press.

LODDER, J. (1970). *The Yeasts*. Amsterdam, North Holland Publishing Company.

Formation of the Trichospore Appendage in *Stachylina grandispora (Trichomycetes)*

STEPHEN T. MOSS

Department of Biological Sciences,
Portsmouth Polytechnic, Portsmouth, Hampshire, England

Trichomycetes (Duboscq, Léger and Tuzet, 1948) is a class of fungi found as obligate commensals within the digestive tracts or as epizoites of many terrestrial and aquatic mandibulate arthropods. The simple organization of their thalli, asexual reproduction by sporangiospores and sexual reproduction, in the Harpellales, by zygospores constitute the principal criteria for their inclusion in the Zygomycotina (Lichtwardt, 1973*a*). The class contains four orders with seven families: Amoebidiales (Amoebidiacea), Eccrinales (Eccrinaceae, Palavasciaceae, Parataeniellaceae), Asellariales (Asellariaceae), Harpellales (Genistellaceae, Harpellaceae), united by their common ecological niche. Recent fine-structural (Farr and Lichtwardt, 1967; Lichtwardt, 1973*b*; Manier, 1973*a*, *b*; Manier and Coste-Mathiez, 1968; Manier and Grizel, 1972; Moss, 1972, 1975; Reichle and Lichtwardt, 1972; Tuzet and Manier, 1967; Whisler and Fuller, 1968), serological (Sangar *et al.* 1972) and histochemical (Sangar and Dugan, 1973; Trotter and Whisler, 1965; Whisler, 1963) studies have indicated that the class may be polyphyletic; the Asellariales and Harpellales show a close affinity with each other and also with the Kickxellaceae (Mucorales) whilst the affinities of the Amoebidiales and Eccrinales remain uncertain.

Stachylina grandispora Lichtwardt (1972) is a member of the Harpellales, an order characterized by infesting the digestive tracts of freshwater insect larvae and nymphs and the production of unique monosporous sporangia, termed trichospores (Manier and Lichtwardt, 1968). These spores develop exogenously and in basipetal succession, each termino-laterally from a generative cell. Attached to the base of the released trichospore are one to several non-motile appendages the number being characteristic of each genus, produced within the subtending generative cell. Results from light microscope studies have provided few details of appendage formation (Lichtwardt, 1967; Moss, 1970) and it is only in well prepared material and using interference light microscopy that the mature appendage may be resolved. The trichospore appendages of only three species have been examined with the electron microscope (Manier, 1973*a*; Manier and Coste-Mathiez, 1968; Reichle and Licktwardt, 1972). These studies showed that the structure of the mature appendage was different in each species but provided only limited information of appendage ontogeny.

This paper presents the results of an ultrastructural study of appendage formation in *S. grandispora*.

Materials and Methods

Stachylina grandispora has not been cultured axenically and all thalli used in this study were dissected from the living host. Thalli were found attached to the peritrophic membranes lining the larval midguts of several species of Chironomidae. Hosts were collected from Sulham Brook, Berkshire, England (O.S. map grid ref. SU 643754) immediately prior to their dissection. Chironomid larvae were either removed from the surface of debris or sieved out of mud dredged from the stream bed. The larvae were dissected under water and their peritrophic membranes removed. Each membrane lies free within the midgut and is attached to the epithelium of the digestive tract only at its region of formation between the stomodaeum and mesenteron. Circumcisal rupture of the body wall immediately posterior to the head capsule, followed by separation of the head from the body, allowed the peritrophic membrane to be withdrawn from the midgut epithelium. The detached membranes were cleaned of debris by lifting them out of the water anterior end first; fungal thalli remained attached by their holdfasts. Complete removal of debris was impracticable owing to the better results obtained by reducing the time between dissection of the host and fixation of the thalli.

Trials with several fixatives indicated that freshly prepared, aqueous, 1% (w/v) potassium permanganate for 1h at room temperature provided the most acceptable results for this investigation. The peritrophic membranes, with the attached fungal thalli, were slit longitudinally and then immersed in the fixative. Fixed material was washed several times in distilled water, dehydrated in a graded ethanol series (15 min in each of 10% steps) to propylene oxide and embedded in Araldite.

Flat embedment of thalli enabled individual structures to be selected and removed for ultramicrotomy. Ultrathin sections were obtained using a Cambridge Huxley ultramicrotome with glass knives. Sections were floated on to distilled water and mounted on collodion coated, carbon stabilized grids. The use of glass knives and the frequent presence of debris around the thalli resulted in progressive specimen damage in serial sections. Sections were stained in a carbon-dioxide free atmosphere in lead citrate (Reynolds, 1963) for 15 min followed by 30 min in the dark in a saturated solution of uranyl acetate in 50% ethanol. Grids were examined in an Hitachi HS-7S electron microscope at 50 kV.

Observations

Light microscopy

The thallus of *S. grandispora* is unbranched and attached to, but does not penetrate, the peritrophic membrane of its host by a secreted, basal holdfast. Immature thalli are attached to the membrane lining the anterior region of the midgut and sporulating thalli to that lining the posterior region. The increase in thallus maturity along the peritrophic membrane is due to the posterior movement of the membrane from its region of production, combined with the germination and attachment of ingested trichospores to the anterior extremity of the membrane. The mature vegetative thallus is septate with 4–8 uninucleate cells. Trichospores develop basipetally and in a unilateral series each from a generative cell (Fig. 1). It has been shown previously that early in trichospore development the single generative cell nucleus divides mitotically; one nucleus migrates into the trichospore and the second remains within the generative cell (Moss, 1974). Enlargement of the trichospore is concurrent with vacuolation of the generative cell and migration of cytoplasm into the spore. No stage in appendage formation was observed. Immediately prior to trichospore release the mature appendage appears as a solid strand, 12–14 μm long and 1–2 μm wide, which extends from the base of the trichospore into the vacuolated generative cell (Fig. 2). On the released trichospore the appendage strand is seen as a single, fine, long (200–500 μm), non-motile filament folded upon itself many times (Figs 3, 4). In *S. grandispora* a region of the generative cell remains attached to the released trichospore to form the collar.

Electron microscopy

The mature trichospore is separated from the generative cell by a crosswall with a central pore occluded by an electron dense, non-membrane bound plug (Fig. 6). Previous studies have shown this type of septal apparatus to be characteristic of the Harpellales and at least two species of the Asellariales (Lichtwardt, 1973*b*; Manier, 1973*a*, *b*; Manier and Coste-Mathiez, 1968; Moss, 1972, 1975; Reichle and Lichtwardt, 1972). No stage in appendage formation was observed before development of this septum (Fig. 5). The presumptive collar of the released trichospore is indicated by an increase in thickness of the outer wall layer of the generative cell, immediately below the spore, continuous with a similarly thickened wall layer of the trichospore (Figs 5–10). Early stages in appendage formation

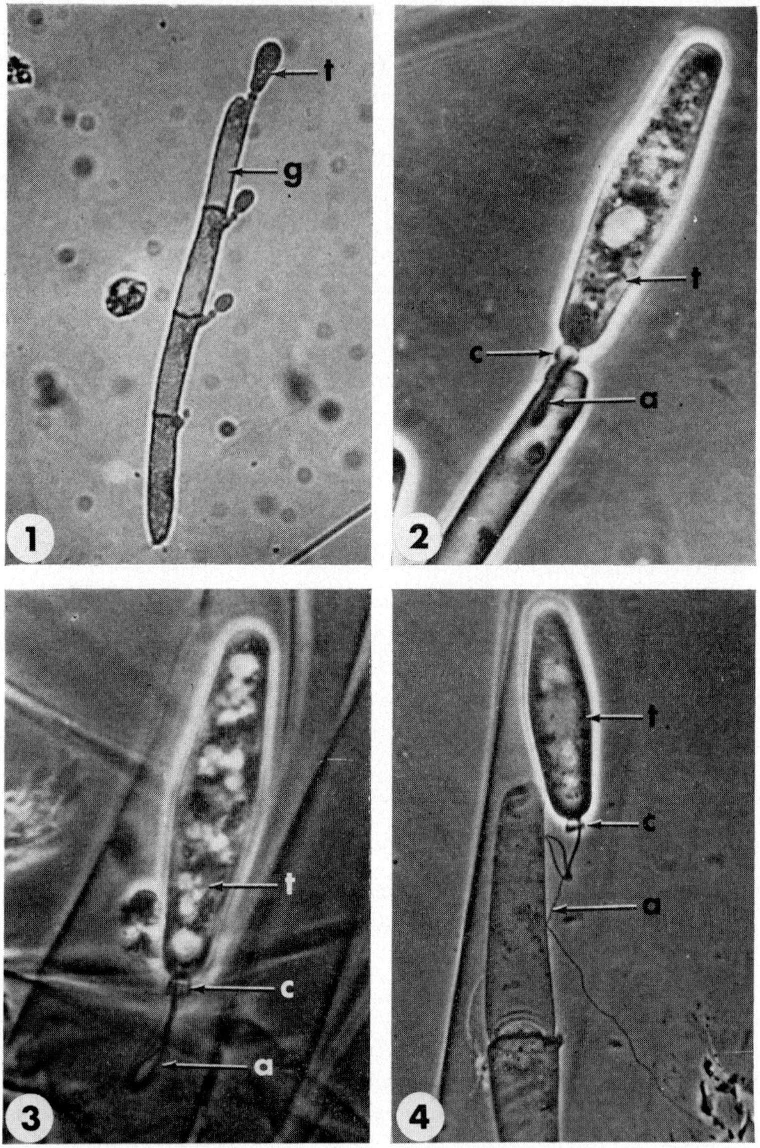

S. grandispora (light micrographs)

FIG. 1. Sporulating thallus showing basipetal development of trichospores (t) from generative cells (g) ($\times$400).

FIG. 2. Apical generative cell with mature trichospore (t) and attached appendage (a) ($\times$880).

FIG. 3. Trichospore (t) immediately after release showing the folded appendage (a) ($\times$880).

FIG. 4. Released trichospore (t) with unfolded appendage (a) attached within the collar (c) ($\times$440).

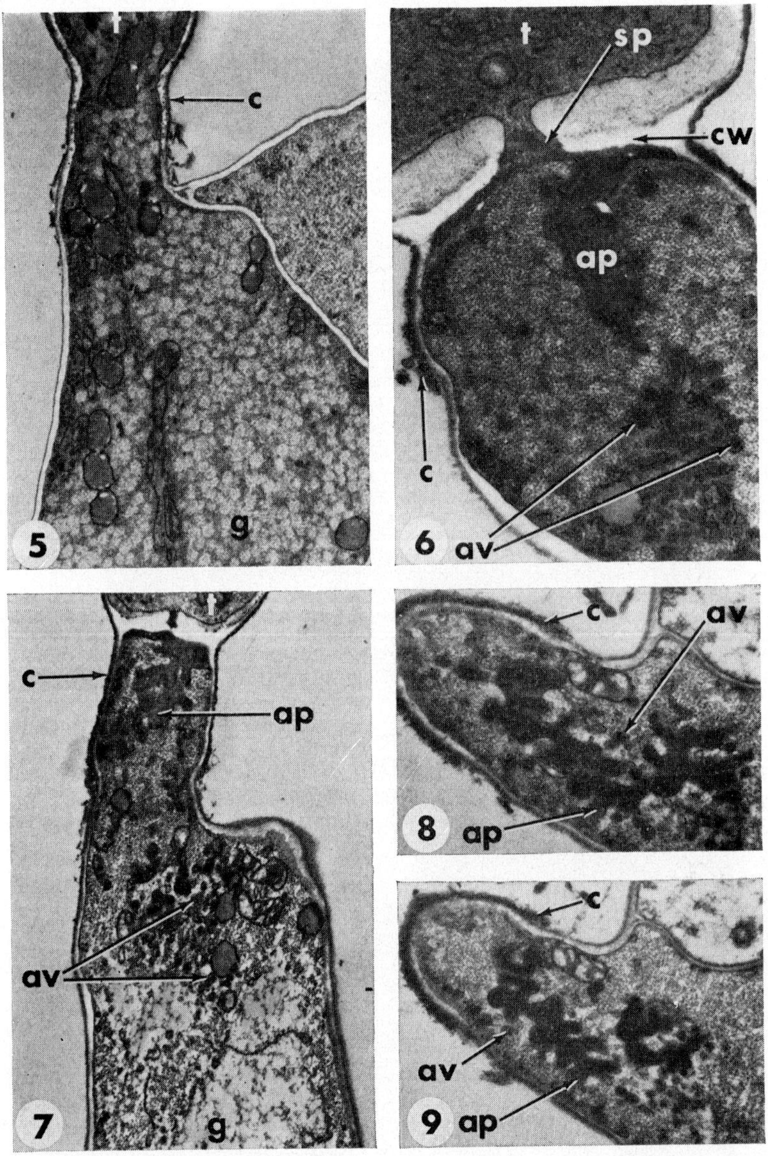

S. grandispora (electron micrographs)

FIG. 5. Longitudinal section of the base of a young trichospore (t) and subtending generative cell (g). Note the thickened wall of the collar (c) and absence of a septum delimiting the spore (×10 400).

FIGS 6, 7. Longitudinal section of a collar (c) with perforate crosswall (cw) and septal plug (sp), deposited electron dense appendage precursor (ap) and appendage vesicles (av). Fig. 6. early in appendage formation (×16 000); Fig. 7. later in appendage formation (×10 400).

FIGS 8, 9. Skipped, serial, longitudinal sections of a collar (c) showing the appendage precursor material (ap) and associated appendage vesicles (av). Areas which appear isolate in single sections are shown to be continuous (×6800).

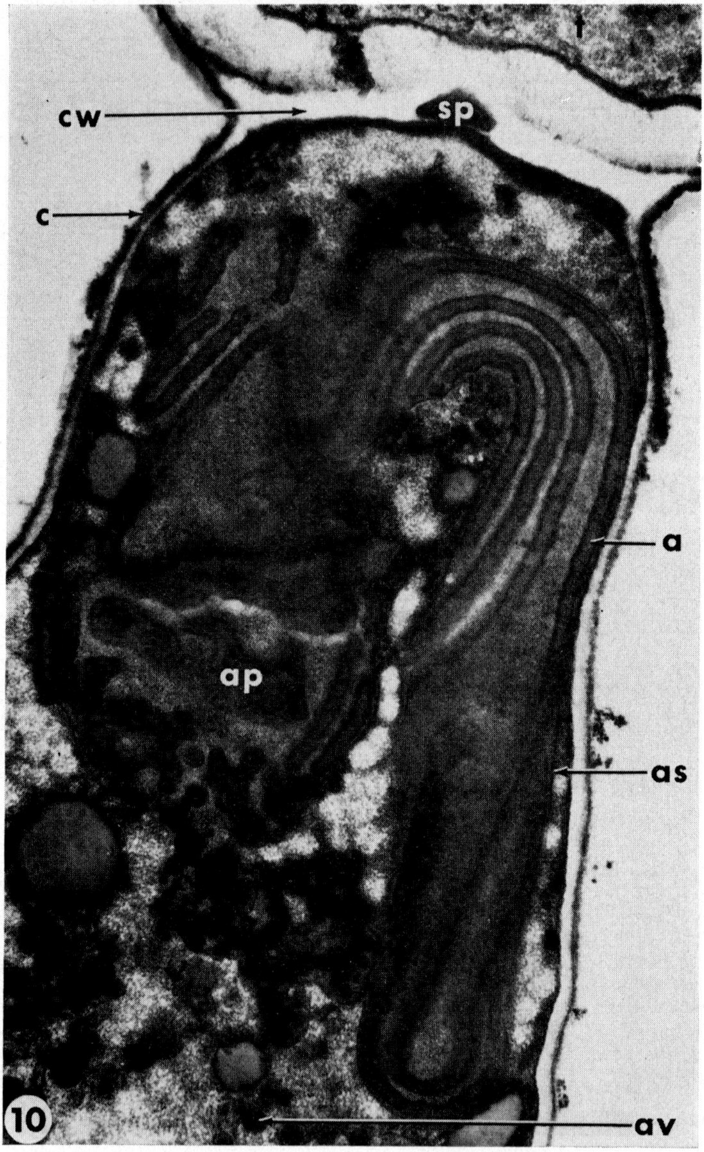

FIG. 10. *S. grandispora* (electron micrograph) Longitudinal section of a collar late in appendage formation showing the forming appendage (a) continuous with the appendage precursor material (ap) deposited by the appendage vesicles (av) ($\times$26 400).

are characterized by the presence of membrane bound vesicles, 100–300 nm in diameter, within the distal region of the generative cell. Each vesicle, termed the appendage vesicle, contains an electron dense core and a less dense peripheral layer. Concurrent with the appearance of appendage vesicles is the development within the collar of an electron dense deposit between the crosswall delimiting the trichospore and an invagination of the plasmalemma (Figs 6 and 7). This deposit contains material identical in electron density to that within the appendage vesicles. Single sections of the generative cell frequently show isolated areas of membrane bound electron dense material. In serial sections these are shown to be continuous with each other (Figs 8 and 9) and also with the deposited material within the invaginated region of the plasmalemma. Many appendage vesicles are continuous with the invaginated plasmalemma and their contents contiguous with zones of similar density contained by the plasmalemma.

At a more advanced stage the deposited electron-dense material is continuous with the appendage (Fig. 10). Differentiation into the appendage is confined to the deposited material closest to the trichospore septum. The formed appendage is enclosed within an elongated second invagination of the plasmalemma, the appendage sac. In longitudinal sections the

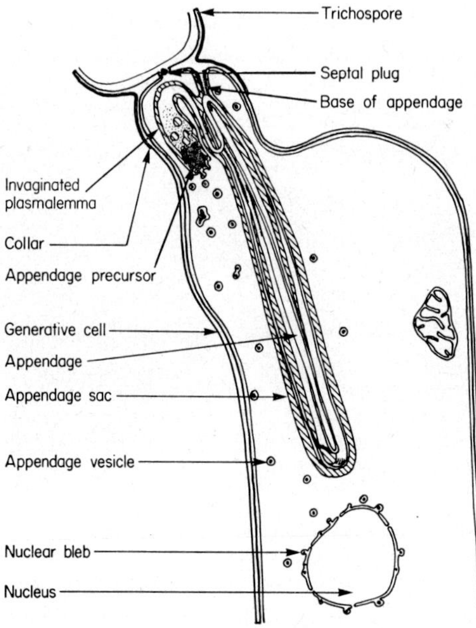

DIAGRAM 1. *S. grandispora:* Interpretation of the sequences in appendage formation.

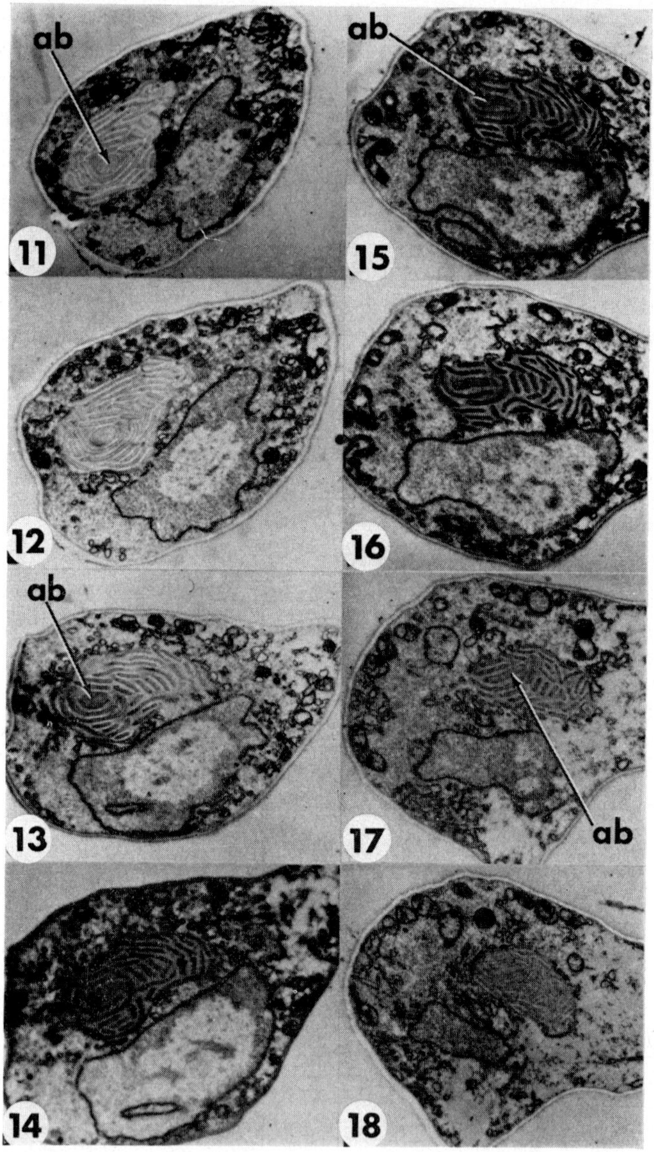

S. grandispora (electron micrographs)

FIGS 11–25. Skipped serial transverse sections of a generative cell and mature, folded appendage within the appendage sac. The eccentric, rod-shaped, basal limb of the appendage (ab, Figs 11–13) becomes central and ribbon-shaped towards the base of the appendage sac (Figs 19–24) where it doubles back to form a parallel, ascending limb (Fig. 25), (Figs 11–22, ×8 400; Figs 23–25, ×22 800).

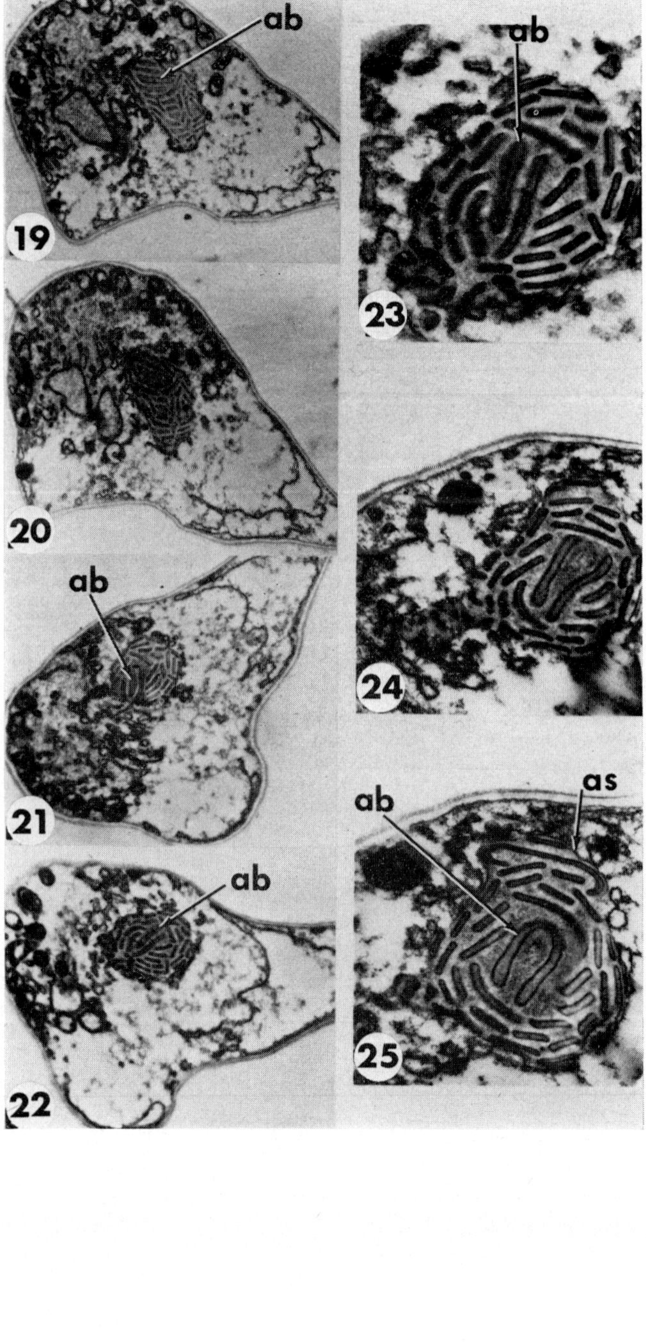

ab
19
20
ab
21
ab
22
ab
23
24
ab
as
25

forming appendage appears as an electron-dense filament folded upon itself several times to form a series of ascending and descending limbs embedded within a less dense matrix. The appendage appears to form from the electron dense core of the deposited material and the matrix from the less dense peripheral layer. At this stage in development the coalesced appendage vesicles, deposited appendage precursor material and formed appendage form an ontogenetic continuum (cf. Fig. 10 with Diagram 1).

During appendage formation vesicles of identical size and electron density to appendage vesicles are associated and continuous with the nuclear membrane of the single generative cell nucleus. Dilated regions of the perinuclear space contain many electron dense deposits similar to those within the appendage vesicles. The occasional presence of appendage vesicles in the nucleoplasm supports the contention that the perinuclear space and nuclear membrane are sites of vesicle formation (Fig. 26).

Serial sections of generative cells containing mature appendages facilitated the construction of a three dimensional model of the unreleased appendage (Figs 11–25; Diagram 2). Longitudinal sections proved unsatisfactory for this purpose owing to the relatively few limbs of the appendage obtained in any one section, the obliqueness of those that were

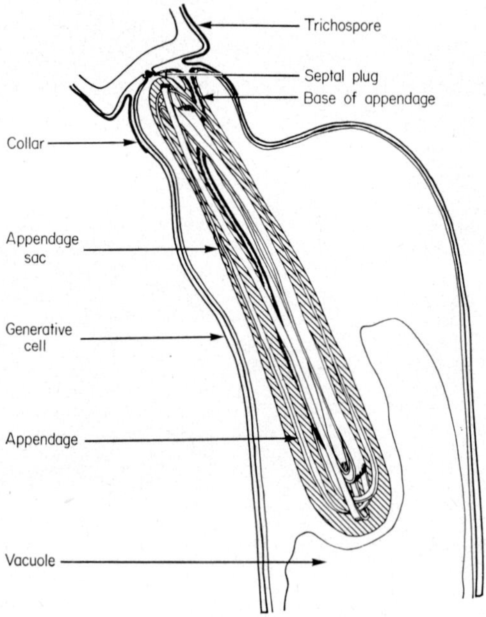

DIAGRAM 2. *S. grandispora:* The mature appendage within the appendage sac.

obtained and the frequent curvature of the appendage sac. Serial transverse sections of the generative cell allowed the entire appendage to be followed. In addition, longitudinal sections through the collar of the unreleased (Fig. 28) and released (Fig. 29) trichospores enabled the region of appendage attached to the trichospore to be identified and correlated with similar configurations of the appendage present in transverse sections of the appendage sac. Sections of the released trichospore show the basal part of the appendage to be 0·3–0–4 μm wide, composed of a homogeneous amorphous material and contiguous with the trichospore at a point on the septum periphery where the collar and crosswall join (Fig. 29). This corresponds with light microscope observations of the released trichospore and appendage (Figs 3, 4). Longitudinal sections of the collar before trichospore release contain many transverse sections of the appendage, 400–600 nm wide and 30–70 nm thick, retained within the appendage sac. Lateral to the appendage bundle but within the appendage sac is a section of appendage, distinguished by its circular appearance, identical in structure and dimensions to that contiguous with the trichospore (Fig. 28). In serial sections (Figs 11–18; Figs 19–25) this apparently rod-shaped initial descending limb of the appendage may be followed towards the base of the appendage sac where it doubles back upon itself to form an ascending ribbon-shaped limb (Figs 21–25 and 27). A similar folding of the appendage occurs in the collar. The skipped serial sections indicate that the mature appendage is arranged within the appendage sac as a number of longitudinally orientated ascending and descending limbs, resembling in arrangement a hank of rope. It may be that the rod-shaped limb of the appendage provides a rigid axis around which the remaining length of appendage is organized.

The trichospore is released from the generative cell below the thickened wall of the collar and the folded appendage is withdrawn from the generative cell. The stage at which the appendage sac disappears was not determined but observations indicate that the appendage unfolds immediately after withdrawal from the generative cell.

Discussion

The single appendage in *S. grandispora* if formed extra cellularly from products of intracellular metabolism transported to the site of appendage formation in membrane bound vesicles. The fine structural observations may be integrated into the following sequence of events to explain appendage formation, summarized in Diagrams 1 and 2.

1 Appendage precursor material either present or synthesized within the perinuclear space is incorporated into appendage vesicles blebbed-off the nuclear membrane.

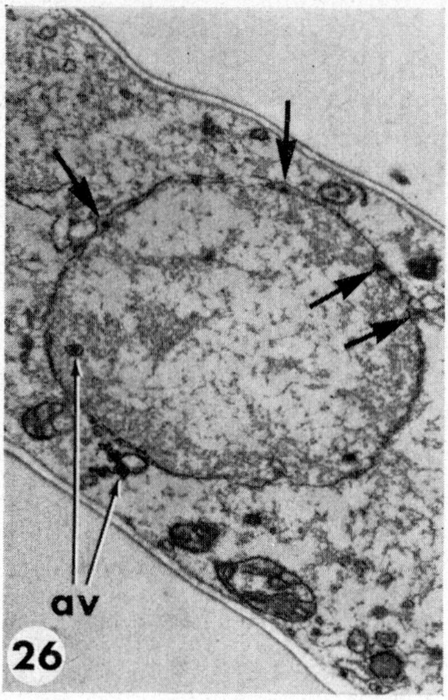

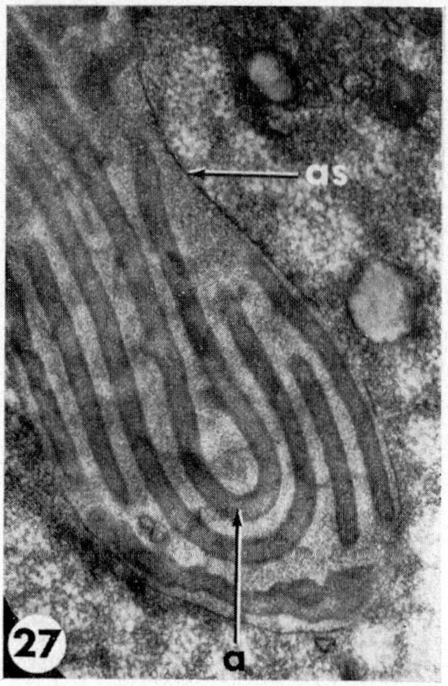

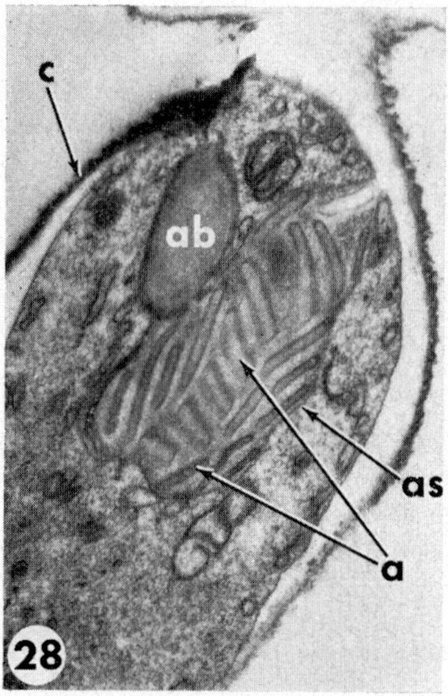

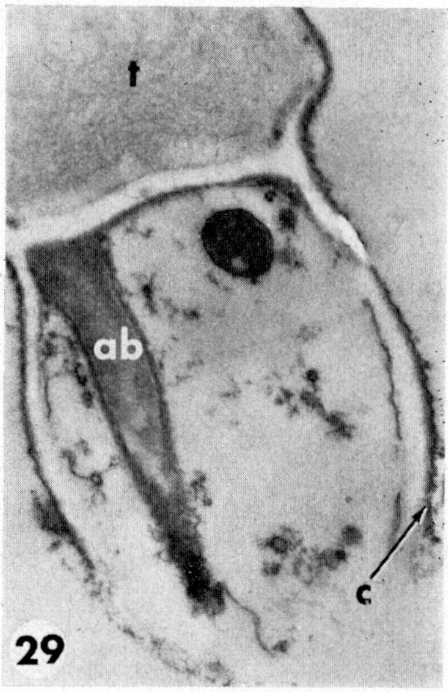

S. grandispora (electron micrographs)

FIG. 26. Generative cell nucleus with dilated regions of the perinuclear space (arrows) each containing an electron dense deposit. Note appendage vesicles (av) within the nucleoplasm and surrounding cytoplasm ($\times$16 000).

FIG. 27. Longitudinal section through the base of an appendage sac (as) showing the folded regions of the appendage (a) ($\times$35 200).

FIG. 28. Longitudinal section of a collar (c) and contained appendage sac (as). The limbs of the appendage (a) are sectioned transversely and/or obliquely owing to the folding of the appendage in this region. Note the eccentric position of the rod-shaped basal limb (ab) of the appendage ($\times$23 200).

FIG. 29. Longitudinal section of the collar (c) of a released trichospore (t) showing appendage (ab) attachment ($\times$23 200).

2 Appendage vesicles migrate to the trichospore septum where they coalesce with the plasmalemma and deposit appendage precursor material on to the crosswall.
3 Reorganization of the appendage precursor produces a single, continuous appendage contiguous with the trichospore septum and retained within an invagination of the plasmalemma, the appendage sac.

This hypothesis proposes that the perinuclear space is a precursor pool. The products of synthesis being transferred in vesicles to a specific site on the plasmalemma. A similar sequence of events occurs in some Algae and zoosporic fungi (Mastigomycotina) where mastigonemes, produced in the perinuclear space and/or endoplasmic reticulum, are transported in vesicles to the cell surface (Bland and Amerson, 1973; Bouck, 1969; Dodge and Crawford, 1971; Heath, Greenwood and Griffiths, 1970; Leadbeater, 1971; Leedale, Leadbeater and Massalski, 1970). Other authors have shown that in several members of the Mastigomycotina vesicles derived from the perinuclear space and outer nuclear membrane transport material to the proximal cisternae of Golgi dictyosomes (Bracker *et al.*, 1971; Grove, Bracker and Morré, 1970). Secretory vesicles formed by the distal cisternae of these dictyosomes may then migrate to the hyphal apex where they coalesce with the plasmalemma and deposit material for cell wall formation (Bartnicki-Garcia, 1973; Brown *et al.*, 1970; Grove and Bracker, 1970; Grove, Bracker and Morré, 1970; Heath, Gay and Greenwood, 1971; McClure, Park and Robinson, 1968). It is suggested that in *S. grandispora* where no Golgi apparatus has been found, appendage precursor material produced in the perinuclear space is transported directly to the cell surface without further elaboration and membrane transformation by a Golgi apparatus. The biochemical nature of the appendage precursor and mature appendage have not been characterized in *S. grandispora* or in any other member of the Harpellales.

The fine structure of trichospore appendages has been previously described for three species of the Harpellales. Reichle and Lichtwardt (1972) showed the four trichospore appendages in *Harpella melusinae* Léger and Duboscq to form outside the plasmalemma adjacent to the longitudinal wall of the generative cell and to have a periodic structure of alternating light and dark bands. Manier (1973*a*) described a similar disposition for the two appendages in *Genistella ramosa* Léger and Gauthier but in their substructure they show closer similarity to the appendages in *S. grandispora*. Further variation in appendage substructure exists in *Smittium mucronatum* Manier and Mathiez (Manier and Coste-Mathiez, 1968) and *Smittium* sp. (unpublished data) where the single appendage is comprised, in transverse section, of five to seven electron-dense, concentric layers. There is also diversity between genera in the stage of trichospore maturation at which the appendages form. In *G. ramosa* the appendages form prior to trichospore initiation whereas in *S. grandispora* and possibly also in *H. melusinae* they develop late in spore maturation. However, the reported close spatial association between vesicles in the generative cells and regions of appendage formation in *H. melusinae* and *G. ramosa* suggests that a similar sequence of events to that reported in *S. grandispora* may occur in other Harpellales.

Clearly the trichospore appendages in all those Harpellales studied show close similarity in that they develop extracellularly from products of intracellular metabolism and are ornamentations of the trichospore wall. It is considered that the spore appendages in the Harpellales aid in the subsequent ingestion of trichospores by feeding hosts. Similarly three aquatic genera of the Eccrinales (Trichomycetes) possess appendaged spores. However in the Eccrinales the appendages are extensions of the sporangiospore wall and not ornamentations of the sporangium wall as in the Harpellales. The presence of appendaged spores in these two morphologically dissimilar orders of the Trichomycetes may be due to convergence rather than phylogenetic affinity.

This work formed part of a thesis for the degree of Ph.D. in the University of Reading.

References

BARTNICKI-GARCIA, S. (1973). Fundamental aspects of hyphal morphogenesis. In *Microbial Differentiation,* 23rd Symposium of the Society for General Microbiology. (Ashworth, J. M. & Smith, J. E., eds) London: Cambridge University Press.

BLAND, C. E. & AMERSON, H. V. (1973). Electron microscopy of zoosporogenesis in the marine phycomycete *Lagenidium callinectes* Couch. *Arch. Mikrobiol.*, **94**, 47.

BOUCK, G. B. (1969). The origin, structure and attachment of flagellar hairs in *Fucus* and *Ascophyllum* antherozoids. *J. Cell Biol.*, **40,** 446.

BRACKER, C. E., GROVE, S. N., HEINTZ, C. E. & MOORE, D. J. (1971). Continuity between endomembrane components in hyphae of *Pythium* spp. *Cytobioll.* **4**, 1.

BROWN, R. M., FRANKE, W. W., KLEINIG, H., FALK, H. & SITTE, P. (1970). Scale formation in chrysophycean algae. I. Cellulosic and noncellulosic wall components made by the Golgi apparatus, *J. Cell Biol.*, **45**, 246.

DODGE, J. D. & CRAWFORD, R. M. (1971). Fine structure of the dinoflagellate *Oxyrrhis marina*. II. The flagellar system. *Protistologica,* **7,** 399.

DUBOSCQ, O., LÉGER, L. & TUZET, O. (1948). Contribution à la connaissance des Eccrinides: les Trichomycètes. *Archs Zool. exp. gén.,* **86,** 29.

FARR, D. F. & LICHTWARDT, R. W. (1967). Some cultural and ultrastructural aspects of *Smittium culisetae* (Trichomycetes) from mosquito larvae. *Mycologia,* 59, 172.

Grove, S. N. & Bracker, C. E. (1970). Protoplasmic organization of hyphal tips among fungi: vesicles and Spitzenkörper. *J. Bact.*, **104,** 989.

Grove, S. N., Bracker, C. E. & Morré, D. J. (1970). An ultrastructural basis for hyphal tip growth in *Pythium ultimum*. *Amer. J. Bot.*, **57,** 245.

Heath, I. B., Gay, J. L. & Greenwood, A. D. (1971). Cell wall formation in the Saprolegniales: cytoplasmic vesicles underlying developing walls. *J. gen. Microbiol.*, **65,** 225.

Heath, I. B., Greenwood, A. D. & Griffiths, H. B. (1970). The origin of flimmer in *Saprolegnia, Dictyuchus, Synura,* and *Cryptomonas*. *J. Cell Sci.*, **7,** 445.

Leadbeater, B. S. C. (1971). The intracellular origin of flagellar hairs in the dinoflagellate *Woloszynskia micra* Leadbeater and Dodge, *J. Cell Sci.*, **9,** 443.

Leedale, G. F., Leadbeater, B. S. C. & Massalski, A. (1970). The intracellular origin of flagellar hairs in the Chrysophyceae and Xanthophyceae. *J. Cell Sci.*, **6,** 701.

Lichtwardt, R. W. (1967). Zygospores and spore appendages of *Harpella* (Trichomycetes) from larvae of Simuliidae. *Mycologia*, **59,** 482.

Lichtwardt, R. W. (1972). Undescribed genera and species of Harpellales (Trichomycetes) from the guts of aquatic insects. *Mycologia*, **64,** 167.

Lichtwardt, R. W. (1973a). Trichomycetes. In *The Fungi*, Vol. 4B. (Ainsworth, G. C. Sparrow, F. K. & Sussmann, A. S., eds) New York: Academic Press.

Lichtwardt, R. W. (1973*b*). The trichomycetes: what are their relationships? *Mycologia*, **65,** 1.

McClure, W. K., Park, D. & Robinson, P. M. (1968). Apical organization in the somatic hyphae of fungi. *J. gen. Microbiol.*, **50,** 177.

Manier, J.-F. (1973*a*). L'ultrastructure de la trichospore de *Genistella ramosa* Léger et Gauthier, Trichomycète Harpellale parasite du rectum des larves de *Baetis rhodani* Pict. *C. r. hebd. Séanc. Acad. Sci., Paris*, **276,** 2159.

Manier, J.-F. (1973*b*). Quelques aspects ultrastructuraux du Trichomycète Asellariale, *Asellaria ligiae* Tuzet et Manier, 1950 ex Manier, 1968. *C. r. hebd. Séanc. Acad. Sci., Paris*, **276,** 3429.

Manier, J.-F. & Coste-Mathiez, F. (1968). L'ultrastructure du filament de la spore de *Smittium mucronatum* Manier, Mathiez 1965 (Trichomycète, Harpellale) *C. r. hebd. Séanc. Acad. Sci., Paris*, **266,** 341.

Manier, J.-F. & Grizel, H. (1972). L'ultrastructure de l'enveloppe et du "pavillon" des Trichomycètes Eccrinales. *C. r. hebd. Séanc. Acad. Sci., Paris*, **274,** 1159.

Manier, J.-F. & Lichtwardt, R. W. (1968). Révision de la systématique des Trichomycètes. *Annls Sci. nat.* (*Bot.*), *Sér.* **9,** 519.

Moss, S. T. (1970). Trichomycetes inhabiting the digestive tract of *Simulium equinum* larvae. *Trans. Br. mycol. Soc.*, **54,** 1.

Moss, S. T. (1972). Occurrence, cell structure and taxonomy of the Trichomycetes, with special reference to electron microscope studies of *Stachylina*. Ph.D. Thesis, University of Reading.

Moss, S. T. (1974). A note on the nuclear cytology of *Stachylina grandispora* (Trichomycetes, Harpellales). *Mycologia*, **66,** 173.

Moss, S. T. (1975). Septal structure in the Trichomycetes with special reference to *Astreptonema gammari* (Eccrinales). *Trans. Br. mycol. Soc.*, **65,** 115.

Reichle, R. E. & Lichtwardt, R. W. (1972). Fine structure of the Trichomycete, *Harpella melusinae*, from black-fly guts. *Arch. Mikrobiol.*, **81,** 103.

Reynolds. E. S. (1963). The use of lead citrate at high pH as an electron-opaque stain in electron microscopy. *J. Cell Biol.*, **17**, 208.

Sangar, V. K. & Dugan, P. R. (1973). Chemical composition of the cell wall of *Smittium culisetae* (Trichomycetes). *Mycologia*, **65**, 421.

Sangar, V. K., Lichtwardt, R. W., Kirsch, J. A. W. & Lester, R. N. (1972). Immunological studies on the fungal genus *Smittium* (Trichomycetes). *Mycologia*, **64**, 342.

Trotter, M. J. & Whisler, H. C. (1965). Chemical composition of the cell wall of *Amoebidium parasiticum*. *Can. J. Bot.*, **43**, 869.

Tuzet, O. & Manier, J.-F. (1967). *Enterobryus oxidi* Lichtwardt, Trichomycète Eccrinale parasite du Myriapode Diplopode *Oxidus gracilis* (Koch) (cycle, ultrastructure). *Protistologica*, **3**, 413.

Whisler, H. C. (1963). Observations on some new and unusual enterophilous Phycomycetes. *Can. J. Bot.*, **41**, 887.

Whisler, H. C. & Fuller, M. S. (1968). Preliminary observations on the holdfast of *Amoebidium parasiticum*. *Mycologia*, **60**, 1068.

Ultrastructural Characteristics of Dinoflagellates – The Red-Tide Algae

JOHN D. DODGE

Department of Botany, Birkbeck College, Malet Street, London, England

Dinoflagellates are present in most lakes and ponds and constitute the second most important component of the marine phytoplankton. At times, when conditions have been suitable, massive growth of these organisms gives rise to red-tides which may directly or indirectly cause the death of other aquatic organisms. Some dinoflagellates produce toxins which, because of their concentration in the tissues of filter-feeding molluscs, can prove dangerous to man. Thus there is a need to be able to distinguish dinoflagellates from other aquatic organisms such as protozoans and algae. By light microscopy it is possible in many cases to do this because of the typical form of the cell but there are dinoflagellates which do not have such obviously distinctive features. Using the electron microscope, work over the past 15 years (summarized in Dodge, 1971*b*) has shown a number of unique features which could be used to identify a member of this group in, say, the tissues of an animal, as a symbiont in an invertebrate, as a component of an embedded plankton sample or as detritus on a filter. The present paper will concentrate on these distinctive features.

Methods

In this work all the standard transmission electron microscope methods have been used including metal shadowing for the flagella and trichocysts, replication for the flagella and theca, positive staining for the trichocysts and embedding and sectioning for all structures. In addition, scanning electron microscopy has been used in studies of the theca.

Observations

General structure of the cell

Each cell normally bears two flagella which are of two distinct types. In many cases these are inserted into the cell at the base of the flagellar canals situated in roughly the midpoint of the ventral side. In some genera they are inserted in the anterior end. The cell is covered by a complex covering termed the theca or amphiesma which is perforated to allow the discharge of trichocysts. Internally (see Fig. 1) the cell contains a single nucleus with prominent chromosomes, and one or more pyrenoids which may be of various types (see Dodge and Crawford, 1971). There may be one or many chloroplasts although some dinoflagellates lack them completely. The remaining space in the cell is occupied by the pusules which are associated with the flagellar canals, by mitochondria, Golgi bodies, micro-bodies, endoplasmic reticulum, starch grains, lipid droplets and possibly by vacuoles.

The flagella

One of the flagella (the longitudinal or posterior flagellum) is more or less normal in structure consisting of a 9+2 axoneme and possibly bearing fine hairs on its tip region. The second, or transverse, flagellum is quite distinctive (Fig. 2). It consists of a 9+2 axoneme which is drawn into the form of a helix by the tension on a proteinaceous thread termed the striated strand which runs from base to tip of the flagellum. Both axoneme and striated strand are bounded by a common membrane sheath which also bears a unilateral array of long fine hairs (Leadbeater and Dodge, 1967). This flagellum beats with a helical wave. No other algae have been found with flagella quite of this type although one flagellum of *Pedinella* (Chrysophyceae) (Swale, 1969) is similar. In general the arrangement of a helical structure in apposition to a straight one is analogous to the construction of the spirochaete bacteria where the body forms the straight structure and the fibrous bundle is in the form of a helix.

The pusule

Most dinoflagellates possess an internal vesiculate structure which discharges to the exterior by way of the flagellar canals. These pusules consist of invaginated plasmalemma which becomes associated in special vesicles or sacks with vacuolar membrane to give a double-membrane wall (Figs 3–4). The pusule is thus surrounded by cell vacuole. The arrangement of the vesicles, collecting chambers and discharge tubules is very varied (see Dodge, 1972) and is generally most complex in freshwater dinoflagellates. The function of the pusule is probably that of osmo-regulation.

Trichocysts

Most dinoflagellates possess ejectile organelles, termed trichocysts, which after stimulation are shot out of the cell through pores in the theca. The

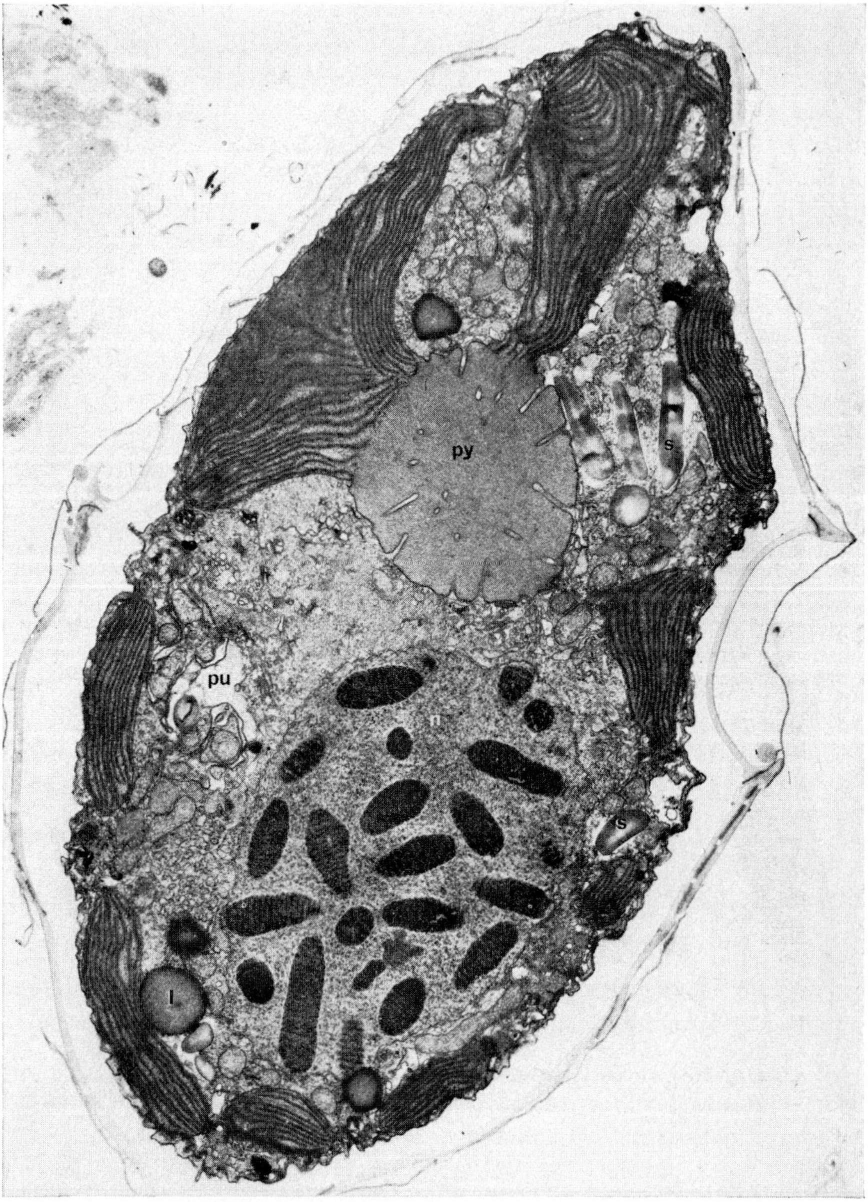

FIG. 1. A longitudinal section of the dinoflagellate *Heterocapsa triquetra*. The theca consisting of plates and membranes has partly become detached from the cell. The nucleus (n) is situated in the lower half and a large pyrenoid (py) is supported by several branches of the peripheral chloroplast reticulum. Part of a pusule (pu) is seen, as are starch grains (s) and lipid globules (1) ($\times$7000).

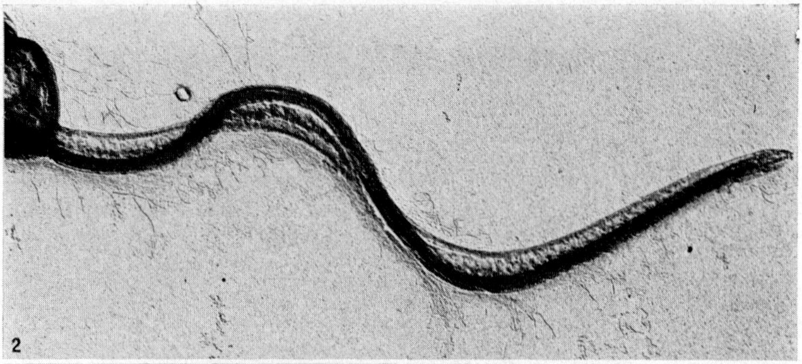

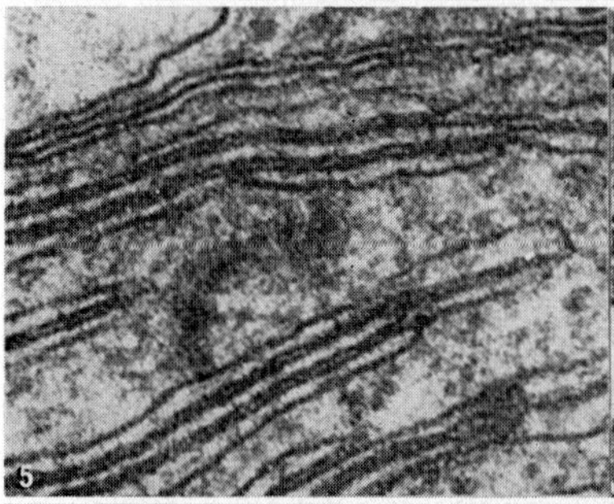

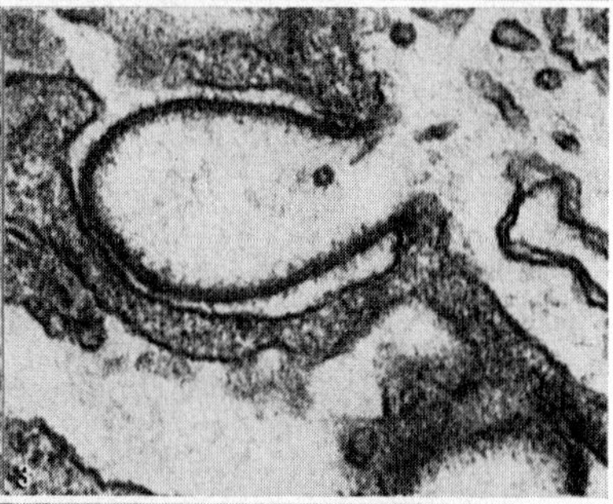

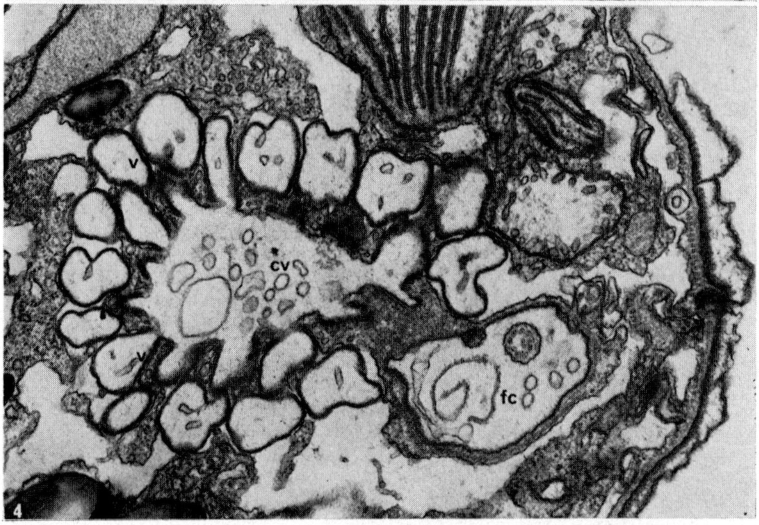

FIG. 2. A shadowed preparation to show the transverse flagellum with the helical axoneme, the straighter striated strand, the flagellar sheath and the fine hairs (×6480).

FIG. 3. A small portion of the pusule of *Gymnodinium nelsonii* showing the vesicle lined by plasma-membrane and sheathed by a jacket-like extension of the cell vacuole (×51 300) (After Dodge, 1972).

FIG. 4. A section of the pusule of *Amphidinium herdmanii* showing numerous vesicles surrounding the collecting vacuole (cv) and to the right the flagellar canal (fc). (×18 900).

FIG. 5. A small part of the chloroplast of *Aureodinium* to show the triple-membrane chloroplast envelope and the thylakoids which are generally stacked in threes (×94 500).

discharged proteinaceous threads can readily be observed in whole mounts which have been shadowed (Fig. 6) or stained with a heavy metal salt (Fig. 7), and they have a regular periodic cross-banding. The repeat distance appears to vary with the species and also with the method of preparation. The mature undischarged trichocyst (Fig. 8) consists of a narrow neck region which is attached to the theca and a wider body portion, the whole being surrounded by a single membrane. The trichocysts are more frequently observed in transverse and oblique sections when the body portion has a square or diamond shape (Fig. 9). In favourable sections it can be seen that the protein sub-units are regularly packed in the form of a cubic lattice. The processes involved in discharge are not understood but presumably they involve the hydration and swelling of the trichocyst body.

The cell covering or theca

The unique feature of the dinoflagellate theca is that it consists of a single entire outer membrane, which must be the functional plasma-membrane, beneath which is situated a single layer of flattened vesicles. The vesicles may appear quite empty as in the genera *Oxyrrhis* and *Amphidinium*, they may contain a very thin layer of material (Fig. 10) which would not seem to justify the term "plate", they may contain thin featureless plates or they might contain thick ornamented plates (Fig. 11) (See Dodge and Crawford, 1970*a*). In all cases there are normally microtubules present immediately beneath the theca. With light microscopy or with scanning electron microscopy (Fig. 12) the shape of the thecal plates and their varied ornamentation can be studied but these techniques give no clue to the fundamental structure of the theca and its relationship to the cell. For this sectioned material is necessary.

The nucleus

The nuclear envelope and nucleolus appear to be quite normal and show no unusual features except in a few species where the envelope is vesiculated. The chromosomes, however, are quite unique in appearance, composition and structure. In a normal section of an interphase nucleus the chromosomes can be observed as numerous rod-shaped bodies which have a banded structure (Figs 13 and 14). Tests with enzymes such as DNAase have shown that the chromosomes consist of DNA fibrils without any appreciable amount of histone protein. Many workers have attempted to describe the way in which the fibrils are arranged together to form the chromosome and the latest attempt (Haarpala and Soyer, 1973) suggests that the individual fibrils are in the form of circles with the chromosome consisting of a large

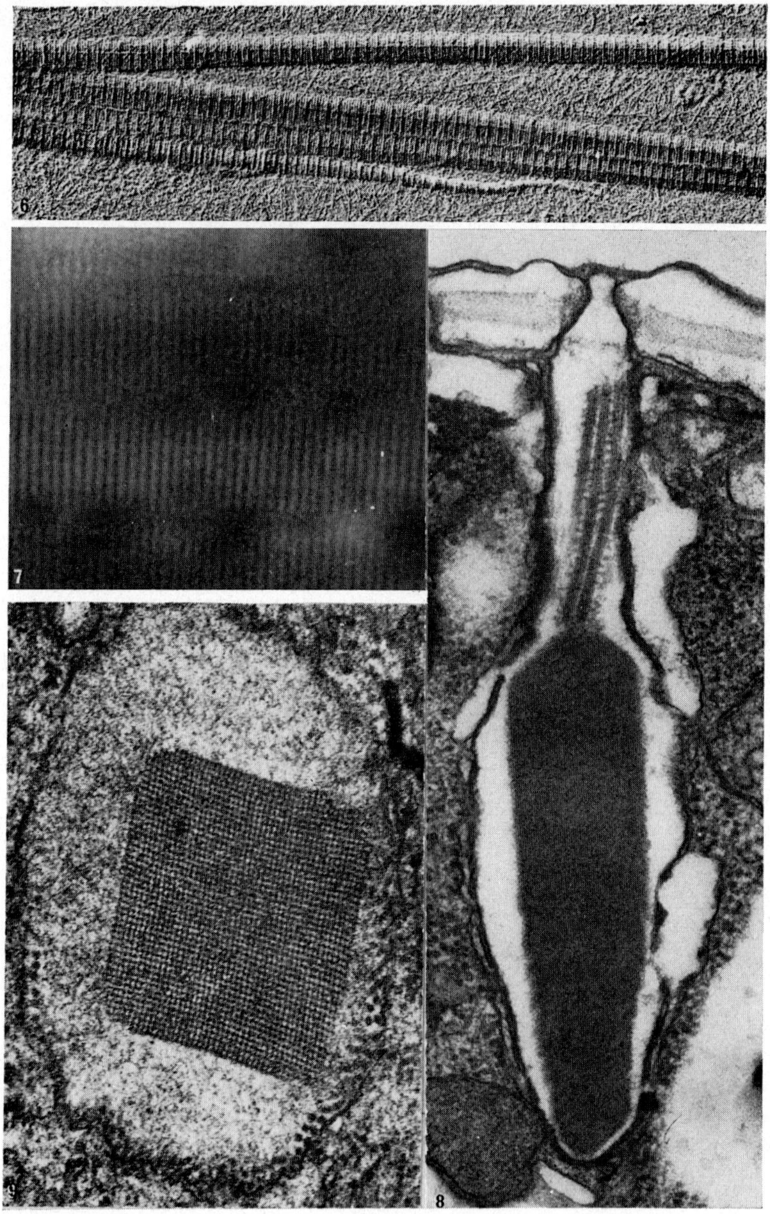

FIGS 6–9. The trichocysts. Figure 6 shows shadowed trichocysts (×18 900) and Fig. 7 those stained with PTA (×113 400). Figure 8 shows a longitudinal section of a mature trichocyst in *Woloszynskia coronata* where the neck, attached to the cell theca, and the body section, are quite distinct (×53 100). Figure 9 shows a cross-section of the body of an undischarged trichocyst of *Prorocentrum* in which the packing of 49×49 sub-units is clearly seen (×61 200). (After Dodge, 1973).

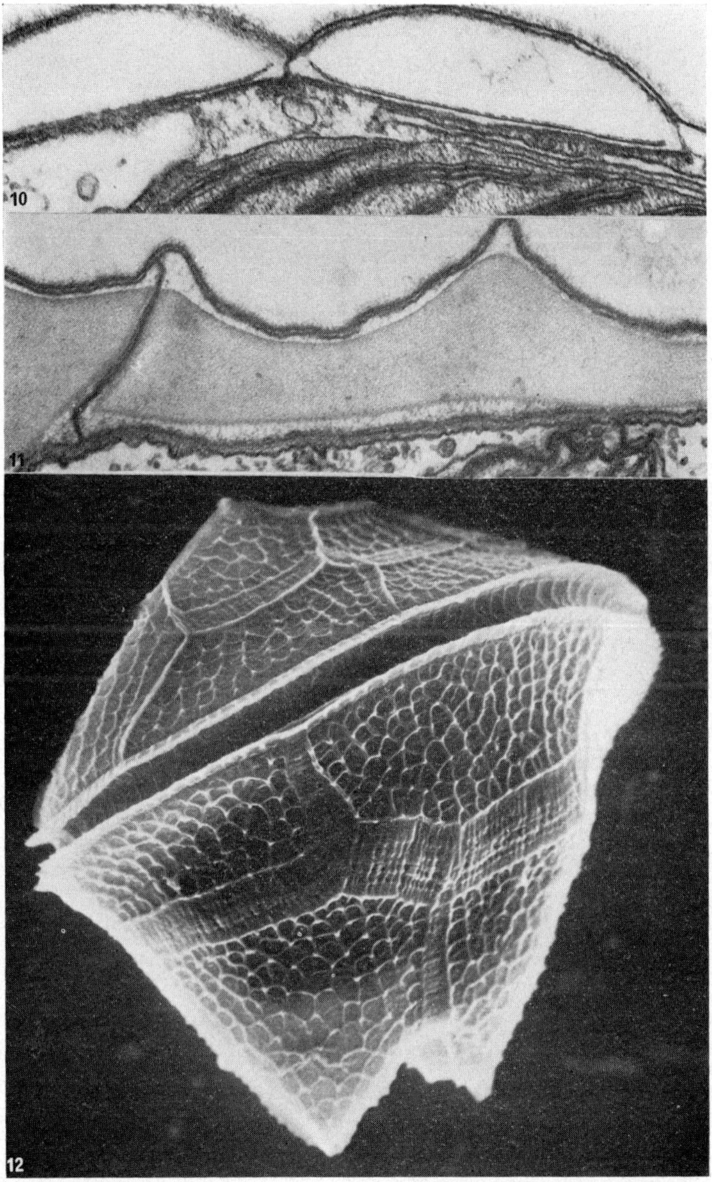

FIGS 10–12. The theca. Figure 10 shows a vertical section of the theca of *Gymnodinium simplex* where the thecal vesicles beneath the plasma-membrane contain very thin structures (×49 500). (After Dodge, 1974). Figure 11 shows a vertical section through the thecal membranes and thick ridged plate of *Ceratium hirundinella*. A suture or junction between two plates is seen at the left. (× 38 700). (After Dodge and Crawford, 1970*b*). Figure 12 is a scanning electron micrograph showing the dorsal view of *Peridinium leonis*. Note the prominent transverse groove or girdle in which the transverse flagellum lies, the junctions between the thecal plates, and the reticulations or ridges over their surfaces (×1350) (After Dodge, 1973).

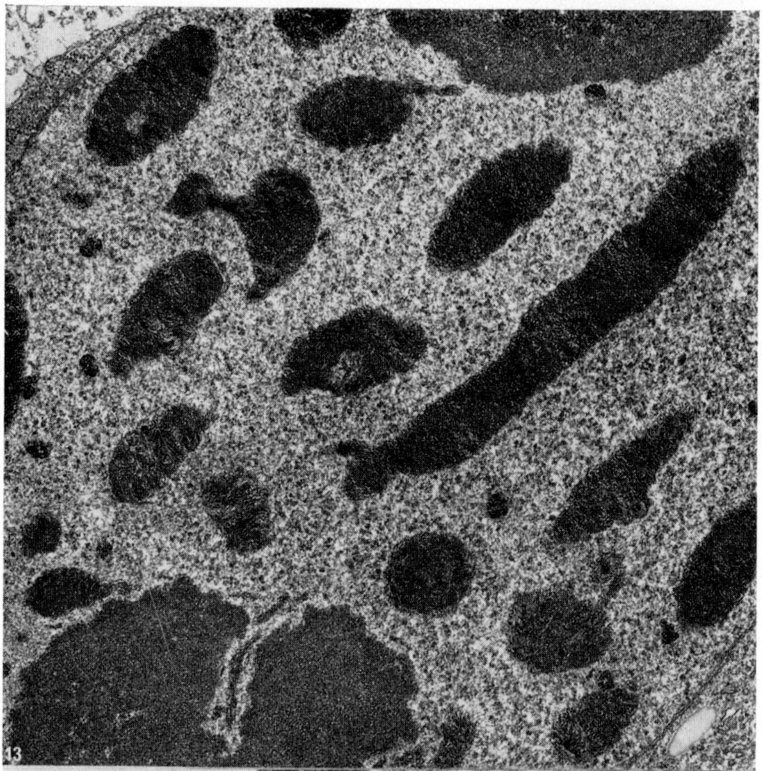

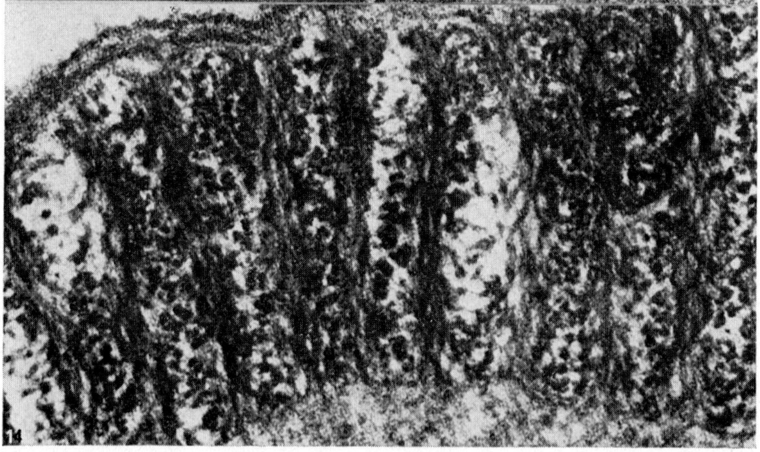

FIGS 13 and 14. The nucleus. Figure 13 shows a section through the nucleus of *Glenodinium foliaceum* with parts of two nucleoli, at top and bottom, and chromosomes sectioned in various planes (×18 900) (After Dodge, 1971*a*). Figure 14 shows at high magnification a small portion of a longitudinal section of a chromosome of *Amphidinium herdmanii*. Note that some of the DNA fibrils are sectioned transversely and others longitudinally (×113 400).

number of the circular fibres which are twisted together much as in a skein of wool. The absence of histone and the circular form of the DNA fibrils would suggest a close similarity of structure with that in the prokaryotic bacteria. This was reinforced by earlier suggestions that the mitosis was mediated by the nuclear envelope rather than by a spindle. Recent studies (Oakley and Dodge, 1974) have shown that the microtubules which penetrate through channels in the dividing nucleus have connections with the chromosomes across the nuclear envelope. Thus, the mitosis in using microtubule traction has more similarity with the general principles of mitosis in eukaryotic organisms than seemed possible.

The chloroplasts

In many respects the chloroplasts of dinoflagellates are very similar to those of many other classes of algae for they contain lamellae which each generally consist of three appressed thylakoids (Fig. 5). This type of structure can be found, for example, in the Chrysophyceae, Haptophyceae, Xanthophyceae, Bacillariophyceae and Euglenophyceae. In the dinoflagellates the unusual feature is the presence of a chloroplast envelope which consists of three membranes in place of the normal two (Fig. 5) (Dodge, 1968). This feature is shared only with the Euglenophyceae in which class the chloroplasts contain chlorophyll a and b whereas in the dinoflagellates chlorophyll a and c are found. So, once again, although the structures are similar there would seem to be no close relationship between these two groups.

Conclusion

The dinoflagellate algae (Dinophyceae or Pyrrophyta) or protozoa (Dinoflagellida) are a distinctive and unique group of organisms which can readily be distinguished from other micro-organisms on the basis of their ultrastructure, in particular the structure of the transverse flagellum, the trichocysts, the pusule, the nucleus and the chloroplast envelope.

References

DODGE, J. D. (1968). The fine structure of chloroplasts and pyrenoids in some marine dinoflagellates. *J. Cell Sci.* **3,** 41.

DODGE, J. D. (1971*a*). A dinoflagellate with both a mesocaryotic and a eucaryotic nucleus. I. Fine structure of the nuclei. *Protoplasma* **73,** 145.

DODGE, J. D. (1971*b*). Fine structure of the Pyrrophyta. *Bot. Rev.* **37,** 481.

DODGE, J. D. (1972). The ultrastructure of the dinoflagellate pusule: A unique osmo-regulatory organelle. *Protoplasma* **75,** 285.

DODGE J. D. (1973). *The Fine Structure of Algal Cells* London and New York. Academic Press

DODGE, J. D. (1974). A redescription of the dinoflagellate *Gymnodinium simplex* with the aid of electron microscopy. *J. mar. biol. Ass. U.K.* **54,** 171.

DODGE, J. D. & CRAWFORD, R. M. (1970*a*). A survey of thecal fine structure in the Dinophyceae. *Bot. J. Linn. Soc.* **63,** 53.

DODGE, J. D. & CRAWFORD, R. M. (1970*b*). The morphology and fine structure of *Ceratium hirundinella* (Dinophyceae). *J. Phycol.* **6**, 137.

DODGE, J. D. & CRAWFORD, R. M. (1971). A fine-structural survey of dinoflagellate pyrenoids and food-reserves. *Bot. J. Linn. Soc.* **64,** 105.

HAARPALA, O.K. & SOYER, M.O. (1973). Structure of dinoflagellate chromosomes. *Nature, New Biol.* **244,** 195.

LEADBEATER, B. & DODGE, J. D. (1967). An electron microscope study of dinoflagellate flagella. *J. gen. Microbiol.* **46,** 305.

OAKLEY, B. R. & DODGE, J. D. (1974). Kinetochores associated with the nuclear envelope in the mitosis of a dinoflagellate. *J. Cell Biol.* **64,** 322.

SWALE, E. M. F. (1969). A study of the nannoplankton flagellate *Pedinella hexacostata* by light and electron microscopy. *Br. phycol. J.* **4,** 65.

Fibrous Structures in Zooflagellate Protozoa

B. P. EYDEN

Zoology Department, Glasgow University, Glasgow, Scotland

Introduction

A feature distinguishing protozoa, and to a lesser extent micro-algae, from other eukaryotic protists is their possession of organelles called *fibres* which constitute an integral part of their interphase structure. Electron microscopy has shown that fibres are composed of elements which are below the resolution of the light microscope and which Pitelka (1969) has termed *fibrils*. Based on the nature of their fibrillar substructure fibres may be designated as *microtubular fibres*, made up of *microtubules*; i.e., long, hollow, cylindrical structures *c*. 25 nm in external diameter with a 5 nm thick wall consisting of 13 tubulin-dimer protofilaments (Sleigh, 1973) or *filamentous fibres*, composed of long, straight, nonhollow fibrils (varying in diameter from *c*. 5 nm to *c*. 20 nm) whose relationship to the more clearly defined classes of filaments in various metazoan cells (see, for example, Goldman *et al*., 1973) is not yet clear.

Pitelka (1969) defined a fibre as a structure visible in the light microscope but the term is employed here to mean any ordered assembly of fibrils whether or not this assembly is visible by light microscopy. Fibres take the form of sheets, ribbons or rods, possess regular fibril-alignments, and are thought to function in support and anisometric growth, undulatory movements and transport of membrane, vesicles and particles. These roles will be discussed in relation to the structure of fibres in zooflagellate protozoa, i.e., non-photosynthetic flagellates feeding by pinocytosis and/or phagocytosis. By possessing clearly defined, well-ordered and frequently large fibrous organelles, these organisms are favourable objects for studying motility, morphogenetic and cytoskeletal functions in terms of ordered assemblies of fibrils in non-metazoan systems. The emphasis in this account is on permanently organized interphase organelles, for although ordered microtubular assemblies form part of the mechanism for nuclear genome segregation in nearly all eukaryotes, these assemblies are transitory and do not usually persist in an organized state into interphase. The examples of fibres given here are taken from the following groups of zooflagellates; fibrous organelles are found in phytoflagellates (Pitelka, 1969) but they seldom approach the complexity of those of zooflagellates:

1 *Kinetoplastids*, which typically have a single mitochondrion containing a mass of DNA called the kinetoplast; the group includes insect-borne parasites of vertebrate blood and tissues, e.g., trypanosomes and leishmanias, but also free-living forms e.g., *Bodo* spp.
2 *Trichomonads*—these mainly endozoic commensals can produce pathological symptoms in vertebrates including Man, e.g., *Trichomonas vaginalis*
3 *Diplomonads* which are principally harmless gut commensals of amphibians and mammals (e.g., *Spironucleus* spp.), include free-living forms (e.g., *Trepomonas agilis*); *Giardia* is a frequent causative agent of human intestinal disorders
4 *Retortomonads*, which are harmless and widespread gut-commensals (e.g., *Chilomastix*)
5 *Oxymonads*—exclusively endocommensal, these typically possess undulating axostyles (e.g. *Monocercomonoides, Saccinobaculus*).

The structure of fibres in zooflagellates omitted from the present treatment (hypermastigids, opalinids) has been described by Grimstone and Gibbons (1966), Pitelka (1969) and Hollande and Carruette-Valentin (1970, 1971).

The Flagellum

The chief component of the flagellum is the *axoneme* which consists of 9 peripheral microtubule-doublets and 2 central single microtubules (see Warner, 1974). Most eukaryotic flagella including those of zooflagellates (Fig. 1, a–c) have this "9+2" configuration. The principal current theory of the mechanism of flagellar movement emphasizes the sliding of microtubules, mediated by the reversible interaction of the myosin-like *dynein arms* of the A-tubule with the B-tubule of the adjacent doublet (Fig 1a; Satir, 1974; Sleigh and Pitelka, 1974). Inter-doublet *nexin* arms, if they are physically distinct from the dynein arms, may help to maintain the geometric arrangement of axonemal microtubules but, if so, they must be labile linkages in order not to interfere with microtubule-sliding (Warner, 1974). Arms projecting from the microtubule cylinder are present in many microtubular fibres implicated in motility and cytoskeletal function (see below).

External ornamentation of the flagellar membrane which is usually lacking in eukaryotic flagella other than those of micro-organisms, may modify flagellar movement. The extent to which the inconspicuous

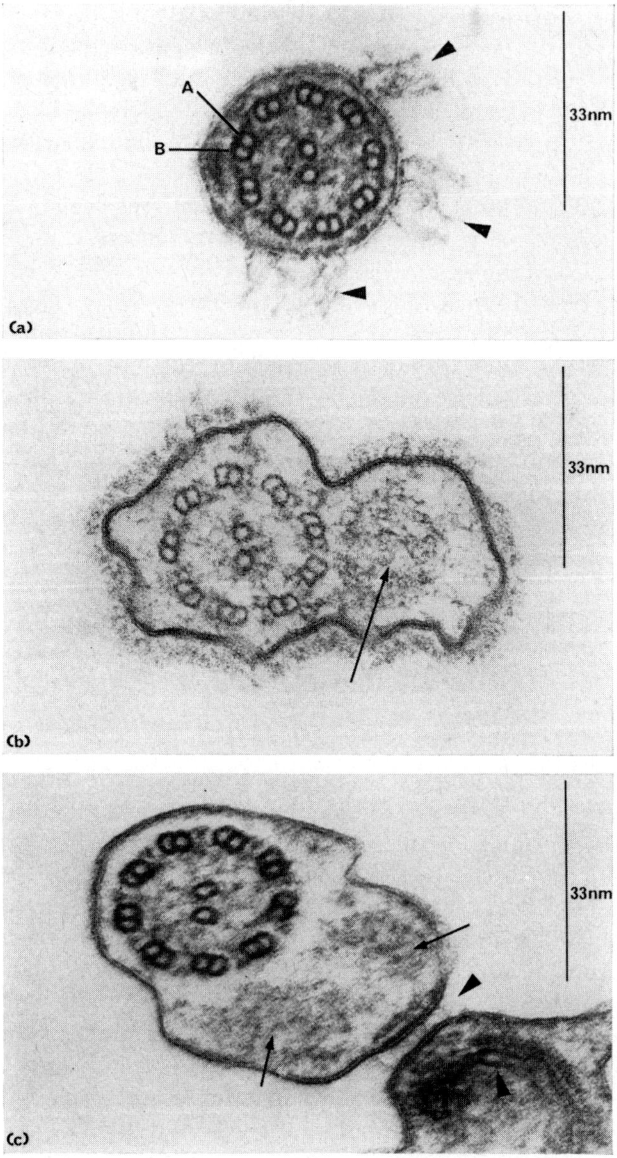

FIG. 1. Flagellar structure in: (a) *Trepomonas agilis*-arrowheads, tomentum; A, complete A-subtubule; B, incomplete B-subtubule; (b) *Bodo designis*-arrow, paraxial rod; (c) *Tritrichomonas* sp. from mouse caecum-arrows, paraxial rod; arrowheads, elements of junctional complex.

tomentum of *Trepomonas agilis* (Fig. 1a) might modify flagellar activity is uncertain, but the more organized filamentous mastigonemes appear to reverse the direction of fluid movement over the flagellum surface and so also of cell body movement in e.g. chrysomonads (Holwill, 1974). Whether this function applies to the mastigonemes of bodonid flagella (Hitchen, 1974) is not known.

In the flagella of kinetoplastids and in the recurrent flagellum of trichomonads the axoneme is accompanied by a *paraxial rod* (Fig. 1, b and c) characterized by regular fibril-alignments (Hollande and Valentin, 1969; Vickerman, 1969; Daniel, Mattern and Honigberg, 1971). The function of the rod is uncertain and even the 3-dimensional conformation of the constituent fibrils is unclear, but it is generally thought to strengthen the flagellum as a propulsive organelle and so oppose more effectively the viscous forces of the ambient fluid. The attachment of the flagellum to the body surface to form an undulating membrane, a feature not uncommon in zooflagellates, is mediated by zonular (Daniel, Mattern and Honigberg, 1971) or macular (Vickerman, 1974) desmosomes; both types of junctional complex include a filamentous component. More unusually, a microtubular fibre appears to mediate attachment as in *Monocercomonoides* (Brugerolle and Joyon, 1973).

Microtubular Fibres

Single row fibres

Fibres consisting of a single row of microtubules take the form of sheets or ribbons. Sheet-like fibres are commonly located beneath the plasmalemma and are most highly developed in kinetoplastids; e.g., *Trypanosoma* (Fig. 2, a and b), *Leishmania* (Vickerman, 1974), *Crithidia* (Brooker, 1972), *Bodo* (Brooker, 1971) and retortomonads e.g., *Chilomastix* (Brugerolle, 1973). In these flagellates almost the entire surface is reinforced by closely spaced anisotropically orientated microtubules pursuing a longitudinal-spiral course (Fig. 2b). Arm-like projections arising from the microtubules, sometimes making contact with adjacent tubules, are probably important in the functioning of these fibres. For example, the microtubular corselet of kinetoplastids probably maintains the elongate shape displayed especially by trypanosomes, and inter-tubule links, possibly of a nexin-like nature may add to the cohesive strength of these cytoskeletal elements. Alternatively, the surface deformations which these flagellates exhibit could be mediated by a parallel microtubule-sliding similar in mechanism to flagellar movement by means of inter-tubule links of a dynein-like nature.

In *Chilomastix* at least 2 morphologically different inter-microtubule inks or arms are present. In the large cytostome (Brugerolle, 1973) inter-

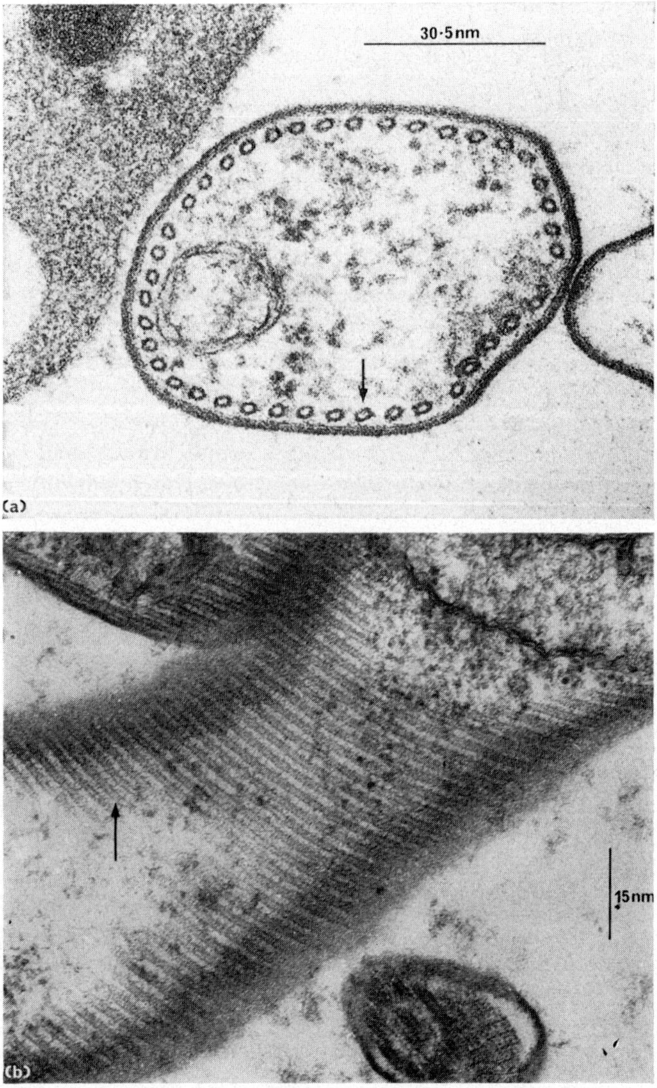

FIG. 2. Subplasmalemmar microtubular fibre in bloodstream *Trypanosoma brucei*: (a) cross-section; (b) tangential section. Arrow, microtubule.

tubule links are sigmoid (Figs 4a, 7a) while inter-tubule links over the rest of the body are linear (Fig. 7b), similar morphologically and possibly functionally, to those of trypanosomes. The former, in making consistent inter-tubule contact, may provide a greater degree of cohesion-promoting activity than the latter, and might be required to maintain the physical contours of a rigid, permanent cytostome.

Inter-tubule arms are clearly demonstrable in bodonids. *Bodo designis* has anterior subplasmalemmar microtubules each with 2 fine filamentous arms (Fig. 7c). Although contact between adjacent tubules is not always evident, the possible artifactual distortion through fixation of a true contact in the living flagellate must be considered, and it is likely that these microtubule links are responsible to some extent for cytoskeletal function in the maintenance of specific surface contours.

Microtubules with arm-like projections developed to an extreme degree are found in the ventral or striated disk of *Giardia* (Friend, 1966). The flagellate's attachment to the duodenal surface is thought to be due to the activity of 2 ventral flagella which displace liquid from a shallow ventral cavity formed by lip-like body surface extensions (which partially surround the disk) and the disk itself, thereby generating an adhesive suction pressure (Holberton, 1973). The closely spaced microtubules of the disk, anchored in the cytoplasm by large electron-dense ribbons (Fig. 7f), themselves connected to adjacent ribbons by fine bridging filaments, confer the property of structural rigidity considered necessary for attachment. The substructural organization of this fibre argues convincingly in favour of a cytoskeletal role.

In addition to the role of support in the maintenance of form microtubules appear to promote the development of anisometric form in cell division and in other morphogenetic processes accompanying life-cycle changes. Cell movements resulting in the formation of specific and characteristic shape during division and, e.g. excystment, probably reflect the growth of microtubules by polymerization of tubulin from a cytoplasmic pool on to specific microtubule-organizing centres. An important example of morphogenetic change relating to microtubules is the transformation of bloodstream trypanosomes to tsetse-fly mid-gut or culture forms. This is marked by a posterior outgrowth of the trypanosome body (see Brown, Evans and Vickerman, 1973) due probably to the functioning of preexisting microtubules as organizing centres for the polymerization of tubulin. An alternative mechanism however, is the sliding into position of preformed microtubules which could be mediated by dynein-like arms.

Subplasmalemmar fibres less extensive than those described above in the form of sheets are exemplified by the ribbon-like fibres in diplomonads such as *Trepomonas agilis*. This flagellate possesses oral grooves for the

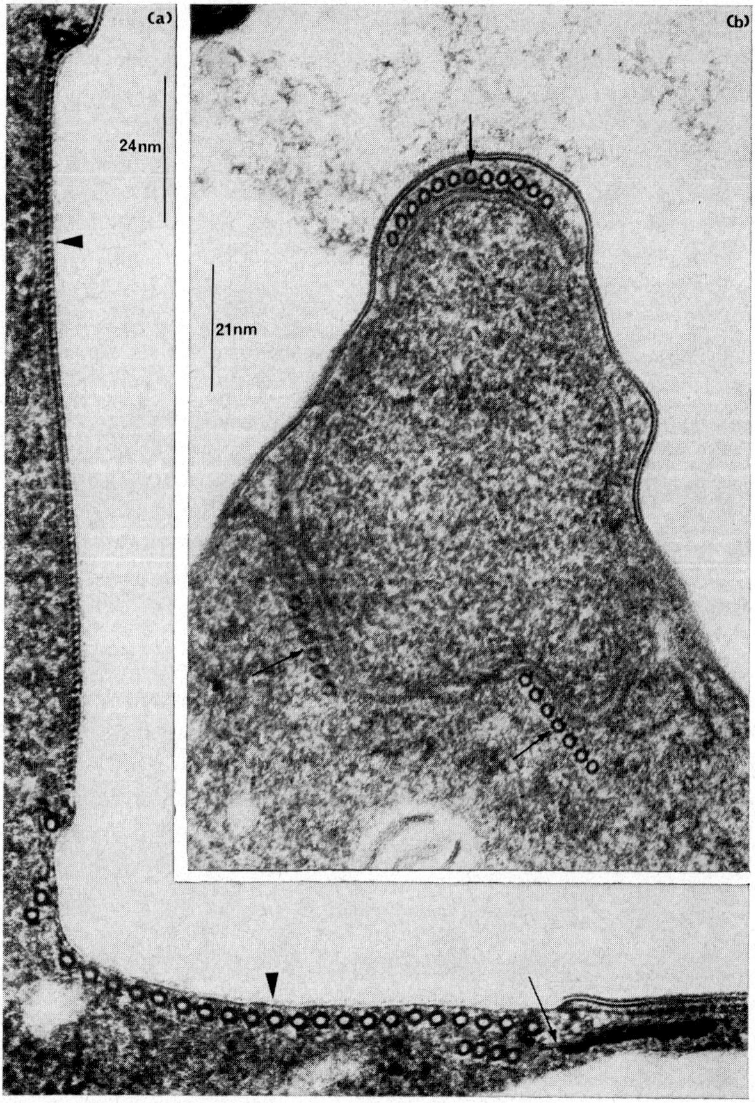

FIG. 3. Microtubular and filamentous fibres in *Trepomonas agilis*: (a) fibres of oral groove-small arrowhead, filamentous lattice; large arrowhead, main oral groove microtubular fibre in cross-section; arrow, cross-sectioned filamentous striated ribbon; (b) perinuclear microtubular fibres (arrows).

ingestion of food particles (Eyden and Vickerman, 1975) and the membrane of each groove is underlain both by filamentous and by microtubular fibres. Most of the tubules of the main oral groove microtubular fibre possess small arms conferring angular profiles in cross-sections (Figs 3a and 7d). The basal bodies of *T. agilis* from which this fibre originates also provide a nucleating site for other microtubular ribbons. These partially envelope the 2 anterior, conspicuously pyriform nuclei (Fig. 3b) like those adjacent to the elongate nuclei of *Spironucleus* (Brugerolle, Joyon and Oktem, 1973). In trophic individuals the oral and perinuclear ribbons seem to maintain oral groove and nuclear shape. These unusual nuclear shapes contrast with the ovoid to spherical nuclei of other diplomonads lacking juxtanuclear microtubules. The changes in flagellate form during division and morphogenesis (e.g., in excystment) may be due to growth or regression of microtubules by tubulin polymerization or depolymerization.

Orientated movement of surface membrane (e.g., in phagocytosis) and of cytoplasmic membranous vesicles may be mediated by dynein-like microtubule arms (Bardele, 1973; Porter, 1973). In *T. agilis* arm-bearing microtubules lie along the pathways of cyclosis of food vacuoles and the arms on both oral groove and subnuclear (Fig. 7e) microtubules may be active in this example of cytoplasmic flow.

Trichomonads possess more deep-seated microtubule rows, one cap-like and anterior (the pelta), the other (the axostyle) passing the length of the body as a partially rolled up cylindrical fibre (see Hollande and Valentin, 1968; Fig. 4b). Its microtubules possessing short arms and fine bridging filaments (Mattern, Honigberg and Daniel, 1973), are likely to have a cytoskeletal role as well as being responsible for the production of anisometric form: trichomonad axostyles appear to lack undulatory properties.

Rod-like microtubular fibres

Several ribbons of microtubules grouped together form a discrete rod-like structure. Such a fibre is found in *Bodo designis* (Fig. 4c) accompanying the tube-like cytopharynx used in food ingestion, along its entire length. Cross-sections show its triangular profile to comprise 4–6 rows, the last row consisting of a single tubule thereby providing the fibre with between 10 and 21 tubules. The hexagonally packed microtubules are linked by 2 kinds of connective. One is clearly defined, as thick as the tubule wall and connects tubules in the same row; the other stains less intensely and connects tubules in different rows. The fibre may be concerned with the maintenance or development of cell shape. Alternatively, the entire fibre may be capable of a movement which could extend the cytostomal aperture to permit ingestion of food particles. The prominent inter-tubule links, possibly with

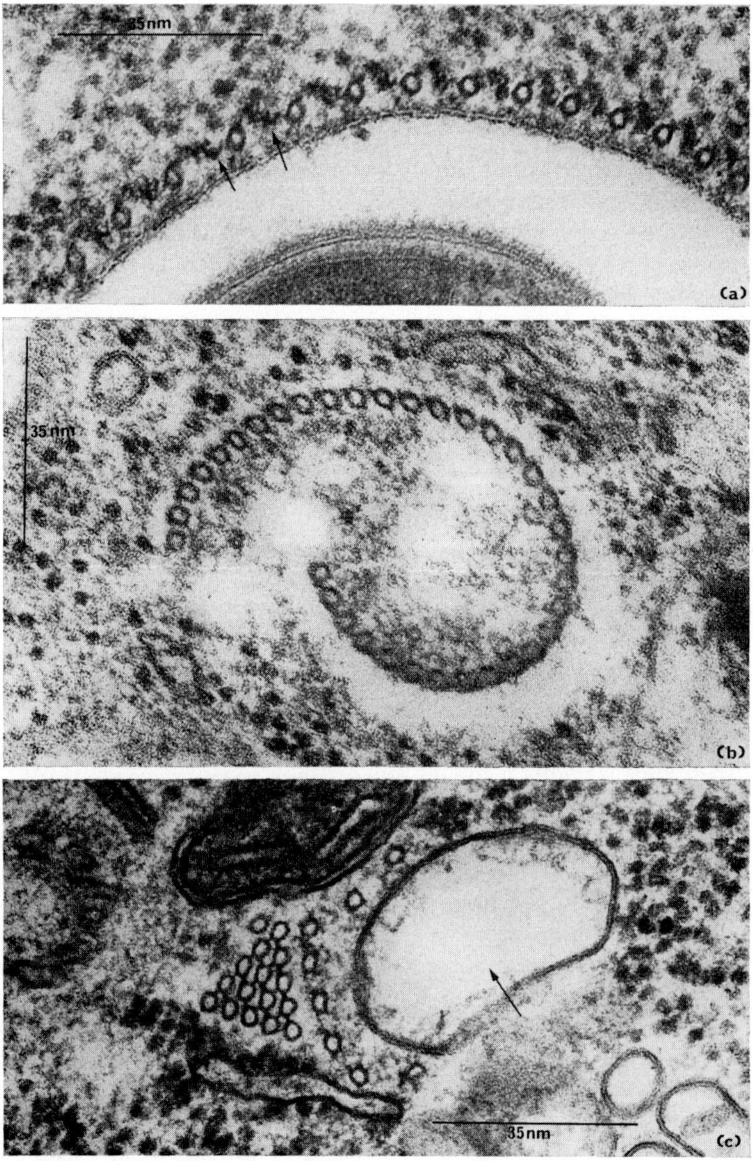

FIG. 4. (a) Subplasmalemmar inter-microtubule connectives (arrows) in cytostome-cytopharynx of *Chilomastix intestinalis*; (b) cross-sectioned axostyle in *Tritrichomonas* sp.; (c) Microtubular rod and arm-bearing microtubules adjacent to the cytropharynx (arrow) of *Bodo designis*.

nexin-like characteristics, within single rows may indicate relative movement of entire rows mediated by the less conspicuous connectives in a dynein-arm type of mechanism. The extent to which such possible movements relate to cytostome-extension, or movement of ingested particles down the cytopharyngeal lumen is not yet clear; but the single row of arm-bearing microtubules lying close to the cytopharyngeal wall may (Fig. 4c) also play a part in this process since they are found in other bodonids having a tube-like cytopharynx but lacking the rod-like fibre. Much larger rods concerned with the feeding process having a similar organization of hexagonally packed inter-linked tubules are conspicuous in other protozoan groups, e.g., euglenids (Mignot, 1966; Nisbet, 1974) and ciliates (e.g., Tucker, 1972; Bardele, 1973).

Fibres built on the same basic plan as that of *Bodo designis*, i.e., consisting of closely applied single-row microtubule-ribbons form the large undulating axostyles of oxymonads. Comprising thousands of tubules, the axostyles are clearly visible in the light microscope. The microtubules within each of the numerous rows are linked by regularly positioned connectives while less regularly placed arms make contact between microtubules in different rows (McIntosh, Ogata and Landis, 1973). These fine structural observations coupled with analyses of extracted axostyles (Sleigh and Pitelka, 1974) showing the presence of tubulin, dynein and nexin make the extension of the sliding filament mechanism for flagellar undulations to the sliding of ribbons in undulating axostyles an attractive proposition. In undulating axostyles nexin could be responsible for the cohesion of microtubules in a single row thus allowing a row to move as a unit.

Filamentous Fibres

Like microtubular fibres, fibres composed of filamentous fibrils take the form of sheets, ribbons or rods, all of which possess ordered fibril-alignments which confer a periodic appearance by electron microscopy. Sheet-like fibres are not conspicuous in lower flagellate groups although they have been described from more complex zooflagellates; e.g., *Trepomonas agilis*, which has a fibrous sheet lying immediately beneath the oral groove membrane (Eyden and Vickerman, 1975). This lattice comprises 2 kinds of fibril; 6–10 nm diameter fibrils are crossed at 90° by thicker 15 nm fibrils (Fig. 5c). Fibres with similar orthogonal fibril-alignments have been described by Brugerolle and Joyon (1973) in *Monocercomonoides*, and by Balamuth and Bradbury (see Pitelka, 1969) in the rhizomastigid *Tetramitus*. In each of these flagellates the fibre acts as a connective between basal bodies (thereby differing from the trepomonad subplasmalemmar lattice); the sizes of fibrils in these fibres is not known.

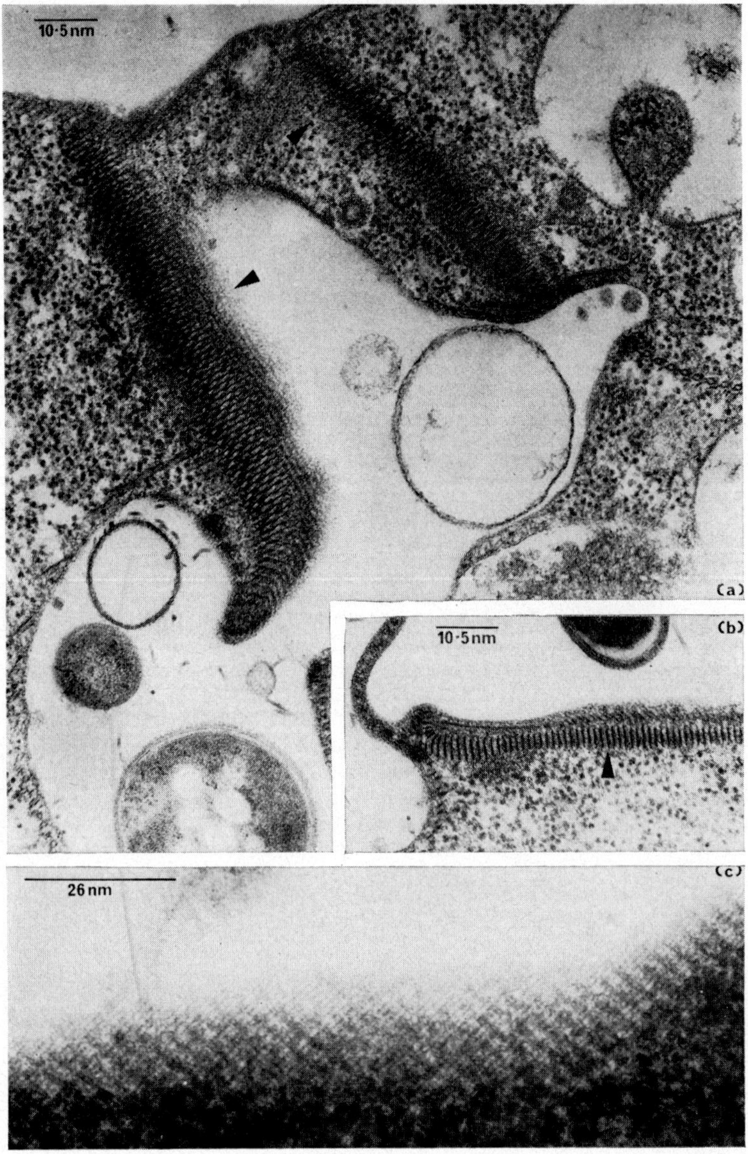

FIG. 5. (a) and (b) right peristomal fibre (arrowheads) in *Chilomastix intestinalis*; (c) filamentous lattice of *Trepomonas agilis* showing thick and thin filaments.

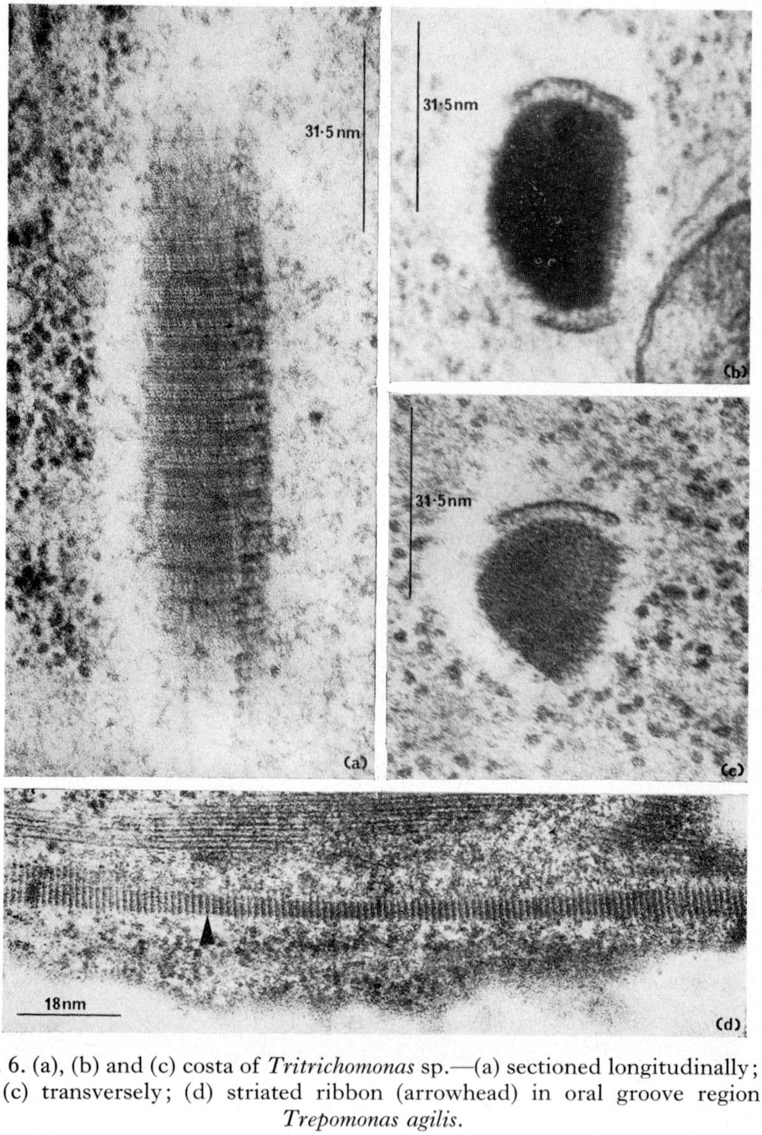

FIG. 6. (a), (b) and (c) costa of *Tritrichomonas* sp.—(a) sectioned longitudinally; (b) and (c) transversely; (d) striated ribbon (arrowhead) in oral groove region of *Trepomonas agilis*.

Fibres showing a greater complexity of fibril-arrangement than the foregoing are exemplified by the striated plates of *Giardia* and the right peristomal fibre of *Chilomastix*. In *Giardia* the peripheral rim-like extension of the body (the ventrolateral flange) is occupied by a pair of fibres called striated plates (Holberton, 1973). Each plate is a filamentous fibre possessing both thin and thick (18 nm) filaments. The right peristomal fibre of *Chilomastix* (Brugerolle, 1973) is a thick ribbon lying against the wall of the cytostome-cytopharynx (Fig. 5, a and b). Several fibril alignments are evident, the most clearly defined being of 6 nm filaments.

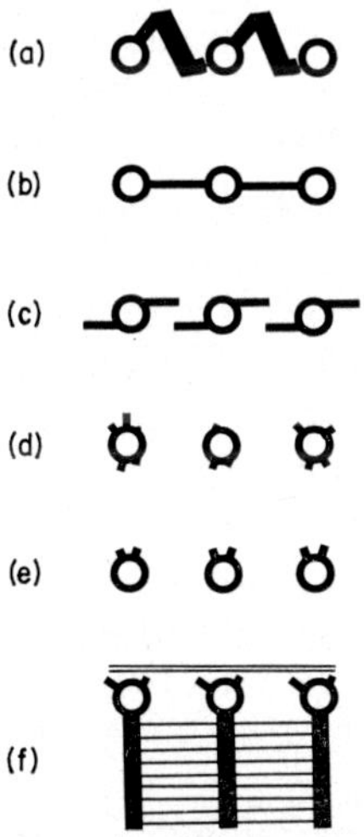

FIG. 7. Arm-like structures of microtubules: (a) and (b) *Chilomastix intestinalis*; (c) *Bodo designis*; (d) and (e) *Trepomonas agilis*; (f) *Giardia muris*.

Another category of filamentous fibre characterized by regularly spaced striations orthogonal to the longitudinal axis of the fibre are common in flagellates (see Pitelka, 1969) and are here described as cross-striated fibres. These are well-developed in diplomonads and trichomonads for example, in *Trepomonas* basal-body-derived cross-striated ribbons lie beneath the oral groove membranes (Fig. 6d). The costa, a rod-like fibre (Fig. 6, a–c), typical of trichomonads shows a more complex set of striation patterns (Fig. 6a). Both the trepomonad fibre and the trichomonad costa reveal fine longitudinal filaments in high resolution electron micrographs which in the latter appear to have a zig-zag pattern (Honigberg, Mattern and Daniel, 1968).

In assigning functions to filamentous fibres the rigidity that might be expected from regular alignments of protein filaments indicates a role in support, like many microtubular fibres. In this context the ultrastructural similarity of cross-striated fibres with collagen is noteworthy. The trichomonad costa in particular bears a strong resemblance to collagen fibres which constitute the main supporting fibrous elements in animal tissues.

While it is thought that peripheral beds of somewhat isotropically orientated microfilaments are responsible for morphogenetic cell movements in developing metazoan tissues (e.g., Spooner, 1973), the extent to which filamentous fibres in zooflagellates provide morphogenetic direction during cell division and other life-cycle changes is not yet known. Nor is there yet any evidence to suggest a role for zooflagellate filamentous fibres in cytoplasmic movements mediated by a contractile mechanism of cortical microfilaments as in tissue culture cells (Wessells, Spooner and Luduena, 1973; Goldman *et al.*, 1973) and amoebae (Wohlfarth-Bottermann, 1974). Several filamentous organelles possess fibrils similar in size to skeletal muscle actin and myosin but nothing is known of any biochemical similarity. Contractility has however been proposed as a function for the striated plates of *Giardia* (Friend, 1966) largely on the basis of the presence and regular alignments of thin as well as thick filaments. Holberton (1973) in discussing these structures in relation to muscle cell structure dismissed any evidence for contractility. The demonstration of actomyosin-like proteins would provide support for such an hypothesis, but so far no studies have shown the presence of actin- or myosin-like components in fibrillar zooflagellate organelles.

Acknowledgements

I thank Professor K. Vickerman for helpful criticism during the preparation of this work. For kind permission to use Figs 13, 16, 19 and 23 in Eyden and Vickerman (1975) reproduced here as Figs 3b, 3a, 6d and 5c, from Vol. 22, *Journal of Protozoology*, I wish to thank Professor B. M. Honigberg (editor).

References

BARDELE, C. F. (1973). Structure, biochemistry and function of microtubules. *Cytobiologie*, **7**, 442.

BROWN, R. C., EVANS, D. A. & VICKERMAN, K. (1973). Changes in oxidative metabolism and ultrastructure accompanying differentiation of the mitochondrion in *Trypanosoma brucei*. *Int. J. Parasit.*, **3**, 691.

BROOKER, B. E. (1971). Fine structure of *Bodo saltans* and *Bodo caudatus* (Zoomastigophora: Protozoa) and their affinities with the Trypanosomatidae. *Bull. Brit. Mus. (Nat. Hist.) Zool.*, **22**, 89.

BROOKER, B. E. (1972). Modifications in the arrangement of the pellicular microtubules of *Crithidia fasciculata* in the gut of *Anopheles gambiae*. *Z. Parasitenk.*, **40,** 271

BRUGEROLLE, G. (1973). Etude ultrastructurale du trophozoite et du kyste chez le genre *Chilomastix* Alexeieff 1910 (Zoomastigophorea, Retortomonadida Grassé 1952). *J. Protozool.*, **20.**

BRUGEROLLE, G. & JOYON, L. (1973). Sur la structure et la position systématique du genre *Monocercomonoides* (Travis 1932). *Protistologica,* **9,** 71.

BRUGEROLLE, G., JOYON, L. & OKTEM, N. (1973). Contribution a l'étude cytologique et phylétique des Diplozoaires Zoomastigophorea, Diplozoa Dangeard 1910. II. Etude ultrastructurale du genre *Spironucleus* (Lavier 1936). *Protistologica,* **9,** 495.

DANIEL, W. A., MATTERN, C. F. T. & HONIGBERG, B. M. (1971). Fine structure of the mastigont system of *Tritrichomonas muris* (Grassi). *J. Protozool.*, **18,** 575

EYDEN, B. P. & VICKERMAN, K. (1975). Ultrastructure and vacuolar movements in the free-living diplomonad *Trepomonas agilis* Klebs. *J. Protozool.*, **54,** 66.

FRIEND, D. S. (1966). The fine structure of *Giardia muris*. *J. Cell Biol.*, **29,** 317.

GOLDMAN, R. D., BERG, G., BUSHNELL, A., CHANG, C.–M., DICKERMAN, L., HOPKINS, N., MILLER, M. L., POLLACK, R. & WANG, E. (1973). Fibrillar systems in cell motility. In *Locomotion in tissue cells* (Porter, R. & Fitzsimons, D. W., eds) Amsterdam: ASP.

GRIMSTONE, A. V. & GIBBONS, I. R. (1966). The fine structure of the centriolar apparatus and associated structures in the complex flagellates *Trichonympha* and *Pseudotrichonympha*. *Phil. Trans. Roy. Soc. Lond.* B, **250,** 215.

HITCHEN, E. T. (1974). The fine structure of the colonial kinetoplastid flagellate *Cephalothamnion cyclopum* Stein. *J. Protozool.*, **21,** 221.

HOLBERTON, D. V. (1973). Fine structure of the ventral disk apparatus and the mechanism of attachment in the flagellate *Giardia muris*. *J. Cell Sci.*, **13,** 11.

HOLLANDE, A. & VALENTIN, J. (1968). Morphologie infrastructurale de *Trichomonas* (*Trichomitopsis* Kofoid et Swezy 1919) *termopsidis*, parasite intestinal de *Termopsis angusticollis* Walk. Critique de la notion de centrosome chez les polymastigines. *Protistologica,* **4,** 127.

HOLLANDE, A. & VALENTIN, J. (1969). La cinétide et ses dépendances dans le genre *Macrotrichomonas* Grassi. Considérations générales sur la sous-famille des Macrotrichomonadinae. *Protistologica,* **5,** 335.

HOLLANDE, A. & CARRUETTE-VALENTIN, J. (1970). Interprétation générale des structures rostrales des Hypermastigines et modalités de la pleuromitose chez les Flagellés du genre *Trichonympha*. *C. r. Acad. Sci., Paris,* **270,** 1476.

HOLLANDE, A. & CARRUETTE-VALENTIN, J. (1971). Les atractophores, l'induction du fuseau et la division cellulaire chez les Hypermastigines. Etude infrastructural et revision systématique des trichonymphines et des spirotrichonymphines. *Protistologica,* **7,** 5.

HOLWILL, M. E. J. (1974). Hydrodynamic aspects of ciliary and flagellar movement. In *Cilia and Flagella* (Sleigh, M. A., ed.) London & New York: Academic Press.

HONIGBERG, B. M., MATTERN, C. F. T. & DANIEL, W. A. (1968). Structure of *Pentatrichomonas hominis* (Davaine) as revealed by electron microscopy. *J. Protozool.*, **15,** 419.

MATTERN, C. F. T., HONIGBERG, B. M. & DANIEL, W. A. (1973). Fine-structural changes associated with pseudocyst formation in *Trichomitus batrachorum*. *J. Protozool.*, **20**, 222.

MCINTOSH, J. R., OGATA, E. S. & LANDIS, S. C. (1973). The axostyle of *Saccinobaculus*. I. Structure of the organism and its microtubule bundle. *J. Cell Biol.*, **56**, 304.

MIGNOT, J.–P. (1966). Structure et ultrastructure de quelques Euglénamonadines. *Protistologica*, **2**, 51.

NISBET, B. (1974). An ultrastructural study of the feeding apparatus of *Peranema trichophorum*. *J. Protozool.*, **21**, 39.

PITELKA, D. R. (1969). Fibrillar systems in Protozoa. In *Research in Protozoology* Vol. 3 (Chen, T.–T., ed.) 279 Oxford: Pergamon Press.

PORTER, K. R. (1973). Microtubules in intracellular locomotion. In *Locomotion of tissue cells* (Porter, R. & Fitzsimons, D. W., eds) Amsterdam: ASP.

SATIR, P. (1974). The present status of the sliding microtubule model of ciliary motion. In *Cilia and Flagella* (Sleigh, M. A., ed.) London & New York: Academic Press.

SLEIGH, M. A. (1973). Cell Motility. In *Cell Biology in Medicine* (Bittar, E. E. ed.) New York: Wiley.

SLEIGH, M. A. & PITELKA, D. R. (1974). Processes of contractility in Protozoa. In *Actualités Protozoologiques* (Puytorac, P. de, & Grain, J., eds) Clermont-Ferrand: Couty.

SPOONER, B. S. (1973). Microfilaments, cell shape changes, and morphogenesis of salivary epithelium. *Amer. Zool.*, **13**, 1007

TUCKER, J. B. (1972). Microtubule-arms and propulsion of food particles inside a large feeding organelle in the ciliate *Phascolodon vorticella*. *J. Cell Sci.*, **10**, 883.

VICKERMAN, K. (1969). On the surface coat and flagellar adhesion in trypanosomes. *J. Cell Sci.*, **5, 163**.

VICKERMAN, K. (1974). The ultrastructure of pathogenic flagellates. In *Trypanosomiasis and Leishmaniasis with special reference to Chagas' disease* (Porter, R. & Knight, J. eds) Amsterdam: ASP.

WARNER, F. D. (1974). The fine structure of the ciliary and flagellar axoneme. In *Cilia and Flagella* (Sleigh, M. A., ed). London & New York: Academic Press.

WESSELS, N. K., SPOONER, B. S. & LUDUEÑA, M. A. (1973). Surface movements, microfilaments and cell locomotion. In *Locomotion in tissue cells* Porter, R. & Fitzsimons, D. W., eds) Amsterdam: ASP.

WOHLFARTH-BOTTERMANN, K. E. (1974). Ameboid movements. In *Actualités Protozoologiques* (Puytorac, P. de & Grain, J. eds) Clermont-Ferrand: Couty.

Extruded Bodies on Erythrocytes

PHYLLIS PEASE, JANICE TALLACK AND
ROBERTA BARTLETT

Department of Bacteriology, University of Birmingham, Birmingham, England

Electron microscopy of mammalian reticulocytes has shown (Seelig, 1972) that protuberant bodies on their surfaces may be associated with the extrusion of discarded nuclear apparatus. Somewhat similar bodies containing discarded Golgi apparatus etc. have been shown in the process of maturation of the nucleated erythrocyte of the trout (Sekhon and Beams, 1969). It has also been suggested that such bodies may be associated with bacterial L-form infection of erythrocytes (Tedeschi *et al.*, 1967; Pease, 1970).

Studies in this laboratory, on the white rat and the New Zealand Black mouse, have shown the occurrence of comparable bodies on the erythrocytes of these animals.

Methods and Materials

Animals

Material used was from strains of New Zealand Black mice and white rats, maintained in this Department.

Embedding

Blood was collected in heparin (0. 1–0.2 mg/ml), usually withdrawn from the heart at death, and embedded by a modification of the method of Zetterqvist (1956). The erythrocytes were washed and centrifuged three times in phosphate buffered saline, placed in 3% isotonic buffer overnight at 4°, washed in saline and emulsified in 2–3 drops of molten ionagar. When solidified, this was cut into cubes less than 1 mm and suspended in 5% uranyl acetate in saline for two hours. The cubes were then dehydrated in graded alcohols, finishing with propylene oxide, left overnight in 1:1 propylene oxide and Taab resin, loosely capped to allow evaporation. They were then transferred to pure Taab resin, dispensed singly in gelatin capsules and polymerized for 48 h at 60°. Sections were cut on a Huxley microtome and picked up on formvar coated grids.

Staining

Grids with attached sections were floated, section downwards, on drops of 2% uranyl acetate in alcohol for 15 min and washed in distilled water. They were further stained, in the same manner, with lead citrate, washing before and after with 0.02 M. sodium hydroxide, and finally, distilled water (Reynolds, 1963). Sections were examined in an AEI 801A microscope.

Results

Sections of erythrocytes of the NZB mouse (Fig. 1) and the white rat (Fig. 2) showed numerous bodies, *c.* 0.5μm in diameter, attached to the surface, singly or in groups. These contained smaller granules; *c.* 0.1μm in diameter,

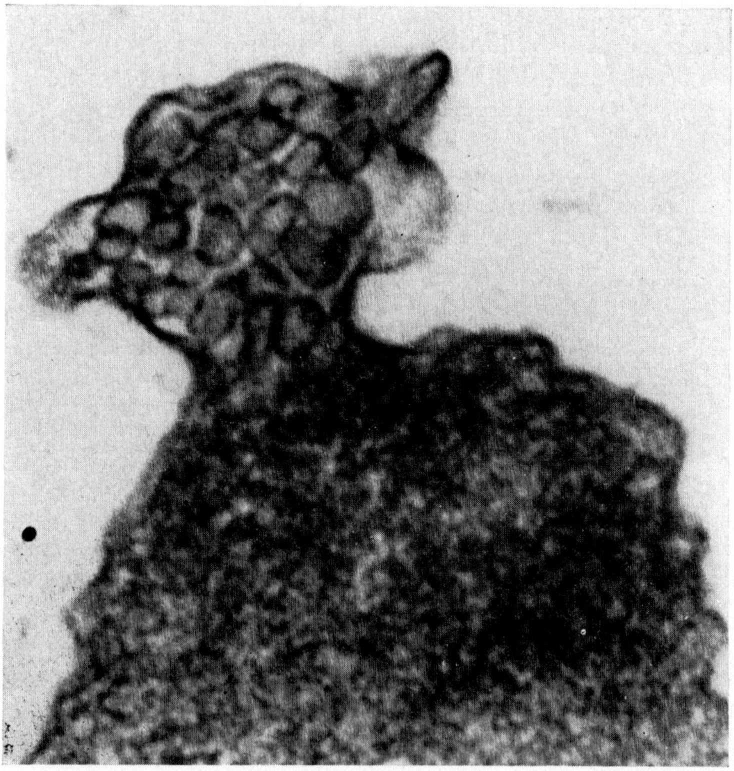

FIG. 1. Extruded body on erythrocyte of NZB mouse, containing numerous, small granules. Electron micrograph of section; stained uranyl acetate and lead citrate ($\times$140 000).

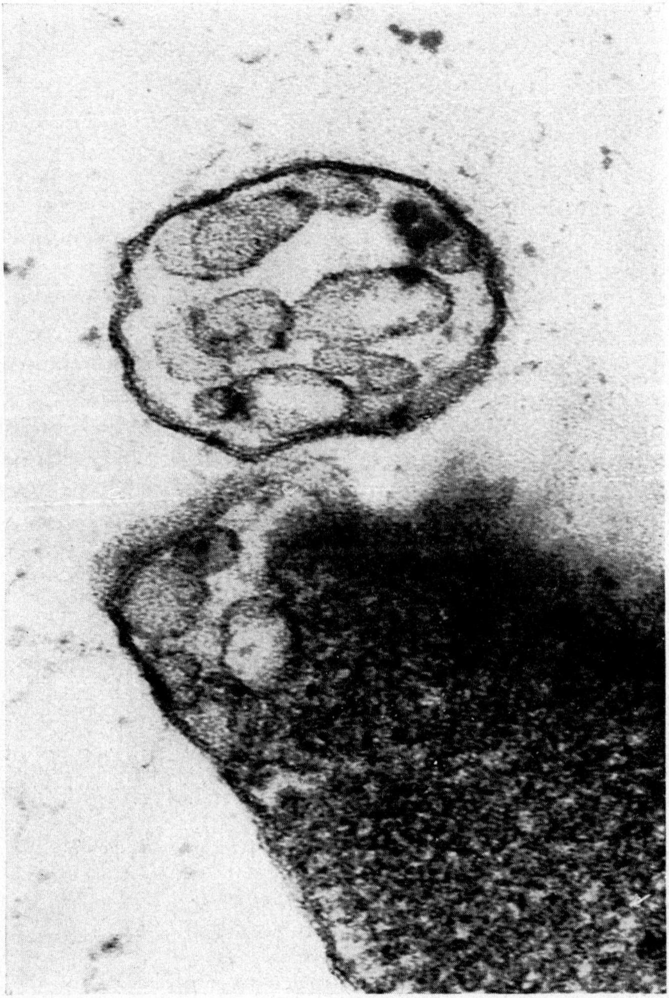

FIG. 2. Extruded body on erythrocyte of white rat, containing numerous, small granules. Electron micrograph of section; stained uranyl acetate and lead citrate (×140 000).

but rather larger in the rat than the mouse, mainly because the former were elliptical in outline, the latter, roughly circular. These appearances were distinct from those found in maturing erythrocytes, and the number of bodies, apparently greater.

The actual numbers however, either of the bodies or the included granules, cannot be accurately estimated in sections.

Discussion

Although the investigation of this problem is at an early stage, and positive conclusions cannot be drawn from the available evidence, it appears possible that the extruded bodies seen on the surface of the erythrocytes of rodents, possibly infected with *Haemobartonella muris*, represent the same bodies demonstrable by microscopic methods in classical descriptions of such infections (Weinman and Ristic, 1968). This permits the speculation, whether this may be a general excretory process through the surface of erythrocytes and whether the small granules, apparently in process of elimination, represent elementary bodies of the actual infective organism. Work now in progress is directed towards the solution of this problem.

References

PEASE, P. E. (1970). Morphological appearances of a bacterial L-form growing in association with the erythrocytes of arthritic subjects. *Ann. Rheum. Dis.*, **29,** 439.

REYNOLDS, E. S. (1963). The use of lead citrate at high pH as an electron opaque stain in electron microscopy. *J. Cell Biol.*, **17,** 208.

SEELIG, L. (1972). Surface multivesicular structures associated with maturing erythrocytes in rats. *Z. Zellforsch. mikrosk. Anat.*, **133,** 181.

SEKHON, S. S. & BEAMS, H. W. (1969). Fine structures of developing trout erythrocytes and thrombocytes with special reference to the marginal band and the cytoplasmic organelles. *Am. J. Anat.*, **125,** 353.

TEDESCHI, G. G., AMICI, D., MURRI, O. & PAPARELLI, M. (1967). Aspetti in microscopia ottica ed elettronica del sangue di pazienti splenectomizzati o affetti da ittero emolitico. *Haematologica,* **52,** 57.

WEINMAN, D. & RISTIC, M. (1968). Infectious Blood Diseases of Man and Animals. New York & London: Academic Press

ZETTERQVIST, H. (1956). The ultrastructural organization of the columnar absorbing cells of the mouse jejunum. An electron microscopic study including some experiments regarding the problem of fixation and an investigation of Vitamin A deficiency. *Thesis*. Karolinska Instituet, Stockholm, p. 83.

Unstable L-forms of Micrococci in Human Platelets

G. G. Tedeschi, D. Amici, I. Santarelli,
M. Paparelli and C. Vitali

Institute of General Physiology, University of Camerino, Italy

The presence of micro-organisms, variously identified as mycoplasmas, stable or unstable L-forms, or other cell-wall-deficient forms of bacteria in the circulating blood of human subjects has been described in Chattman, Mattman and Mattman (1969), Pease (1969), Tedeschi, Amici and Paparelli (1969) and Pohlod, Mattman and Tunstall (1972).

Isolates from human subjects, initially identified as mycoplasmas, were subsequently identified as L-forms of micrococci (Kelton, Gentry and Ludwig, 1960). Following the results of previous researches in this and other laboratories (Dmochowski *et al.* 1967; Tedeschi, Paparelli and Amici, 1968; Fajardo, 1973), suggesting that platelets might carry microbial forms, the present investigation was undertaken.

Materials and Methods

Platelets from 1000 human subjects, normal and pathological, and from the dog, rabbit, guinea-pig, rat, mouse, calf, horse and sheep were examined.

Suspensions of platelets in plasma or serum were incubated at 37° in Krebs-Ringer solution, PPLO or nutrient broth (Difco), or on nutrient agar.

The uptake of ^{44}C-thymidine (Radiochemical Centre, Amersham) was measured by the methods of Tedeschi *et al.* (1969).

Results

An uptake of ^{14}C-thymidine beyond that explicable by the synthesis of mitochondrial DNA was detected in platelets of human beings and dogs, but not of the other animals. This uptake required the presence of factors that were provided by plasma or serum from any species examined. When incubation was conducted in the presence of oxytetracycline the incorporation was inhibted. Provisionally, the uptake of thymidine is attributed to the metabolic activity of bacterial L-forms.

Suspensions of human platelets incubated in broth or on agar in the presence of plasma or serum, showed appearances interpreted as the growth and multiplication of L-forms derived from the platelets, especially from those which were strongly basophilic and undergoing lysis. On suitable media, L-type colonies (Fig. 1) developed. From these incubation products,

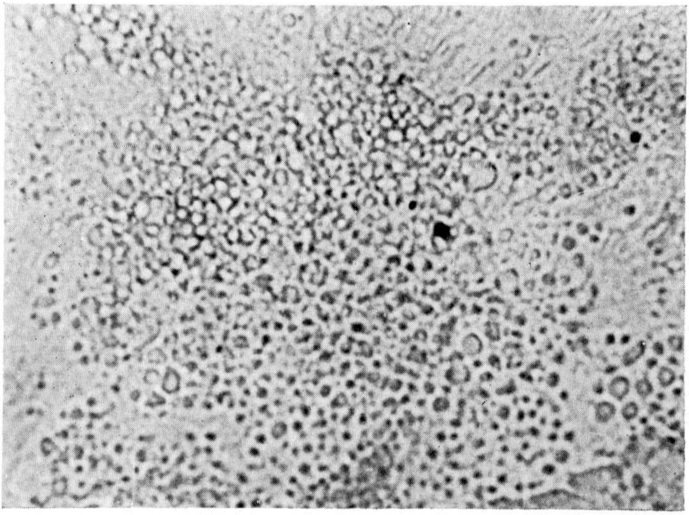

FIG. 1. Plasma containing numerous platelets incubated on PPLO agar (Difco) with penicillin 100 units/ml. Groups of particles are set free and give rise to colonies resembling those of L-forms (Phase contrast ×2000).

micro-organisms provisionally considered to be micrococci could sometimes be subcultured, but not in all cases. Both L-forms and conventional bacterial forms were strictly aerobic.

Discussion

The evolution of these forms, which are not explicable as normal structures in platelets, could be detected at various stages by electron microscopy. They were observed to originate from granules with a high density, bounded by a double-layered membrane, and from granules not at first morphologically distinguishable from alpha-granules (Figs 2–8).

These presumed L-forms in the platelets represent a phenomenon distinct from the platelet-bacterial interaction which has been described as

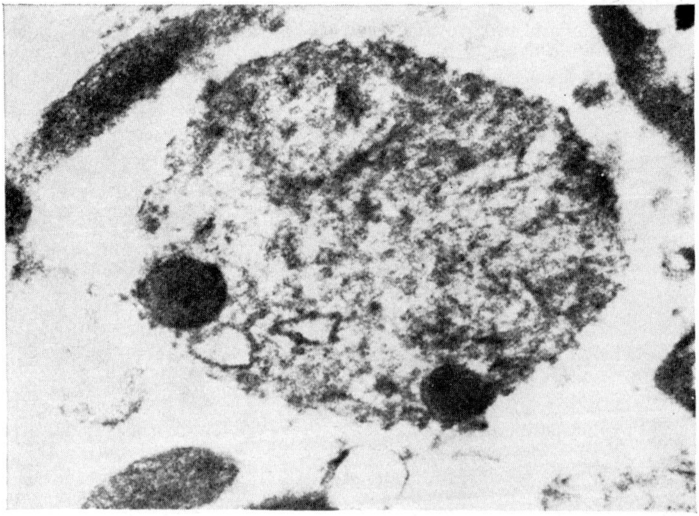

FIG. 2. Circulating blood. Some structurally altered platelets carry strongly osmiophilic granules bounded by a well defined double-layered membrane. (×53 000).

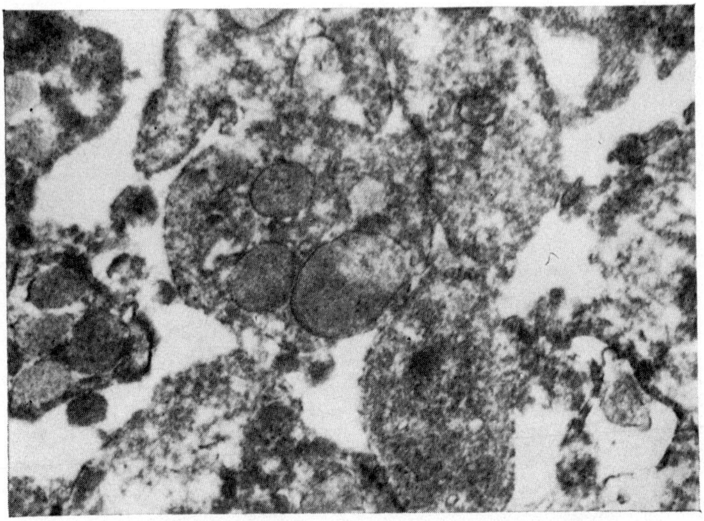

FIG. 3. After 48 h incubation of plasma, some granules, initially indistinguishable from alpha-granules, increase in size (×28 500).

FIGS 4, 5, 6, & 7. The enlarged granules are set free into the medium; their profile is modified, they grow and multiply like bacterial L-forms; often they become surrounded by a cell wall and look like L-forms reverting to "conventional" bacterial forms (Fig. 4 ×61 000, Fig. 5 ×48 000, Fig. 6 ×33 500, Fig. 7 ×49 000.)

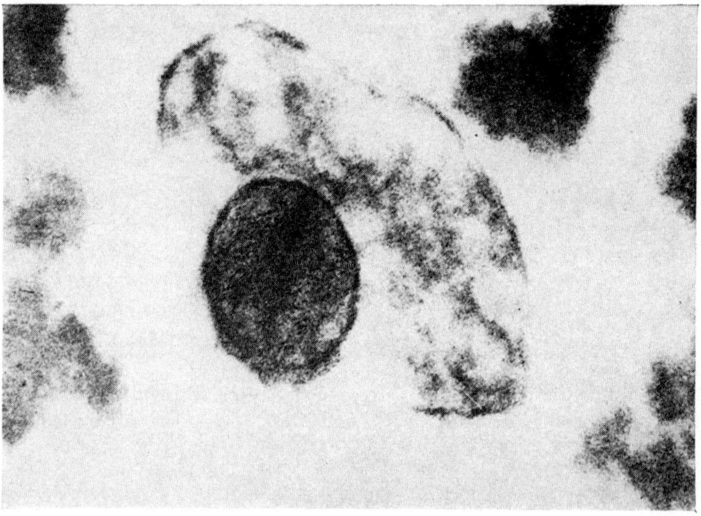

FIG. 4.

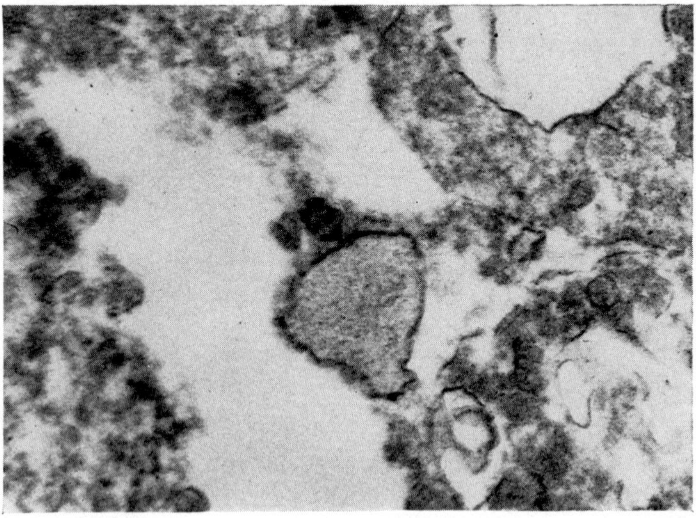

FIG. 5.

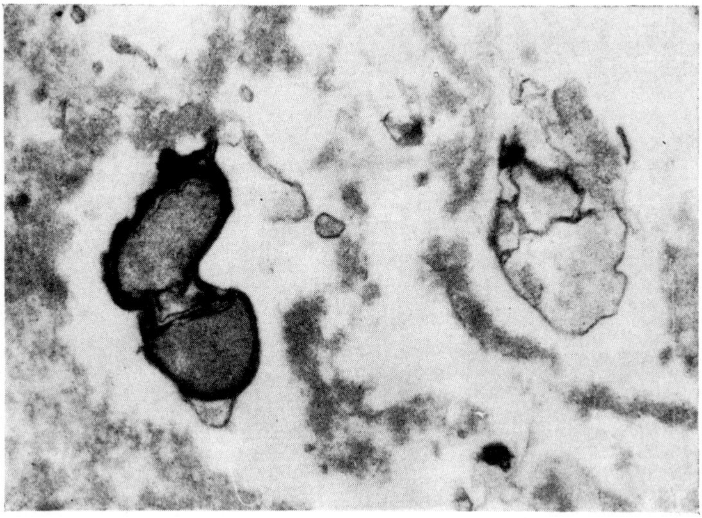

FIG. 6.

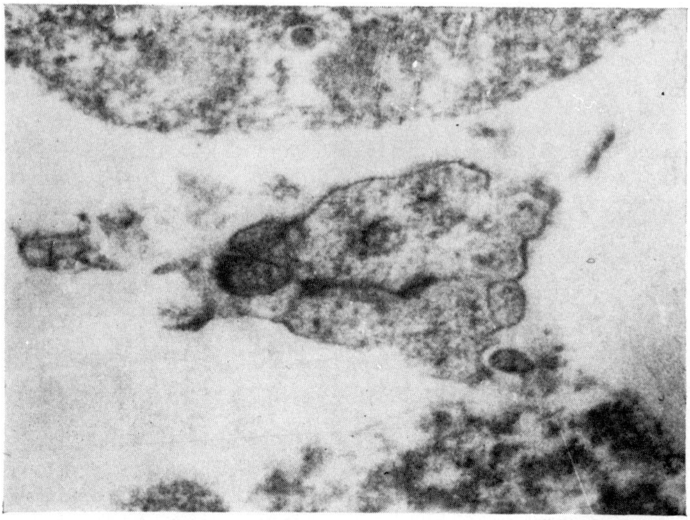

FIG. 7.

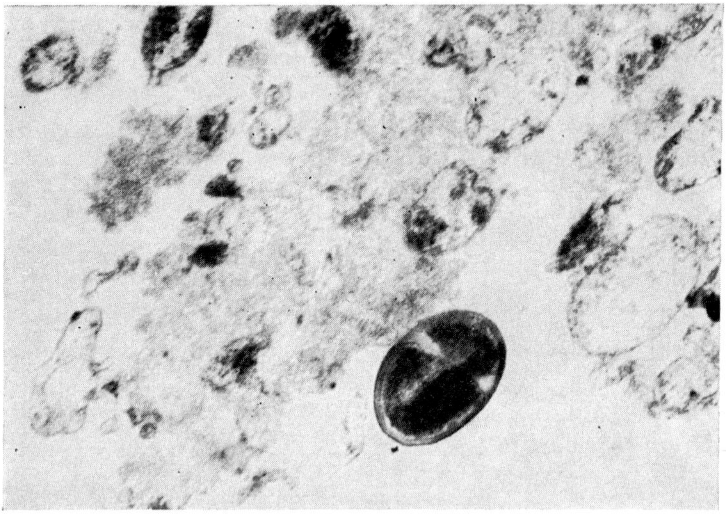

FIG 8. In some cases conventional bacterial forms resembling micrococci, and apparently originating from L forms carried by the platelets, multiply free in the medium but cannot readily be subcultured. Some smaller, incomplete forms are seen in the opposite corner. (×23 800).

a possible mechanism of clearance of bacteria from the circulation (Clawson and White, 1971). However, we cannot neglect the possibility that bacteria released into the circulation or the culture medium may proceed to a sequence of events comparable with the clearing processes.

References

CHATTMAN, M. S., MATTMAN, L. H. & MATTMAN, P. E. (1969). L-forms in blood cultures demonstrated by nucleic acid fluorescence. *Am. J. Clin. Path.*, **51,** 41.

CLAWSON, C. C. & WHITE, J. G. (1971). *In vitro* reactions of Platelets and Bacteria. *J. Reticuloendothelial Soc.*, **9,** 621.

DMOCHOWSKI, L., DREYER, D. A., GREY, C. E., HALES, R., LANGFORD, P. L., PIPES, F., RECHER, L., SEMAN, G., SHIVELY, J. A., SHULLENBERGER, C. C., SINKOVICS, J. H., TAYLOR, H. G., TESSMER, C. F. & YUMOTO, T. (1967). Studies on the submicroscopic morphology of structures resembling mycoplasma and virus particles in mice and men. *Ann. N.Y. Acad. Sci.*, **143,** art. 1, 578.

FAJARDO, L. F. (1973) Malarial parasites in mammalian platelets. *Nature, Lond.*, **243**, 298–299.

KELTON, W. H., GENTRY, R. F. & LUDWIG, E. H. (1960). Derivation of Gram-positive cocci from pleuropneumonia like organisms. *Ann. N. Y. Acad. Sci.*, **79,** art. 10, 410.

PEASE, P. E. (1969). Bacterial L-forms in the blood and joint fluids of arthritic subjects. *Ann. Rheum. Dis.*, **28**, 270.

POHLOD, D. I., MATTMAN, L. H., & TUNSTALL, L. (1972). Structures suggesting cell-wall-deficient forms detected in circulating erythrocytes by fluorochrome staining. *Appl. Microbiol.*, **23**, 262.

TEDESCHI, G. G., PAPARELLI, M. & AMICI, D. (1968). Aspetti morfologici e istochimici di materiale particolato ottenibile in seguito a incubazione di plasma umano su terreno all'agar. *Boll. Soc. It. Biol. Sper.*, **44**, 610.

TEDESCHI, G. G., AMICI, D. & PAPARELLI, M. (1969). Incorporation of nucleosides and aminoacids in human erythrocyte suspensions: possible relation with a diffuse infection of mycoplasma or bacteria in the L-form. *Nature, Lond.*, **222**, 1285.

Subject Index